Plants of Central Asia

Volume 12

Plants of Central Asia

Plant Collections from China and Mongolia

Volume 12

Nymphaeaceae—Ceratophyllaceae
Ranunculaceae—Berberidaceae
Menispermaceae

A.E. Borodina-Grabovskaya
V.I. Grubov
M.A. Mikhailova

CRC Press
Taylor & Francis Group
Boca Raton London New York

CRC Press is an imprint of the
Taylor & Francis Group, an **informa** business
A SCIENCE PUBLISHERS BOOK

First published 2007 by Science Publishers

Published 2019 CRC Press
Taylor & Francis Group
6000 Broken Sound Parkway NW, Suite 300
Boca Raton, FL 33487-2742

ISBN-13: 978-1-57808-441-8 (hbk) (Vol 12)
ISBN-13: 978-1-57808-062-5 (Set)

Visit the Taylor & Francis Web site at
http://www.taylorandfrancis.com

and the CRC Press Web site at
http://www.crcpress.com

Library of Congress Cataloging-in-Publication Data

Rasteniia Tsentral'noi Azii. English
Plants of Central Asia: plant collections from China and Mongolia
[editor-in-chief, V.I. Grubov].
p. cm.
Research based on the collections of the V.L. Komarov Botanical Institute.
Includes bibliographical references
Contents: v. 12. Nymphaeaceae—Ceratophyllaceae—Ranunculaceae—Berberidaceae—Menispermaceae
ISBN 978-1-57808-441-8 (vol. 12)
I. Botany--Asia, Central. I. Grubov, V.I., II. Botanicheskii institut im. V.L. Komarova. III. Title.
QK374, R23613 2002
581.958--dc21 99-36729
CIP

Translation of : Rasteniya Tsentral' noi Azii, vol. 12, 2001
St.-Petersburg Chemical-Pharmaceutical Academy Press.

ANNOTATION

PLANTS OF CENTRAL ASIA. From the materials of the V.L. Komarov Botanical Institute, Russian Academy of Sciences, vol. 12: Nymphaeaceae, Ceratophyllaceae, Ranunculaceae, Berberidaceae, Menispermaceae. Compilers: A.E. Borodina-Grabovskaya, V.I. Grubov and M.A. Mikhailova. 2001. Izdatel'stvo sankt-peterburgskoi gosudarstvennoi khimiko-farmatsevticheskoi akademii (State Chemical-Pharmaceutical Academy), St.-Petersburg.

This twelfth volume of the illustrated lists of plants of Central Asia (within the People's Republics of China and Mongolia) continues the description of flowering plants and covers the treatment of families Nymphaeaceae, Ceratophyllaceae, Ranunculaceae, Berberidaceae, Menispermaceae. Keys are provided for the identification of genera and species. For each species are given references to nomenclature, information on habitat and geographic distribution.

Ill. : 8 plates, 4 maps of distribution ranges.

V.I. Grubov
Editor-in-Chief

PREFACE

This volume of "Plants of Central Asia" covers 5 families: water lilies, Nymphaeaceae (2 genera, 4 species); hornworts, Ceratophyllaceae (1, 1); buttercups, Ranunculaceae (25, 237); barberries, Barberidaceae (4, 25); and moonseeds, Menispermaceae (1, 1). These 5 families together constitute 33 genera and 268 species. To these should be added 65 species of Ranunculaceae and 4 species of Berberidaceae inhabiting Kyrgyzstan, Kazakhstan and Eastern Pamir falling in the former Soviet part of Central Asia. The total number of species of these families would then rise to 337.

The largest genera treated in this volume are *Delphinium* (54 species) and *Ranunculus* (46 species). In terms of species abundance, the important genera are *Aconitum* (27) and *Berberis* (22). Genera comprising 14-16 species each are *Anemone* (16), *Clematis* (16) and *Thalictrum* (14); 8 species each *Pulsatilla*, *Aquilegia* and *Trollius*; 2-5 species each *Adonis* (5), *Atragene* (5), *Batrachium* (5), *Paeonia* (4), *Halerpestes* (4), *Caltha* (3), *Callianthemum* (3), *Cimicifuga* (2), *Paraquilegia* (2), *Nuphar* (2) and *Nymphaea* (2); genera represented by single species are: *Actaea*, *Ceratocephala*, *Circaeaster*, *Hepatica*, *Leptopyrum*, *Oxygraphis*, *Paropyrum*, as well as *Ceratophyllum*, *Gymnospermium*, *Leontice*, *Podophyllum* and *Menispermum*.

Many species have been cited only on the strength of literature sources since authentic specimens were not available to us. The number of such species is particularly large in genera *Delphinium*, *Ranunculus* and *Aconitum*.

The most extensively discussed family is that of buttercups, following in this respect the author of Flory Kitaya (Flora of China) (Fl. Sin. 1979, 1980). Genus *Paeonia* falls in subfamily Paeonioideae Hutch. while monotypic subendemic Chinese genus *Circaeaster* Maxim. has been included in tribe Anemoneae DC. We did not recognise the independent status of some genera (*Anemonastrum*, *Hegemone* and *Thacla*) while *Atragene* has been accepted as an independent genus.

In Central Asia, there are 28 endemic species of Ranunculaceae (11% of total number); these include 12 species of *Delphinium*, 7 of *Ranunculus*, 3 of *Aconitum*, 2 of *Anemone*, and 1 each of *Atragene*, *Clematis*, *Circaeaster* and *Adonis*. These endemic species are not confined to any one particular part of Central Asia but are found in Tibet, specially in Southern Tibet (many of *Delphinium*), Qinghai, Ordos and Alashan. *Clematis fruticosa* is endemic in Mongolia and East. Kashgar. Apart from this, there are 4 endemic species of barberries in Central Asia (16% of total number of Berberidaceae). The overall endemism covers 13% of families discussed in this volume.

New combinations have been adopted in this volume in genera *Anemone*, *Atragene* and *Ranunculus*; it also cities some species that are new to the Chinese territory: *Delphinium inconspicuum*, *Pulsatilla flavescens*, *Ranunculus gobicus*, *R. krasnovii*, *R. severzovii* and others.

Buttercups represent a boreal family. Geographically, buttercups of Central Asia exhibit closest affinity to the flora of Southern Siberia (about a third of total number of species) and Far East. Some species are distributed predominantly in boreal Eurasia and even North America. Their affinity to flora of Eastern Asia as well as Himalayas is significant. Floristic affinity to Middle and Fore Asia is extremely feeble. On the contrary, barberries of Sinkiang tend toward Mid. Asia and those of Qinghai toward Eastern Asia. Many species of *Berberis* and *Podophyllum emodi* Wall. represent remnants of Tertiary forest flora of Eastern Asia.

Boreal species of Southern Siberia and Eastern Asia could naturally penetrate deep into Central Asia through the surrounding mountain systems and produce there their own derivatives, i.e. local endemics.

Chorologically, the disjunctive disribution range of *Circaeaster agrestis* Maxim. is interesting. This tender annual plant with rather unclear taxonomic affinity was unexpectedly detected recently at 2 places in Eastern Tien Shan (Sinkiang): in forests on northern slopes in Bogdo-Ula southward of Urumchi town and north-west in Usu (Wusu) town region in Tomurte mountains (northern spur of Irenkhabirga mountain range). Another enclave of this species has been known for long in Kumaon forests of Western Himalayas where it inhabits among mosses on rocks (see Fl. Brit. India, 5 (1886) 100-101). Since this plant is very small and poorly visible, it could easily be overlooked by collectors, and its range could in fact be far more extensive. Moreover, it is easily propagated by animals as single-seeded small free follicles (1-4) of *Circaeaster* are covered with sticky uncinate bristles. One way or the other, its present disjunctive range suggests past direct territorial contact of eastern Tibet with Eastern Tien Shan through Himalayas and Tien Shan finds of this species represent relicts.

Most Ranunculaceae are plants of moist habitats: they are found in meadows (often solonetzic and marshy), in mountain forests and their fringes, on banks of brooks and rivers (as well as on water), around springs, scrubs, wet pebble beds and talus, in clayey and sandy soil, mainly on mountain massifs and often at considerable (up to 5000 m) altitudes.

However, some of them—their number is much too small to be significant—have found their niches in the arid environment of dry steppes and barren lands and occupied a prominent place in the vegetative cover.

Foremost among them should be mentioned *Clematis songorica* Bge. Its large and bright green bushes are covered profusely in white flowers, often grow massively to form thickets along margins of solonchaks, flanks and floors of gorges, sand fringes in barren Gobi desert (Junggar to Alashan) and brighten the landscape. *C. fruticosa* Turcz. and *C. tangutica* (Maxim.) Korsh. are also widely distributed in conical hillocks (mud cones) and low mountains along barren and steppe slopes and gorges. *Pulsatilla bungeana* C.A. Mey and *P. turczaninovii* Kryl. et Serg. play a significant role in arid steppes, quite often forming a prominent feature. The common, although less visible component of steppes, *Thalictrum foetidum* L. and *Th. squarrosum* Steph. ex Willd., *Trollius asiaticus* L. and *T. altaicus* C.A. Mey. ("warm weather plants") as well as *Thalictrum petaloideum* L., as in Siberia, impart colour to meadows in forest and alpine belts of Mongolian and Gobi Altay. In the grass cover of forests and coastal groves, *Paeonia anomala* L. as well as species of *Aconitum* L. and *Delphinium* L. strike the eye. *Paeonia veitchii* Lynch, *Aquilegia ecalcarata* Maxim., *Trollius pumilus* D. Don and others play an analogous role in Nanshan. Among barberries, *Berberis heteropoda* Schrenk, *B. dasystachya* Maxim., B. nummularia Bge. and *B. diaphana* Maxim. play a prominent role in tugai vegetation of Junggar, Kashgar and Qinghai.

The members of families of water lilies, hornworts and moonseeds are extremely few and rare and play no perceptible role either in the flora or vegetative cover of Central Asia.

* * *

For this volume, O.I. Starikova translated the Chinese references and herbarium labels. Artist O.V. Zaitseva prepared the drawings for plates while A.E. Borodina-Grabovskaya plotted the maps of distribution ranges and compiled the indexes.

V.I. Grubov
A.E. Borodina-Grabovskaya

CONTENTS

CONTENTS

TAXONOMY

SPECIAL ABBREVIATIONS

Abbreviations of Names of Collectors

Bar.	—	V.I. Baranov
Bud.	—	A.L. Budantsev
Ching	—	R.C. Ching
Chu	—	C.N. Chu
Czet.	—	S.S. Czetyrkin
Divn.	—	D.A. Divnogorskaya
Fed.	—	B.A. Fedtschenko (Fedczenko)
Fet.	—	A.M. Fetisov
Glag.	—	S.A. Glagolev
Golubk.	—	N.S. Golubkova
Grombch.	—	B.L. Grombchevski
Grub.	—	V.I. Grubov
Gr.-Grzh.	—	G.E. Grum-Grzhimailo
Gub.	—	I.A. Gubanov
Gur.	—	N.P. Guricheva
Gus.	—	V.A. Gusev
Ik.-Gal.	—	N.P. and V.A. Ikonnikov-Galitzkij
Isach.	—	E.A. Isachenko (also known as E.A. Volkova)
Ivan.	—	A.F. Ivanov
Kal.	—	A.V. Kalinina
Kam.	—	R.V. Kamelin
Karam.	—	Z.V. Karamysheva
Klem.	—	E.N. Klements
Knorr.	—	O.E. Knorring
Kom.	—	V.L. Komarov
Krasch.	—	I.M. Krascheninnikov
Krasn.	—	A.N. Krasnov
Kryl.	—	P.N. Krylov
Kuan	—	K.C. Kuan
Lad.	—	V.F. Ladygin
Lavr.	—	E.M. Lavrenko

Lee — A.R. Lee
Lips. — V.I. Lipsky
Litw. — D.I. Litwinow
Lom. — A.M. Lomonossov
Merzb. — G. Merzbacher
Mois. — V.S. Moiseenko
Muld. — A.A. Muldashev
Pal. — I.V. Palibin
Petr. — M.P. Petrov
Pias. — P.Ya. Piassezki
Pob. — E.G. Pobedimova
Pop. — M.G. Popov
Pot. — G.N. Potanin
Przew. — N.M. Przewalsky
Rachk. — E.I. Rachkovskaya
Reg. A. — A. Regel
Rob. — V.I. Roborowsky
Sap. — V.V. Sapozhnikov
Schischk. — B.K. Schischkin
Serp. — V.I. Serpukhov
Sold. — V.V. Soldatov
Sum. — I.Yu. Sumerina
Tug. — A.Ya. Tugarinov
Ulzij. — N. Ulzijkhutag
Volk. — E.A. Volkova (also known as E.A. Isachenko)
Wang — K.S. Wang
Yuan' — Yi-f. Yuan
Yun. — A.A. Yunatov
Zam. — B.M. Zamatkinov

Abbreviations of Names of Herbaria

AA — Herbarium of the Botanical Institute of the Academy of Sciences of Kazakhstan, Alma Ata
B — Botanisches Museum, Berlin-Dahlem
BM — British Museum of Natural History, London
BP — Departement Botanique du Musée d'Histoire Naturelle (Botanical Department of the Hungarian History Museum), Budapest
BR — Herbarium du Jardin botanique de l'État, Brussels, Belgium
CAL — Central National Herbarium, Botanical Survey of India, Calcutta (Kolkata)

E	—	Royal Botanic Garden, Edinburgh
G	—	Conservatoire et Jardin botanique, Genève (Geneva)
HIMC	—	Biology Department, University of Inner Mongolia, Hohhot
IFP	—	Institute of Forestry and Pedology, Academia Sinica, Shenyang, Liaoning
IMD	—	Institute of Medicinal Plant Development, Chinese Academy of Medical Sciences, Beijing
K	—	The Herbarium and Library, Royal Botanic Gardens, Kew, Richmond, Surrey, London
KW	—	Herbarium, M.G. Kholodny Institute of Botany, National Academy of Sciences of Ukraine, Kiev, Ukraina (Ukraine)
LE	—	Herbarium, V.L. Komarov Botanical Institute of the Russian Academy of Sciences, St.-Petersburg
Linn.	—	Herbarium, The Linnean Society of London, London
LIVU	—	City of Liverpool Museum, Liverpool
M	—	Botanische Staatssammlung, München (Munich)
MPU	—	Institut de Botanique, Université de Montpellier, France
MW	—	Herbarium of the Moscow State University, Moscow
NWFC	—	Dendrological Herbarium, Department of Forest Resources Protection, Northwest Forestry College, Yangling, Xianyang, Shaanxi
NWTC	—	Institute of Botany, Northwest Normal University, Lanzhou
O	—	Universitéts Botanisk Museum, Oslo
OXF	—	Fielding Herbarium, Druce Herbarium, Department of Botany, Oxford, Great Britain
P	—	Museum National d'Histoire Naturelle, Laboratoire de Phanerogamie, Paris
PE	—	Herbarium, Institute of Botany, Academia Sinica, Beijing
PR	—	Botanical Department of National Museum, Praha
S	—	Section for Botany, Swedish Museum of Natural History, Stockholm
TK	—	P.N. Krylov Herbarium of State University, Tomsk, Russia
UPS	—	Institute of Systematic Botany, University of Uppsala, Sweden
W	—	Naturhistorisches Museum, Botanische Abteilung, Wien (Vienna)
WU	—	Botanisches Institut und Botanischer Garten der Universität Wien (Vienna)
XJA	—	Xinjiang August 1st Agricultural College, Urumqi (Urumchi)
XJFA	—	Xinjiang Academy of Forestry Sciences, Urumqi (Urumchi)

Family 45. **NYMPHAEACEAE** Salisb.

1. Flowers yellow, with 5 sepals and many tiny, narrow petals ....1. **Nuphar** Smith.
+ Flowers white, with 4 sepals and many tiny large broad petals ....2. **Nymphaea** L.

1. **NUPHAR** Smith

in Sibth. et Smith, Fl. Graec. Prodr. 1 (1809) 361

1. Flowers 4-6 cm in diam.; stigma concave, with 10-20 rays not reaching up to stigma margin..............................1. **N. lutea** (L.) Smith.
+ Flowers 1.5-3 cm in diam.; stigma domed, with 8-10 rays running beyond stigma margin.................................2. **N. pumila** (Timm) DC.

1. **N. lutea** (L.) Smith in Sibth. et Smith, Fl. Graec. Prodr. 1 (1809) 361; Kryl. Fl. Zap. Sib. 5 (1931) 1110; Kom. in Fl. SSSR, 7 (1937) 13; Fl. Kazakhst. 4 (1961) 6; Fl. Sin. 27 (1979) 14; Claves pl. Xinjiang, 2 (1983) 267. —*Nymphaea lutea* L. Sp. pl. (1753) 510. —**Ic.**: Fl. Kazakhst. 4, Plate 1, fig. 3.

Described from Europe. Type in London (Linn.).

In oxbow-lakes.

IIA. Junggar: ***Cis-Alt.*** ("meander of Irtysh river"—Claves pl. Xinjiang, l.c.).

General distribution: Aralo-Casp.; Europe, Caucasus, West. Sib., East. Sib., Mediterr., Balk.-Asia Minor (Asia Minor), Caucasus, North America.

2. **N. pumila** (Timm) DC. Syst. 2 (1821) 61; Kryl. Fl. Zap. Sib. 5 (1931) 1111; Kom. in Fl. SSSR, 7 (1937) 13; Fl. Kazakhst. 4 (1961) 7; Abdullaeva in Opred. rast. Sr. Azil [Key to Plants of Mid. Asia] 3 (1972) 138; Fl. Sin. 27 (1979) 13, cum auct. bas. Hoffm.; Grub. Opred. rast. Mong. [Key to Plants of Mongolia] (1982) 106; Fl. Intramong. ed. 2, 2 (1990) 419; Claves pl. Xinjiang. 2 (1983) 268; Gub. Konsp. Fl. Vneshn. Mong. [Conspectus of Flora of Inner Mongolia] (1996) 47. —*Nymphaea lutea* L. var. *pumila* Timm, Mag. Naturk. Mecklenb. 2 (1792) 256. —**Ic.**: Fl. Kazakhst. 4, Plate 1, fig. 4; Fl. Sin. 27, tab. 5.

Described from Europe. Type in Geneva (G) ?

In lakes, oxbow-lakes, riverine backwaters, streams.

IA. Mongolia: ***Depr. Lakes*** ("Khara-Nur, Khara-Us-Nur"—Grub. l.c.).

IIA. Junggar: ***Cis-Alt.*** ("meander of Irtysh river"—Claves pl. Xinjiang, l.c.).

General distribution: Fore Balkh., Jung.-Tarb. (Zaisan basin); Europe, West. Sib., East. Sib., Far East, Nor. Mong. (Fore Hubs., Hang.), China (Altay, Dunbei, Nor. and others), Japan, North America.

2. **Nymphaea** L.

Sp. pl. (1753) 510

1. Flower cross-section 8-12 cm. Petals 15-25, gradually transforming into stamens. Stigma with 6-15 (20), more often 12-15, rays. ....1. **N. candida** C. Presl.

\+ Flower cross-section 4-5, rarely 9 cm. Petals 10-12, differing sharply from stamens. Stigma with 7-10 rays............2. **N. tetragona** Georgi.

1. **N. candida** C. Presl in J. et C. Presl, Del. Prag. (1822) 224; Kryl. Fl. Zap. Sib. 5 (1931) 1108; Kom. in Fl. SSSR, 7 (1937) 8; Fl. Kazakhst. 4 (1961) 5; Abdullaeva in Opred. rast. Sr. Azii [Key to Plants of Mid. Asia] 3 (1972) 137; Fl. Sin. 27 (1979) 9; Claves pl. Xinjiang, 2 (1983) 267; Gub. Konsp. Fl. Vneshn. Mong. [Conspectus of Flora of Outer Mongolia] (1996) 47. —**Ic.**: Fl. SSSR, 7, Plate 1, fig. 2: Fl. Kazakhst. 4, Plate 1, fig. 1.

Described from Austria. Type in Praha (PR).

In lakes and riverine backwaters.

IA. Mongolia: ***Depr. Lakes*** (Gub. l.c.).

IB. Kashgar: ***Nor.*** (Kuruk-tag mountain range, Kurlya river, near Bash-Akin picket, in water along fringe of rapid stream, Aug. 24, 1929—Pop.), ***East.*** (south. slope of Karashar basin, south-west portion of Bagrashkul' lake, Aug. 15, 1958—E.M. Muratov and N.T. Kuznetsov).

IIA. Junggar: ***Cis-Alt.*** ("meander of Irtysh river"—Claves pl. Xinjiang, l.c.), ***Tien Shan*** (Savan district, from Paotai to Shaomyn'tsza, along Manas river, No. 1592, June 29, 1957—Kuan).

General distribution: Fore Balkh., Jung.-Tarb., Nor. and East. Tien Shan; Europe, West. Sib., East. Sib., Caucasus (West. Transcaucasus), Mid. Asia (Kyzylkums, deltas of Syr Darya and Amu Darya rivers), Nor. Mong. (Hang.).

2. **N. tetragona** Georgi, Bemerk. Reise Russ. Reich. 1 (1775) 220; Forbes and Hemsl. Index Fl. Sin. 1 (1886) 33; Kryl. Fl. Zap. Sib. 5 (1931) 1109; Kom. in Fl. SSSR, 7 (1937) 11; Fl. Kazakhst. 4 (1961) 6; Abdullaeva in Opred. rast. Sr. Azii [Key to Plants of Mid. Asia] 3 (1972) 138; Fl. Sin. 27 (1979) 9; Grub. Opred. rast. Mong. [Key to Plants of Mongolia] (1982) 106; Claves pl. Xinjiang, 2 (1983) 267; Fl. Xizang. 2 (1985) 2; Fl. Intramong. ed. 2, 2 (1990) 417; Gub. Konsp. Fl. Vneshn. Mong. [Conspectus of Flora of Outer Mongolia] (1996) 47. —**Ic.**: Fl. Kazakhst. 4, Plate 1, fig. 1; Fl. SSSR, 7, Plate 1, fig. 1; Fl. Intramong. ed. 2, 2, tab. 186, fig. 1-3; Fl. Sin. 27, tab. 4.

Described from East. Siberia. Type not found.

In lakes, oxbow-lakes and riverine backwaters.

IA. Mongolia: ***Depr. Lakes*** ("Khara- Nur, Khara-Us-Nur"—Grub. l.c.).

IIA. Junggar: ***Cis-Alt.*** ("meander of Irtysh river"—Claves pl. Xinjiang, l.c.).

General distribution: Fore Balkh., Jung.-Tarb. (Zaisan basin); Europe, West. Sib., East. Sib., Far East, Nor. Mong. (Fore Hubs., Hang.), China, Korean peninsula, Japan, Indo-Mal., Nor. Amer.

Family 46. **CERATOPHYLLACEAE** S.F. Gray

1. **Ceratophyllum** L.

Sp. pl. (1753) 992; id. Gen. pl. ed. 5 (1754) 428

1. **C. demersum** L. Sp. pl. (1753) 992; Kryl. Fl. Zap. Sib. 5 (1931) 1113; Kuzeneva in Fl. SSSR, 7 (1937) 15; Fl. Kirgiz. 6 (1955) 12; Fl. Kazakhst. 4 (1961) 7; Ikonnik. Opred. rast. Pamira [Key to Plants of Pamir] (1963) 111; Abdullaeva in Opred. rast. Sr. Azii [Key to Plants of Mid. Asia] 3 (1972) 138; Fl. Sin. 27 (1979) 16; Grub. Opred. rast. Mong. [Key to Plants of Mongolia] (1982) 106; Claves pl. Xinjiang, 2 (1983) 268; Fl. Xizang. 2 (1985) 2; Fl. Intramong. ed. 2, 2 (1990) 420; Gub. Konsp. Fl. Vneshn. Mong. [Conspectus of Flora of Outer Mongolia) (1996) 47. —**Ic.:** Fl. SSSR, 7, Plate 2, fig. 2; Fl. Sin. 27, fig. 6; Fl. Intramong. ed. 2, 2, tab. 167, fig. 4-6.

Described from Europe. Type in London (Linn.).

In lakes, oxbow-lakes, riverine backwaters.

IA. Mongolia: ***Depr. Lakes*** ("Khara-Us-Nur"—Grub. l.c.), ***Alash. Gobi*** (Bayan-Khoto, Tengeri sand, July 26, 1958—Petr.).

IIA. Junggar: ***Jung. Gobi*** (east.: within Mongolia—Gub. l.c., "Shache"—Claves pl. Xinjiang, l.c.).

General distribution: Aralo-Casp., Fore Balkh., Jung.-Tarb., Nor. and Cen. Tien Shan; Europe, Mediterr., Mid. Asia, West. Sib., East. Sib., Far East., Nor. Mong. (Mong.-Daur.), China.

Family 47. **RANUNCULACEAE** Juss.

1. Leaves simple, palmatisected with dichotomous nervation. Fruit, 1-seeded free follicle, single or 2-4 together, covered with uncinate bristles. Highly reduced plant .... 15. **Circaeaster** Maxim. (*C. agrestis* Maxim.).
+ Characteristics different .... 2.
2. Flowers irregular, symmetrical only in one vertical plane, with coloured corolliform, 5-membered perianth .... 3.
+ Flowers regular, actinomorphic, with 2 or more planes of symmetry .... 4.
3. Upper unpaired perianth lobe (sepal) larger than rest and shaped like hood of various forms: falcate, carinate, conical or terete. Nectaries (modified petals) tiny, cowl-like with long claw, largely or completely concealed inside hood, invariably 2 .... 6. **Aconitum** L.
+ Upper unpaired perianth lobe (sepal) large, flat and with hollow spur at base. Nectaries large, open, with blade transforming into

long tubular spur, enclosed in spur of upper perianth lobe........
....7. **Delphinium** L.

4. Leaves opposite, entire, pinnatisected or ternate. Fruit, tiny nut with long plumose arista. Shrubs with erect, procumbent or climbing stems, rarely large erect herbaceous perennials............5.

\+ Leaves alternate; if opposite, plant small, with dichotomously branching stem (*Anemone dichotoma*); if fruit, nutlike with long arista and small plants with radical leaves (*Pulsatilla*). Grasses. ....6.

5. Perianth simple, corolliform, with 4-8 lobes of identical form and colour...13. **Clematis** L.

\+ Perianth double, 4-membered, outer perianth lobes (sepals) differing sharply in form (broader) and generally in colour from inner specialised petaloid staminodes; latter sometimes highly reduced and then sepals petaloid............................14. **Atragene** L.

6. Flowers small, unattractive, with quickly falling green or white perianth lobes, in terminal simple racemose or paniculate inflorescences. Stamens long, considerably longer than perianth. Leaves with 2-4-pinnatisected or ternate blade...............................7.

\+ Flowers relatively large, with brightly coloured perianth and mostly single, rarely in umbelliform or loose corymbose-paniculate inflorescence. Stamens shorter than perianth, non-secreting.......9.

7. Flowers white, in simple oval cluster. Pistil single. Fruit, a berry. ....5. **Actaea** L. (*A. acuminata* Wall. ex Royle).

\+ Flowers yellow-green or canescent, in paniculate, rarely simple long clusters. Pistils 2 or more...8.

8. Flowers with petaloid staminodes. Fruit, many-seeded follicles, 2-4. Leaves 2-3-ternate; leaf lobes dentate all along margin ....4. **Cimicifuga** Wernisch.

\+ Flowers without staminodes. Fruits 3-15, tiny ribbed nuts. Leaves pinnati- or trisected, with fairly distinct ochrea at base; leaf lobes dentate only in upper part....................................25. **Thalictrum** L.

9. Aquatic submerged plants with leaves 2-4-palmatisected into filiform lobes. Flowers white, 0.5-1 cm in diam. Transversely rugose small nuts..19. **Batrachium** (DC.) S.F. Gray.

\+ Land plants, rarely (some *Ranunculus*) aquatic but then with floating stems and leaves. Leaf lobes not filiform........................10.

10. Fruit with tiny nutlike 1-seeded follicle...11.

\+ Fruit with many-seeded follicles (sometimes follicle single) ....19.

11. Perianth simple, all its pinnae of identical or almost identical shape and colour..12.

+ Perianth binary, outer perianth pinnae (sepals) differ sharply in shape, and generally in colour, from inner (nectary petals), sometimes greatly reduced and then sepals petaloid..................14.

12. Radical leaves tripartite, cauline (pinnae of spathe) highly proximated to flower and sepal-like... ....12. **Hepatica** Mill. (*H. falconeri* (Thoms.) Steward).

+ Radical leaves tri- or palmatisected, cauline (pinnae of spathe) far closer to flower or inflorescence..12.

13. Carpels (and pistils) with long plumose arista (style). Spathe multipartite, connate at base..................................11. **Pulsatilla** Mill.

+ Carpels (and pistils) with short beak (style), shorter than carpel itself. Spathe comprising 3 whorled-proximated and mostly pentiolate leaves, rarely bifoliate.............................10. **Anemone** L.

14. Plants with radical, entire or lobate-incised leaves and leafless peduncles..15.

+ Plants with developed leafy stems. Leaves mostly palmati- or pinnatipartite, rarely entire...17.

15. Sepals coriaceous, persisting on fruits. Leaves with entire, undivided blade. High-altitude plant.. ....17. **Oxygraphis** Bge. (*O. glacialis* (Fisch. ex DC.) Bge.).

+ Sepals thin, herbaceous, deciduous. Plant of plains and low-mountain regions...16.

16. Plants with creeping, rooting shoots, growing in damp solonchak soils. Nuts with oblong exserted nerves. Leaves with entire or lobate blade...16. **Halerpestes** Greene.

+ Tiny plant without rooting shoots growing in arid soils in steppes and deserts. Nuts without exserted nerves, with 2 hollow tubercles and elongated corniculate acute beak at base. Leaves palmati-tripartite into entire or 2-3 lobate segments............................ ....20. **Ceratocephala** Moench (*C. testiculata* (Crantz) Bess.).

17. Leaves entire, ternate, palmate or, rarely, pinnatipartite or laciniated. Flowers golden yellow, invariably with 5-membered perianth...18. **Ranunculus** L.

+ Leaves 2-3-pinnatisected. Flowers with many petals (up to 20, rarely 5), white, yellow, light blue or red.......................................18.

18. Leaf lobes orbicular or ovoid, blunt. Petals with nectar cavity at base, white. Nuts glabrous................8. **Callianthemum** C.A. Mey.

+ Leaf lobes linear, cuspidate. Petals without nectar cavity, yellow, red, light blue or white. Nuts tomentose, rarely (in year-old plants) glabrous..9. **Adonis** L.

19 (10). Leaves entire, orbicular or reniform. Perianth simple, yellow or white...2. **Caltha** L.

+ Leaves laciniated. Perianth double..20.

20. Leaves 2-3-trisected..21.
+ Leaves 5-palmatisected, orbicular. Perianth orange, yellow or bluish. Nectary petals narrow, liguliform........................3. **Trollius** L.
21. Nectary petals prominent, with spur turned downward from flower base between coloured sepals; rarely, spur poorly developed and instead with rather small depression..............24. **Aquilegia** L.
+ Nectary petals without spur, tiny, poorly visible or with nectar plate..22.
22. Nectar plate present. Flowers large, 8-15 cm in diam., white, pink or red, with green leaf-shaped sepals. Follicles 2-5, large, with stout rough walls. Large plant, 40-80 cm tall................1. **Paeonia** L.
+ Nectary petals small, poorly visible, 1/3-1/2 of sepals, without spur, funnel-shaped, tubular or concave toward base.................23.
23. Annual, loose-branched plant. Follicles 10 or more (up to 20), free, narrow-lanceolate. Flowers small, 0.5-1 cm in diam., light yellow. ....22. **Leptopyrum** Reichb.
+ Perennial plant..24.
24. Plants compactly caespitose, stems with contracted internodes, rigid, covered in lower part with many amplexicaulous remnants of year-old leaf stalks. Peduncles without leaves, only with 2 bracts, single-flowered. Flowers 1.5-3 cm in diam. Petals light blue, tubular only at base, sessile. Follicles 4-5, mature ones largely erect, coriaceous.....................23. **Paraquilegia** J. Drumm. et Hutch.
+ Plants lax; stems single or few, thin, deflecting, with long internodes, arising from thin horizontal rhizome and sparsely covered in lower part with remnants of year-old leaves. Peduncles leafy in upper part, with 1-3 flowers. Flowers 1.2-1.6 cm in diam. Petals white, tubular for 2/3 of their length from base, short-petiolate. Follicles 2-5, deflected in mature ones, leaf-shaped......... ....21. **Paropyrum** Ulbr. (*P. anemonoides* (Kar. et Kir.) Ulbr.).

Subfamily 1. PAEONIOIDEAE Hutch.

1. **Paeonia** L.

Sp. pl. (1753) 530

1. Leaves ternate or biternate (radical leaves 3-ternate). Leaf lobes entire, broad, sometimes connate at base, with very fine chondroid crenations along margin. Petals white. Nectar plate perceptible, about 5 mm long..3. **P. lactiflora** Pall.
+ Leaf lobes laciniated, entire. Petals pink or purple; nectar plate not more than 2 mm long or almost altogether imperceptible..........2.

2. Flowers terminal, single, large, erect, purple. Follicles glabrous or pubescent..3.
+ Flowers on stem generally few (2-4); flowers very small, nutant, pink. Follicles compactly tomentose....................4. **P. veitchii** Lynch.
3. Plants large, 40-100 cm tall. Lateral leaf lobes on first branch subsessile; only midlobe petiolate. Leaf lobes lanceolate, up to 2.5 cm broad, long-cuspidate at tip. Flowers large, 8-18 cm in diam. Follicles (3) 4-5, glabrous, rarely pilose. Seeds black, with underground growth..1. **P. anomala** L.
+ Plants less large, 20-70 cm tall. Lateral leaf lobes on first branch, as well as midlobe, long-petiolate. Leaf lobes linear or linear-lanceolate, up to 1-1.5 cm broad, generally subobtuse at tip. Flowers less large, 8-12 cm in diam. Follicles 2-3 (4), compactly pubescent, soft with yellow pubescence. Seeds generally light cinnamon or rarely black, with surface growth..............................2. **P. hybrida** Pall.

1. **P. anomala** L. Mant. 2 (1771) 247; Maxim. Fl. Tangut. 1 (1889) 28; id. Enum. pl. Mong. 1 (1889) 31; Huth in Bot. Jahrb. 14 (1892) 267; Stern, Stud. Paeonia (1946) 112; Kryl. Fl. Zap. Sib. 5 (1931) 1116; Schipcz. in Fl. SSSR, 7 (1937) 33; Fl. Kazakhst. 4 (1961) 12; Pakhom. in Opred. rast. Sr. Asii [Key to Plants of Mid. Asia] 3 (1972) 142; Fl. Sin. 27 (1979) 59; Grub. Opred. rast. Mong. [Key to Plants of Mongolia] (1982) 107; Claves pl. Xinjiang, 2 (1983) 273; Gub. Konsp. Fl. Vneshn. Mong. [Conspectus of Flora of Outer Mongolia] (1996) 47. —**Ic.**: Fl. Kazakhst. 4, Plate 2, fig. 11.
Described from Siberia. Type in London (Linn.).
In larch and mixed forests, forest borders, up to upper forest boundary.

IA. Mongolia: ***Khobd.*** (Shur-Ulyastai-Gol river valley, forb-cereal grass larch forest, July 12, 1973—Banzragch et al.), ***Mong. Alt.*** (Gub. l.c.).

IIA. Junggar: ***Cis-Alt.*** (Kandagatai river, among bushes, Sept. 13, 1876—Pot.; upper courses of Kurtu river, June 20, 1903—Gr.-Grzh.; Koktogoi, No. 1803, Aug. 13, 1956—Ching; 23-50 km north of Koktogoi, right bank of Kairta river, Kuidyn valley, forest belt, on southern steppe rocky slope, July 15, 1959, Yun. et al.), ***Tarb.*** (Saur mountain range, southern slope, Karagaitu river valley, right bank of Bain-Tsagan creek valley, larch forest, June 23, 1957—Yun. et al.; Dachen, No. 1629, Aug. 13, 1957—Kuan), ***Jung. Alat.*** (in Toli district, No. 1047, 1283 and 1297, Aug. 6 and 7, 1957—Kuan).

General distribution: Jung.-Tarb. (Saur); Europe (east. East Europe), West. Sib., East. Sib., Nor. Mong. (Fore Hubs., Hent., Hang., Mong.-Daur.), China (North).

Note. Plants with pubescent as well as glabrous fruits are found in Junggar.

2. **P. hybrida** Pall. Fl. Ross. 1, 2 (1788) 94; Kryl. Fl. Zap. Sib. 5 (1931) 1117; Schipcz. in Fl. SSSR, 7 (1937) 34; Fl. Kazakhst. 4 (1961) 13; Pakhom. in Opred. rast. Sr. Azii [Key to Plants of Mid. Asia] 3 (1972) 142; Claves

pl. Xinjiang, 2 (1983) 273. —*P. intermedia* C.A. Mey. in Ledeb. Fl. alt. 2 (1830) 277; Fl. Kirgiz. 6 (1955) 15; Kamelin et al. in Bot. zhurn. 76, 4 (1991) 614; Gub. Konsp. Fl. Vneshn. Mong. [Conspectus of Flora of Outer Mongolia] (1996) 47. —*P. anomala* L. var. *hybrida* (Pall.) Trautv. Bull. Soc. natur. Moscou, 33 (1886) 88; Huth in Bot. Jahrb. 14 (1892) 269. —*P. anomala* L. var. *intermedia* (C.A. Mey) O. et B. Fedtsch. Beih. Bot. Centralbl. 18, 2 (1905) 216; Stern, Stud. Paeonia (1946) 113; Fl. Sin. 27 (1979) 59. —*P. hybrida* L. var. *intermedia* Kryl. Fl. Sib. Occid. 5 (1931) 1117. —*P. anomala* auct. non L. : Maxim. Enum. pl. Mong. 1 (1889) 31, p.p. —**Ic.**: Pallas, tab. 86; Fl. Kazakhst. 4, Plate 2, fig. 12; Fl. Sin. 27, tab. 10.

Described from West. Siberia (vicinity of Barnaul). Type in London (BM).

On exposed mountain slopes, talus margins, forest glades, scrubs, among cereal-forbs steppe vegetation and tall grasses, on steppes, subalpine and alpine meadows and among neve basins.

IA. Mongolia: ***Mong. Alt.*** (Dzhelta river basin, Kurumty-Gol valley, near border, Aug. 9, 1979—Gub. (MW); upper course of Chernyi Irtysh, Sanginin-Gol river, near border with People's Republic of China, 25 km south-east of Dayan-Nurgate, 1800-2100 m, July 16-17; Ėlt-Gol river (upper course of Chernyi Irtysh), 30 km south of Altay settlement, July 22—1988, Kam., Bud., Gub. et al.).

IIA. Junggar: ***Cis-Alt.*** (Altay, No. 2443, Aug. 26, 1956—Ching; 20 km north-west of Shara-Sume, on Kran river, bushy meadow steppe, July 7, 1959—Yun. et al.; in Altay mountains, 1500 m, No. 10567, July 16, 1959—A.R. Lee et al.; Sanginin-Gol river basin (upper course of Chernyi Irtysh), 25 km south-east of Dayan-Nur gate, July 16-17, 1988—Kam., Bud., Gub. et al.), ***Tarb.*** (north-west of Khobsair (= Khobuk-saira) (Obo mountain), 2000 m, No. 10563, June 22, 1959—A.R. Lee et al. (*Paeonia anomala* × *P. hybrida*); around Dachen, No. 1545 and 1620, Aug. 12 and 13, 1957—Kuan), ***Jung. Alat.*** (in Toli region, on slope, No. 2495 and 2611, Aug. 4 and 7, 1957—Kuan), ***Tien Shan*** (Tien Shan [without reference to exact location of find], July 16-19—1876, Pevtsov; on Sodzhan river, around Narin-kara, June 13; at Tsan-ma river sources, 2100 m, in forest, June 20, 1877—Przew.; Dzhagastai, Aug. 8, 1877; Sarybulak on Suidun river, 1200-1800 m, April 24; Almaty gorge around Kul'dzha, 1200-1500 m, May 26—1878, A. Reg.; Burkhantau, June 5, 1878—Fet.; jugi montium Alatau dschungaricus, bei Manass, July 5-17, 1908, Merzb.; "Fukan, Urumchi"—Claves pl. Xinjiang, l.c.).

General distribution: Jung.-Tarb., Nor. and Cent. Tien Shan; Mid. Asia, West. Sib., China (Altay).

3. **P. lactiflora** Pall. Reise, 3 (1776) 286; Stern, Stud. Paeonia (1946) 91; Fl. Sin. 27 (1979) 51; Grub. Opred. rast. Mong. [Key to Plants of Mongolia] (1982) 107; Claves pl. Xinjiang, 2 (1983) 273; Fl. desert Sin. 1 (1985) 507; Fl. Intramong. ed. 2, 2 (1990) 574. —*P. albiflora* Pall. Fl. Ross. 1, 2 (1788) 92; Hance in J. Bot. (London) 20 (1882) 257; Franch. Pl. David. 1 (1884) 23; Forbes and Hemsl. Index Fl. Sin. 1 (1886) 21; Maxim. Enum. pl. Mong. 1 (1889) 31; Huth in Bot. Jahrb. 14 (1892) 265; Finet et Gagnep. in Bull. Soc. Bot. Fr. 51 (1904) 526; Danguy in Bull. Mus. Hist. Nat. (Paris) 19, 7 (1913) 5; Schipcz. in Fl. SSSR, 7 (1937) 25; Kitag. Lin. Fl. Mansh. (1939) 220; Grub. Konsp. Fl. MNR [Conspectus of Flora of People's Republic of Mongolia]

(1955) 135; Fang in Acta Phytotax. Sin. 7 (1958) 320; Gub. Konsp. Fl. Vneshn. Mong. [Conspectus of Flora of Outer Mongolia] (1996) 47. —**Ic.**: Pallas, 1788, tab. 84 (sub *P. albiflora*); Fl. Sin. 27, tab. 7; Grub. Plate 47, fig. 219; Fl. Intramong. ed. 2, 2, tab. 234, fig. 1-5.

Described from East. Siberia (Transbaikal). Type in St.-Petersburg (LE).

In meadow steppes, steppes and forb meadows, sparse scrubs.

IA. Mongolia: ***East. Mong.*** (Muni-Ula mountains, June 25, 1871—Przew.; "Khailar" —Danguy, l.c.; nor. slope of Gurban-Chilotai mountains, June 25; Tsagan-Nuru mountains, June 27—1925, V. Kazakevich; Khuntu somon, 35 km east-south-east of Basin-Tsagan somon, Aug. 6, 1949, Yun.; left bank of Khalkhin-Gol river in Derkhin-Tsagan-Obo conical hillock (mud cone) region, Aug. 16, 1970, Grub. et al.).

General distribution: East. Sib. (Transbaikal), Far East (south.), Nor. Mongolia (Mong.-Daur., Cis-Hing.), China (Dunbei, Nor., Nor.-West., Cent., South-West.), Korean peninsula, Japan.

4. **P. veitchii** Lynch in Gard. Chron., ser. 3, 46 (1909) 2; Stern, Stud. Paeonia (1946) 115; Fang in Acta Phytotax. Sin. 7, 4 (1958) 321; Fl. Sin. 27 (1979) 56; Fl. Xizang. 2 (1985) 8. —*P. beresowskii* Kom. in Not. Syst. (Petrop.) 2 (1921) 5. —*P. veitchii* var. *beresowskii* (Kom.) Schipcz. in Not. Syst. (Petrop.) 2 (1921) 46. —*P. anomala* auct. non L.: Maxim. Fl. Tangut. 1 (1889) 28; id. Enum. pl. Mong. 1 (1889) 31. —**Ic.**: Lynch, l.c. fig 1; Stern, l.c. pl. 63, 2; Fl. Sin. 27, tab. 9.

Described from South-West. China. Type as depicted in plate at first description. Plate 1, fig. 1.

In margins, among scrubs and in mountain forests at 2800-4200 m alt.

IA. Mongolia: ***Ordos*** (Ordos, in midcourse of Huang He river, May; Huang He river valley, on rocks, on way to Gashun-Nor lake, May 27—1872, Przew.).

IIIA. Qinghai: ***Nanshan*** (nor. slope toward Tetung river, in river valley, July 4, 1872; on Yusup-Khatyma river, 2700-3000 m, among bushes of forest belt, July 24, 1880—Przew.; vicinity of Dan'ger and Sin'chen towns, 2100-2700 m, Aug. 22, 1901—Lad.; Xining town, 1908—Czet.).

IIIB. Tibet: ***South.*** ("Tszyanda"—Fl. Xizang. l.c.).

General distribution: China (Nor., North-West., South-West.).

Note. This species was described from plants grown from seeds collected in western Sichuan province. It was wrongly attributed to *P. anomala* although it differs from the latter in fewer flowers on stems, compactly tomentose-pubescent fruits, lower and middle cauline leaves on long petioles measuring 1/3-1/2 of leaf length, thus creating impression of less compact plant foliage. The author also points to the following characteristics in his first description: leaves with light green lustrous surface on both sides and distinct exsertion between nerves. *P. veitchii* is the lone species of Central Asian peonies growing in mountains at significant altitudes.

Subfamily 2. **HELLEBOROIDEAE** Hutch.

Tribe 1. **Helleboreae** DC.

2. **Caltha** L.

Sp. pl. (1753) 558

1. Aquatic plants with stems submerged in or ascending from water. Flowers white, small, 8-12 mm in diam. Follicles several, in globose head..2. **C. natans** Pall. ex Georgi.
\+ Marsh or meadow-marsh plants. Flowers yellow, larger, 15-30 mm in diam. Follicles 8-12..2.
2. Flowering shoot without leaves, single-flowered (flower 15-20 mm in diam.), follicles petiolate. Several radical leaves gathered in rosette. Leaf blades 0.5-2 cm broad, compact, usually remaining green on drying, with incumbent basal lobes. High-altitude plant. ....3. **C. scaposa** Hook. f. et Thoms.
\+ Shoot branching 1-2 times and terminating in 2 flowers of 1.8-3 mm in diam. Follicles sessile. Radical rosette not manifest. Leaf blades 5-6 cm broad, qutie thin, darkening on drying, with separated basal lobes.......................1. **C. membranacea** (Turcz.) Schipcz.

1. **C. membranacea** (Turcz.) Schipcz. in Not. Syst. (Petrop.) 2 (1921) 168; Schipcz. in Fl. SSSR, 7 (1937) 41; Kitag. Lin. Fl. Mansh. (1939) 215; Gub. Konsp. Fl. Vneshn. Mong. [Conspectus of Flora of Outer Mongolia] (1996) 49. —*C. palustris* var. *membranacea* Turcz. in Bull. Soc. natur. Moscou, 15 (1842) 62; Fl. Sin. 27 (1979) 63; Fl. Intramong. ed. 2, 2 (1990) 426. —*C. palustris* var. *sibirica* Regel, Pl. Radd. 1 (1861) 53; Maxim. Enum. pl. Mong. 1 (1889) 24; Fl. desert Sin. 1 (1985) 508; Fl. Intramong. ed. 2, l.c.: 426. —*C. palustris* auct. non L.: Smit in Blumea, 21, 1 (1973) 144, p.p.; Fl. Sin. 27 (1979) 61, p.p.; Grub. Opred. rast. Mong. [Key to Plants of Mongolia] (1982) 108 (for Depr. Lakes); Claves pl. Xinjiang, 2 (1983) 274; Fl. desert Sin. l.c. 507; Fl. Xizang. 2 (1985) 8; Gub. l.c. (for Depr. Lakes). —**Ic.** Fl. SSSR, 7, Plate 4, fig. 2; Fl. Sin. 27, tab. 11, fig. 6; Fl. Intramong. ed. 2, 2, fig. 171, 2-3 (sub *C. palustris* var. *sibirica*).

Described from East. Siberia. Type in St.-Petersburg (LE).

In marshy meadows.

IA. Mongolia: ***East. Mong.*** (Khailar town, Nanlin'tun' village, No. 642, June 10, 1951—Ch. Wang; "Khulunbuir, Khailar"—Fl. Intramong. ed. 2, l.c.), ***Depr. Lakes*** (vicinity of Ubsu-Nur lake, Ulangom, June 28, 1979—Pot.).

General distribution: East. Sib., Far East, Nor. Mong., China (Dunbei, Nor.-West.).

Note. This species substitutes *C. sibirica* Tolm. described from Sakhalin and extensively distributed in Far East and Japan. *C. membranacea* and

C. sibirica differ from *C. palustris* L. s. str. in much shorter stature, weak stalk with fewer branches, much smaller flowers and thin membranous leaf blade.

2. **C. natans** Pall. ex Georgi, Bemerk. Reise Russ. Reich. 1 (1755) 222; Kryl. Fl. Zap. Sib. 5 (1931) 1119; Schipcz. in Fl. SSSR, 7 (1937) 36; Kitag. Lin. Fl. Mansh. (1939) 215; Smit in Blumea, 21, 1 (1973) 144; Fl. Sin. 27 (1979) 67; Fl. Intramong. ed. 2, 2 (1990) 424; Gub. Konsp. Fl. Vneshn. Mong. [Conspectus of Flora of Outer Mongolia] (1996) 49. —*Thacla natans* (Pall. ex Georgi) Deyl et Sojak in Acta Mus. Nat. Pragae, ser. B, 26 (1970) 31. —**Ic.**: Fl. Sin. 27, tab. 12, fig. 6-8; Grub. Opred. rast. Mong. [Key to Plants of Mongolia] (1982) Plate 47, fig. 220; Fl. Intramong. ed. 2, 2, tab. 170.

Described from East. Siberia. Type in London (BM).

In meanders, shallow water with nearly stagnant water.

IA. Mongolia: ***Cent. Khalkha*** (in Kerulen bend, 1200-1500 m, Aug. 6-12, 1906—Novitskii).

General distribution: West. and East. Sib., Far East, Nor. Mong., China (Dunbei), Korean peninsula, Nor. America.

3. C. scaposa Hook. f. et Thoms. Fl. Ind. 1 (1855) 40; Huth in Helios, 9 (1892) 101; Smit in Blumea, 21, 1 (1973) 144; Fl. Sin. 27 (1979) 65; Fl. Xizang. 2 (1985) 9. —*C. palustris* L. var. *scaposa* Maxim. Fl. Tangut. 1 (1889) 16; Finet et Gagnep. in Bull. Soc. Bot. Fr. 51 (1904) 390. —*C. scaposa* Hook. f. et. Thoms. var. *pusilla* Huth. l.c. 101. —**Ic.**: Huth, tab. 1, fig. 14; Fl. Sin. 27, tab. 12, fig. 1-5; Fl. Xizang. 2, tab. 2, fig. 1-3.

Described from Himalayas (Sikkim). Type in London (K). Isotypes in Uppsala (UPS) and St.-Petersburg (LE). Plate II, fig. 4.

In small marshy meadows along banks of brooks and around springs in alpine belt, 3200-6000 m alt.

IIIA. Qinghai: ***Amdo*** (Dzhakhar-Dzhargyn mountains, 3200-3500 m, June 24, 1880—Przew.).

IIIB. Tibet: ***Weitzan*** ("along nor. slope of waterdividing mountain range between Huang He and Yangtze rivers, in meadow around brook..., June 12, Yangtze river, July 11"; Nor. Tibet [without date or location of find]—1884, Przew.; south. slope of Amnen-Kor mountain range, 4000-4300 m alt., June 7—1900, Lad.; Yantszytszyan (Goluboi) river, Gudochoki area on Gonchyu river, 3900 m, May 2-3; Senitan area on left bank of Yaluntszyan, tributary of Goluboi river, 3700 m, May 5—1901, Lad.), ***South.*** ("Lhasa, ? Batin, Dintsze"—Fl. Xizang. l.c.).

General distribution: China (Nor.-West., South-West.), Himalayas (west., east.).

Note. Small plants of var. *pusilla* Huth with 3-5 cm tall peduncles, found mainly in Northern Tibet, differ from much larger plants of Nanshan, South-West. China and Himalayas.

3. **Trollius** L.

Sp. pl. (1753) 556

1. Sepals pale-lilac, dingy, persisting on fruits........6. **T. lilacinus** Bge.
+ Sepals yellow or orange, falling before fruiting.................................2.
2. Sepals 5, rarely up to 7, declinate. Nectary petals nearly 1/2 of large prominent sepals. Stem scapiform, usually only with single leaf at base; flowering plants from a few to 20 cm tall; 10-20 cm, or sometimes up to 30 cm tall at fruiting..3.
+ Sepals 11 or 15-20. Stem distinctly leafy; plant often more than 30 cm tall..5.
3. Sepals deciduous, green even when dry. Leaf lobes highly laciniated, extreme lobes quite short, whitish on underside of leaf blade. Flowers 2.5-4.5 cm in diam. Petals somewhat shorter than stamens, spatulate......................................8. **T. ranunculoides** Hemsl.
+ Sepals persisting, turning red or brown, often denticulate along margin of rounded tip..4.
4. Leaf lobes highly emarginate-dentate, teeth mostly very sharp. Flowers 2-3.5 cm in diam.; petals shorter than stamens, more or less cuneate..7. **T. pumilus** D. Don.
+ Leaf lobes more or less short-trilobate with some short broad teeth. Stems mostly very slender downward. Flowers 2-3 cm in diam.; petals shorter than stamens, spatulate-linear......5. **T. farreri** Stapf.
5. Nectary-petals roughly 1.5-2 times longer than sepals and 2-3 times longer than stamens, up to 2.5 cm long, gradually cuspidate upward...3. **T. chinensis** Bge.
+ Nectary-petals not longer than sepals and, if somewhat longer, not cuspidate but rounded at tip...6.
6. Nectary-petals 1.5-2 times longer than stamens and only slightly shorter (occasionally longer) than sepals, nectar cavity located 2-3 mm above base of petals. Style and stigma yellowish..2. **T. asiaticus** L.
+ Nectary-petals as long as stamens or shorter and considerably shorter than sepals, nectar cavity located at about 1 mm height.....7.
7. Style (and stigma) purple, if rugose or subglabrous, somewhat curved at tip. Nectary petals (8) 9-12 mm long, linear, tapering rather often from middle toward tip, longer than pistils or rarely as long. Follicles 12-16 mm long, beak 2.5-3 mm long. Leaf segments narrow-rhomboid, rhomboid or narrow-ovoid, gradually tapering toward tip; lateral lobes and teeth generally narrow, acute or subacute at tip...1. **T. altaicus** C.A. Mey.
+ Style (and stigma) yellowish, with fairly intensely rugose lateral protuberances, erect or curved. Nectary petals (5) 6-8 mm long,

not tapering, spatulate-orbicular at tip, quite often with involuted lateral margins, mostly shorter than pistils. Follicles 10-11 (12) mm long, beak 2 mm long. Leaf segments obovoid, broad-rhomboid or suborbicular, rather shallowly (sometimes only barely) divided into fairly broad lobes, more or less short blunt-toothed on margin..**4. T. dschungaricus** Regel.

1. **T. altaicus** C.A. Mey. in Verz. Pfl. Cauc. (1831) 200, in adnot.; Huth in Helios, 9 (1892) 9; Kryl. Fl. Zap. Sib. 5 (1931) 1122; Schipcz. in Fl. SSSR, 7 (1937) 45; Fl. Kirgiz. 6 (1955) 19; Fl. Kazakhst. 4 (1961) 14; Pakhom. in Opred. rast. Sr. Azii [Key to Plants of Mid. Asia] 3 (1972) 144; Fl. Sin. 27 (1979) 83; Grub. Opred. rast. Mong. [Key to Plants of Mongolia] (1982) 108; Claves pl. Xinjiang, 2 (1983) 275; Fl. Intramong. ed. 2, 2 (1990) 429; Gub. Konsp. Fl. Vneshn. Mong. [Conspectus of Flora of Outer Mongolia] (1996) 51. —**Ic.**: Fl. Kazakhst. 4, Plate 2, fig. 10.

Described from Altay. Type in St.-Petersburg (LE).

In forb meadows, on banks of brooks and ponds, borders of larch forests, terraces and ledges of rocks in alpine belt.

IA. Mongolia: ***Khobd.*** (Gub. l.c.), ***Mong. Alt.*** (Dzhavankul' lake, June 26, 1869—Malevskii; bank of Khara-Nur lake (upper Bulugun river, Sept. 4, 1899—Lad.; sources of Bel. Kobdo river, alpine tundra, June 29, 1906—Sap.; upper Bulugun river, Koltin-Gol valley, on south. slope of Shara-Khamrin-Orai mountain range, 3 km above Kudzhurtu settlement, larch forest on steep northern slope, July 3; upper Buyantu river, Chichirtei-Gol, northern slope of Chichirtei-Ul (opposite Chishirgii-Ul), 2600-2800 m, young larch grove along creek valley with brook of neve basin, July 4; upper Kobdo river, Donyagiin-Khara-Nur lake, east. bank at nor. extremity, on Dayan-Nur road, 2500 m, meadow steppe on west. slope, July 7—1971, Grub. Ulzij. et al.; 10-15 km west of Dayan-Nur post, near Khara-Nur, July 20; Ikh-Dzhargalant river basin (right tributary of Bulgan river), 10-15 km on nor.-west. Bulgan settlement, July 27—1988, Kam., Gub. et al.).

IIA. Junggar: ***Cis-Alt.*** (Kran river, July 24, 1903—Gr.-Grzh.; in Timulbakhan region, No. 1076, July 19, 1959—A.R. Lee et al.; "Altay town"—Claves pl. Xinjiang, l.c.), ***Tarb.*** (nor.-west of Khobsair (Chagan-Obo town), No. 19527, June 22, 1959—A.R. Lee et al.; Dachen, No. 1670, Aug. 14, 1957—Kuan), ***Jung. Alat.*** ([excursion in Borotal and Khorgos river sources], 2100-2400 m, Aug. 1878—A. Reg.; in Toli district, in gorge, No. 1120, Aug. 7, 1957—Kuan), ***Tien Shan*** (Talki, July; Sairam lake, Talkibash mountains [2100 m], July 19; Dzhagastai gorge, 1500-2100 m, Aug. 9; Muzart gorge, 1500-2100 m, lower boundary of forest, Aug. 15—1877, A. Reg.; Burkhan-Tau [West of Kul'dzha], June 6; Urtak-Sary river, west of Sairam, July 20—1878, Fet.; south-east of Ketmen'sk pass, 2400-2700 m [June 20], 1878; tributary of Taldy river, 3050 m, June 21; high mountains on tributaries of Pilyuchi and Dzhargalan rivers, 1800 m, April 24—1879, A. Reg.), ***Jung. Gobi*** (east.: nor. macroslope of Baitag-Bogdo mountain range, forest and alpine belt, Budun-Khargaity-Gol gorge, 15 km east of Baitag-Bogdo post, Aug. 1, 1988—Gub., Kam., Bud. et al.; "west. part of Inner Mongolia province"—Claves pl. Xinjiang, l.c.).

General distribution: Jung.-Tarb., Nor. and Cent. Tien Shan; Mid. Asia, West. Sib. (Altay), China (Altay).

2. **T. asiaticus** L. Sp. pl. (1753) 557; Forbes and Hemsl. Index Fl. Sin. 1 (1886) 17; Maxim. Enum. pl. Mong. 1 (1889) 25, p.p., excl. var. *chinensis;* Huth in Helios, 9 (1892) 8; Finet et Gagnep. in Bull. Soc. Bot. Fr. 51 (1904) 396; Kryl. Fl. Zap. Sib. 5 (1931) 1121; Schipcz. in Fl. SSSR, 7 (1937) 49; Fl. Kazakhst. 4 (1961) 15; Pakhom. in Opred. rast. Sr. Azii [Key to Plants of Mid. Asia] 3 (1972) 144; Fl. Sin. 27 (1979) 85; Grub. Opred. Mong. [Key to Plants of Mongolia] (1982) 108; Claves pl. Xinjiang, 2 (1983) 276; Gub. Konsp. Fl. Vneshn. Mong. [Conspectus of Flora of Outer Mongolia] (1996) 51. —**Ic.**: Grub. Plate 48, fig. 222.

Described from Siberia. Type in London (Linn.).

In grass cover of larch forests, their borders, floodplain meadows and in meadow steppes, and also in alpine meadows protected under rocks.

IA. Mongolia: ***Khobd.*** (Dzussylan [west of Ulangom]; July 13, 1879—Pot.; Turgen' mountain range, Turgen'-Gol valley 7 km beyond estuary, right bank of northern slope, July 17, 1971—Grub., Ulzij. et al.; 30 km nor.-east of Kobdo somon, Umne-Otor-Ula town massif, 3150 m, July 3; 20 km west-south-west of Ulangom, Mukhur-Ulyasty-Gol gorge, 2500 m, south. slope, July 9—1977, Karam. et al.), ***Mong. Alt.*** (nor. slope of Khan-Taishiri, 15 km south-east of Yusun-Bulak, Sept.1, 1948—Grub.; south. slope of Khan-Taishiri mountain range, in upper Shine-Usu (nor. of Khamyun somon), 2380-2450 m, on nor. slope under rocky ridge, June 18, 1971—Grub., Ulzij. et al.; Tsast-Bogdo-Ula town, near Tsetseg somon, July 28, 1975—Zhurba (MW)), ***East. Mong.*** (nor. slope of Muni-Ula mountain, July 6, 1871—Przew.), ***Gobi-Alt.*** (Baga-Bogdo, lower grassy slopes and in upper canyon heads, alt. 2100-2700 m, 1925—Chaney; plateau-like ridge of Ikhe-Bogdo-Ula mountain range in nor.-west. half, 3600 m, alpine mat protected by high rocks, July 29, 1948—Grub.; nor.-east. slope of Ikh-Bogdo mountain range, 3300 m, June 28, 1972—Banzragch et al.; 50 km south-east of Barun-Bayan-Ulan somon, Baga-Bogdo mountains, July 6, 1978—Volk.).

IIA. Junggar: ***Tarb.*** (Saur mountain range, southern slope, valley of Karagaitu river, right bank creek valley of Bain-Tsagan, June 23, 1957—Yun. et al.), ***Jung. Alat.*** (in Toli district, No. 995, Aug. 6; 5 km south-west of Dzhagastai, No. 745, Aug. 8; 10 km north of Chzhaosu, No. 888, Aug. 14—1957, Kuan), ***Tien Shan*** ("Thian schan oriental" —Finet et Gagnep. l.c., "Khami"—Fl. Sin. l.c., Claves pl. Xinjiang, l.c.).

General distribution: Jung.-Tarb., Nor. Tien Shan; Arct. (Asian), Europe (north-east. East. Europe), West. Sib., East. Sib., Nor. Mong., China (Dunbei, Nor., East.), Korean peninsula.

3. **T. chinensis** Bge. Enum. pl. China Bor. (1832) 3 and in Mém. Ac. Sci. St. Pétersb., Sav. Etr. 2 (1835) 77; Franch. Pl. David. 1 (1884) 20; Huth in Helios, 9 (1892) 10; Finet et Gagnep. in Bull Soc. Bot. Fr. 51 (1904) 396; Kitag. Lin. Fl. Mansh. (1939) 228; Fl. Sin. 27 (1979) 86; Fl. Intramong. ed. 2, 2 (1990) 428. —*T. asiaticus* var. *chinensis* (Bge.) Maxim. Enum. pl. Mong. 1 (1889) 25. —**Ic.**: Fl. Sin. 27, tab. 1, fig. 1-6; Fl. Intramong. ed. 2, 2, tab. 172, fig. 1-3.

Described from Nor. China. Type in St.-Petersburg (LE).

In forests and along alpine meadows.

IA. Mongolia: ***East. Mong.*** (Muni-Ula, nor. slope, in forest and alpine meadows, abundant, July 6—Przew.; "Ourotes, No. 2768, July 1866—David"—Finet et Gagnep. l.c.; Khalilar town, meadowy steppe, 1959—Ivan.).

General distribution: Far East, China (Dunbei, Nor., Cent.), Korean peninsula.

4. **T. dschungaricus** Regel in Acta Horti Petrop. 7 (1880) 383; Huth in Helios, 9 (1892) 9; Schipcz. in Fl. SSSR, 7 (1937) 44; Persson in Bot. Not. (Lund) (1938) 283; Fl. Kirgiz. 6 (1955) 16; Fl. Kazakhst. 4 (1961) 14; Pakhom. in Opred. rast. Sr. Azii [Key to Plants of Mid. Asia], 3 (1972) 143; Fl. Sin. 27 (1979) 75; Claves pl. Xinjiang, 2 (1983) 275. —*T. europaeus* var. *songaricus* Regel et Herder in Bull. Soc. natur. Moscou, 18 (1870) 243; Maxim. Enum. pl. Mong. 1 (1889) 24. —**Ic.**: Fl. Kazakhst. 4, Plate 2, fig. 9; Fl. Sin. 27, tab. 14, fig. 8-10.

Described from specimens grown in St.-Petersburg Botanical Garden from seeds collected by A. Regel in Mid. Asia. Type in St.-Petersburg (LE). Plate II, fig. 3.

In wet meadows and other humid habitats in spruce forest belt, along mountain brooks, alpine and subalpine meadows.

IB. Kashgar: ***Nor.*** (Ucha pass, June 18, 1904—Divn.), ***West.*** (Sarykol' mountain range, Bostan-Terek, July 11, 1929—Pop.; in Koshkulat, Upal, No. 266, June 10, 1959—A.R. Lee et al.).

IIA. Junggar: ***Jung. Alat.*** (in Toli district, Albakzin mountain, No. 2581, Aug. 7; 10 km south from Dzhagastai on way to Chzhaosu, No. 695, Aug. 8; Chzhaosu district, between Syata and Ven'tsyuan', No. 3412, Aug. 13; 10 km north of Chzhaosu, No. 895 and 3297; 20 km south of Ven'tsyuan', No. 1507, Aug. 14—1957, Kuan), ***Tien Shan*** (near Tsanma sources, June 17; Mal. Yuldus, June 29—1877, Przew.; Shargsu river, 2000-2400 m, May 26; Khanakhai, Ili river valley, 1500-2000 m, June 16; Ketmen' mountain range, 2400-2700 m, June 19—1878; nor. slope of Iren-khabirga mountain range, 2700 m, Kumdaban, May 29; Kumbel' [pass], 2700-3000 m, May 30; Naryn-Gol brook near Tsagan-Usu, June 10; on Borborogussun river, 2700 m, June 15; Dzhagastai, Chapchal gorge, 2400 m, June 28—1879, A. Reg.; declivitas septentrionalis jugi montium Tianschan orient., Syrt von Karabulak und Kara-dschon, July 30, 1907; Nordliches Kiukönik Tal, June 17-19, 1908—Merzb.; in Savan district, Datszymyao village, No. 1301, July 8; Danyu, nor. slopes, in gorge, No. 1415, July 16; on Koisu river, No. 119, July 17—1957, Kuan; Koisu river valley at its confluence with Ulan-Usu, July 17, 1957—Yun.; 36 km south-east of Nyutsyuan'tsza, No. 157; between Daban' and Dachyu, No. 2015—July 19; Datszymyao village, No. 426, July 22—1957, Kuan; Ardyn-Daban pass separating Mal. Yuldus basin from upper Kunges (Ili), north-western slope, Aug. 3; upper Ili river, nor. foothills of Narat mountain range descending gently into Tsanma valley, Aug. 7—1958, Yun. and Yuan').

IIIB. Tibet: **Pamir** ("Jerzil, 3500 m, July 19, 1930—Persson"—Persson, l.c.; King-Tau mountain range, nor. slope 3-4 km south-east of Kosh-Kulak settlement, spruce-juniper grove, 3500 m, June 10, 1959—Yun.).

General distribution: Jung.-Tarb., Nor. and Cent. Tien Shan; Mid. Asia.

5. **T. farreri** Stapf in Curtis's Bot. Mag. 152 (1928), tab. 9143; Hand.-Mazz. Pl. Sin. (1939) 41; Fl. Sin. 27 (1979) 78; Fl. Xizang. 2 (1985) 11. —**Ic.**: Stapf, l.c.; Fl. Sin. 27, tab. 15, fig. 6-8.

Described from South-West. China. Type as depicted in plate of first description (Stapf, l.c. tab. 9143).

On rubble slopes, grasslands, at 2000-5600 m alt.

IIIA. Qinghai ("east."—Fl. Xizang. l.c.).

IIIB. Tibet: ***Weitzan*** ("Bizhu, Sosyan'"—Fl. Xizang. l.c., "nor.-east. Tibet"—Fl. Sin. l.c.).

General distribution: China (Nor.-West., South-West.).

Note. This species is very close to *T. pumilus* D. Don and *T. ranunculoides* Hemsl.

6. **T. lilacinus** Bge. in Mém. Ac. Sci. St. Pétersb., Sav. Etr. 2 (1835) 555; Maxim. Enum. pl. Mong. 1 (1889) 24; Huth in Helios, 9 (1892) 12; Finet et Gagnep. in Bull. Soc. Bot. Fr. 51 (1904) 394; Kryl. Fl. Zap. Sib. 5 (1931) 1125; Pakhom. in Opred. rast. Sr. Azii [Key to Plants of Mid. Asia] 3 (1972) 145; Fl. Sin. 27 (1979) 87; Claves pl. Xinjiang, 2 (1983) 275. —*Hegemone lilacina* (Bge.) Bge. in Ledeb. Fl. Ross. 1, 1 (1841) 51; Schipcz. in Fl. SSSR, 7 (1937) 53; Fl. Kirgiz. 6 (1955) 19; Fl. Kazakhst. 4 (1961) 16; Kamelin et al. in Bot. zhurn. 74, 9 (1989) 135; Gub. Konsp. Fl. Vneshn. Mong. [Conspectus of Flora of Outer Mongolia] (1996) 50. —**Ic.**: Fl. Sin. 27, tab. 15, fig. 1-2.

Described from Altay. Type in St.-Petersburg (LE).

On rubble mounds and rocky slopes, alpine meadows, banks of snow brooks and periglacial grasslands in upper belt of mountains, at 2700-2850 m alt.

IA. Mongolia: ***Khobd.*** ("Sailyugem mountain range"—Kamelin et al., 1989), ***Mong. Alt.*** (south. bank of Dayan-Nur, July 28, 1974—Zhurba (MW).

IIA. Junggar: ***Cis-Alt.*** (30 km south of Altay settlement, Ëlt-Gol river basin (upper Cher. Irtysh), July 22, 1988—Kam., Gambold et al.), ***Tien Shan*** (picket, 1878, Larionov; upper Taldy river, 2400-2700 m, May 17; Taldy pass, 3350 m, May 20; Dumbedan-Kumdaban, nor. slope of Irenkhabirga mountain range, 2400-2700 m, May 28; on Borborogusun river, 2700 m June 15—1879, A. Reg.; Oberstes Dschanart Tal, June 14-17—1903; Pässe zwischen Kinsu und Kurdai, July 3; circa 3400 m hohes Plateau eines Berges im oberen Koksu Gebiet, July 8-10—1907; Kiukonik Pass Nordseite, June 16; Kungess Tal, May 1-5—1908, Merzb.; Shichan, 2400 m, No. 3563, Oct. 3, 1956—Ching, 6-7 km south of Danyu, No. 465, July 21; 3 km south of Yakou, No. 1666, Aug. 30; Bargat (south of Yakou), No. 1719, Aug. 31; from Nilki to Dzinkho through Yakou, No. 4023, Sept. 1—1957, Kuan; Bogdashan' mountain range near Turfan, No. 5686, June 17; east of Khomot, Taskhan canal, marsh, 3280 m, No. 7101, Aug. 6, 1958—Lee and Chu; Narat mountain range, Dasht pass, Aug. 6, 1958—Yun. and Yuan'; on Narat mountain in Kunges, 3000 m, No. 9124, Aug. 7, 1958—Lee and Chu).

General distribution: Jung.-Tarb., Nor. and Cent. Tien Shan; Mid. Asia (Pam.-Alay, West. Tien Shan), West. Sib. (Altay), East. Sib., Nor. Mong. (Hentey), China (Altay).

7. **T. pumilus** D. Don, Prodr. Fl. Nepal. (1825) 195; Maxim. Fl. Tangut. 1 (1889) 16; Kanitz, Wiss. Erg. Reise Szech. Ostas. 2 (1898) 679; Gr.-Grzh. Opis. put. Zap. Kit. [Report of Expedition in West. China] 3 (1907) 476; Fl. Sin. 27 (1979) 77; Fl. Xizang. 2 (1985) 11. —**Ic.**: Fl. Sin. 27, tab. 14, fig. 4-7; Fl. Xizang. 2, tab. 2, fig. 6-7.

Description from Himalayas. Type in London (BM). Plate II, fig. 2. Map 1.

In alpine forb meadows, small swamps, at 3800-5000 m alt.

IA. Qinghai: ***Nanshan*** (in alpine meadows of North Tetung mountain range, July 5, 1872; same site, often in gorges around mountain streams, June 22—1873; in South Tetung mountain range meadows [early Aug.], 1880—Przew.; "Xining alps, North Tetung mountain range, 3000-4000 m, May-Aug. 1905—Gr.-Grzh."; "in vertice Khamtchuk a Sinig-fu versus"— Kanitz, l.c.), ***Amdo*** (Syan'sibei mountain range, 3400-3700 m, in gorges, May 29; in alpine belt of Mudzhik mountain, June 18—1880, Przew.).

IIIB. Tibet: ***Weitzan*** (on bank of Bychu river, June 15, 1884—Przew.), ***South.*** ('Nan'mulin, Pulan'"—Fl. Xizang. l.c.).

General distribution: China (South-West.), Himalayas (west., east.).

8. **T. ranunculoides** Hemsl. in J. Linn. Soc. (London) Bot. 29 (1893) 301; Fl. Sin. 27 (1979) 81; Fl. Xizang. 2 (1985) 13. —**Ic.**: Fl. Sin. 27, tab. 15, fig. 3-5; Fl. Xizang. 2, tab. 2, fig. 8.

Described from South-West. China (Sichuan). Type in London (K).

In alpine meadows, river banks, spruce groves, 3500-4600 m alt.

IIIA. Qinghai ("south. and east. Qinghai"—Fl. Sin. l.c.).

IIIB. Tibet: ***Weitzan*** ("Tszyali"—Fl. Xizang. l.c.).

General distribution: China (Nor.-West., South-West.).

Tribe 2. **Cimicifugeae** Torr. et Gray

4. **Cimicifuga** Wernisch.

Gen. pl. (1763) 298, 321

1. Stem and petioles covered with fine glandular hairs admixed with simple hairs...2. **C. foetida** L.

\+ Stem (excluding inflorescence) and petioles glabrous or with rare simple hairs. Staminodes forked-bipartite, with 2 hollow anthers. Flowers unisexual, plant dioecious. Leaves 2-3-ternate. Inflorescence, branched raceme...1. **C. dahurica** (Turcz. ex Fisch. et Mey.) Maxim.

1. **C. dahurica** (Turcz. ex Fisch. et Mey.) Maxim. in Mém. Ac. Sci. Sav. Etr., St. Pétersb. 9 (1859) 28; Forbes and Hemsl. Index Fl. Sin. 1 (1886) 21; Maxim. Enum. pl. Mong. 1 (1889) 31; Huth in Bot. Jahrb. 16 (1893) 316, cum auct. Huth; Finet et Gagnep. in Bull. Soc. Bot. Fr. 51 (1904) 522; Kitag. Lin. Fl. Mansh. (1939) 215; Schipcz. in Fl. SSSR, 7 (1937) 83; Fl. Sin. 27 (1979) 102; Fl. Intramong. ed. 2, 2 (1990) 431. —*Actinospora dahurica* Turcz. ex Fisch. et Mey. Index Sem. Horti Petrop. 1 (1835) 21; Franch. Pl. David. 1 (1884) 23. —**Ic.**: Fl. Sin. 27, tab. 20, fig. 1-4; Fl. Intramong. ed. 2, 2, tab. 173.

Described from East. Siberia. Type in St.-Petersburg (LE).

On forest borders, scrubs.

IA. Mongolia: ***East. Mong.*** ("Muni-Ula mountain range [= Ulashan'], David"—Maxim. 1889; "Vnutr. Mongoliya [Inner Mongolia]"—Fl. Sin. l.c.).

General distribution: East. Sib., Far East, Nor. Mong. (Cis.-Hing.), China (Dunbei, Nor., Nor.-West.), Korean peninsula.

2. **C. foetida** L. Syst. natur. ed. 12, 2 (1767) 659; Maxim. Fl. Tangut. 1 (1889) 28; Huth in Bot. Jahrb. 16 (1893) 317, p.p.; Finet et Gagnep. in Bull. Soc. Bot. Fr. 51 (1904) 521; Kryl. Fl. Zap. Sib. 5 (1931) 1131; Schipcz. in Fl. SSSR, 7 (1937) 84; Fl. Kazakhst. 4 (1961) 25; Fl. Sin. 27 (1979) 101; Fl. Xizang. 2 (1985) 15. —**Ic.**: Fl. Sin. 27, tab. 19, fig. 1-4; Fl. Xizang. 2, tab. 4, fig. 1-2.

Described from Siberia. Type in Stockholm (S).

On forest borders and meadows, scrubs, alpine meadows.

IIIA. Qinghai: *Nanshan* (between Choibsen monastery and mountain range south of Tetung river, July 29; nor. slope of mountain range south of Tetung river, Aug. 7 and 11, 1872; same site, lower forest belt, 2300 m, Aug. 7, 1880—Przew.; alpine meadow on Yusyp-Khatyma river, 2700-3000 m, July 24; Chortentan temple, 2100 m, in scrubs, Sept. 1901—Lad.).

IIIB. Tibet: *Weitzan* (Mekong river basin, Tszachu Rv. 3350-4000 m, Oct. 1, 1900—Lad.).

General distribution: West. Sib., East. Sib., Nor. Mong. (Fore Hubs., Hent., Hang., Mong.-Daur.), China (Dunbei, Nor., Nor.-West., Cent., South-West.), Himalayas, Korean peninsula, Japan.

Note. Specimen from Tszachu river was re-identified as *Actaea mairei* (H. Levl.) J. Compton in Taxon, 47, 3 (1997) 622.

5. **Actaea** L.

Sp. pl. (1753) 504

1. **A. acuminata** Wall. ex Royle, Ill. Bot. Himal. Mount. 1 (1839) 57; Kitag. Lin. Fl. Mansh. (1939) 212; Schipcz. in Fl. SSSR, 7 (1937) 80. —*A. asiatica* Hara in J. Jap. Bot. 15 (1939) 313; Fl. Sin. 27 (1979) 103; Fl. Xizang. 2 (1985) 16; Pl. Vasc. Helanshan, (1986) 101; Fl. Intramong. ed. 2, 2 (1990) 436. —*A. spicata* var. *asiatica* (Hara) S.H. Li et Y.H. Huang in Fl. Pl. Herb. Chin. Bor.-Or. 3 (1975) 105. —*A. spicata* auct. non L.: Maxim. Enum. pl. Mong. 1 (1889) 31; Finet et Gagnep. in Bull. Soc. Bot. Fr. 51 (1904) 518. —**Ic.**: Fl. Sin. 27, tab. 21; Fl. Intramong. ed. 2, 2, tab. 175, Fig. 1-5.

Described from Himalayas. Type—? sketch.

In broad-leaved mountain forests.

IA. Mongolia: ***East. Mong.*** ("montagnes d'Oulachan, No. 2693, June 1866—David" —Franchet, l.c.; Maxim. l.c.; "Datsinshan' mountain range"—Fl. Intramong. l.c.).

IIIA. Qinghai ("Qinghai"—Fl. Xizang. l.c.).

General distribution: Far East, China (Dunbei, Nor., Nor.-West., Cent., South-West.), Himalayas, Korean peninsula, Japan.

Note. *A. erythrocarpa* Fisch. is found in Nor. Mongolia.

Tribe 3. **Delphinieae** Warming

6. **Aconitum** L.

Sp. pl. (1753) 532, nom. cons.

1. Annual plant. Perianth segments (sepals) tapered at base into long claws between which are seen ovaries and stamens. Ovaries 8. Leaves finely laciniated (Subgenus 3. *Gymnaconitum*)..23. **A. gymnandrum** Maxim.

\+ Perennial plant. Perianth segments (sepals) sessile. Ovaries and stamens entirely concealed. Ovaries 3-5..2.

2. Plants with vertical particularising rhizome. Flowers with tall cylindrical or conical hood, yellow or canescent violet. Follicles 3 (Subgenus 4. *Lycoctonum*)..3.

\+ Plants with horizontal rhizome or tubers. Flowers with broad keeled, falcate or tall conical hood, violet-blue, light-blue or yellow. Follicles 5-6, rarely 3..5.

3. Flowers small, not more than 10 mm long, yellow, turning brown on drying. Nectary petals very small: spur together with claw not more than 7.5 mm long. Hood 5-8 mm tall. Racemose inflorescence up to 60 cm long. Plant up to m tall. Tien Shan Plant .... 24. **A. apetalum** (Huth) B. Fedtsch.

\+ Flowers, including nectary petals, considerably large...................4.

4. Flowers sulphureous, with yellowish hairs outside and hairy tuft exserting from throat. Leaf blades palmatisected into 3-5 segments down to base; segments divided into narrow-lanceolate, falcate lobes..25. **A. barbatum** Patr. ex Pers.

\+ Flowers white- or grey-violet. Cauline leaves palmatisected not down to base into 3-9 broad-rhomboid lobes.. ....26. **A. leucostomum** Worosch.

5. Rhizome catenate up to 8 cm breadth with conical stem roots growing together beadlike...6.

\+ Bulbs oblong or globose, not joined beadlike..................................10.

6. Some rhizomes (stem roots) not more than 5-8 mm long and broad. Plants 0.5-1 m tall...7.

\+ Some rhizomes (stem roots) large, (8) 10-25 mm or more broad.....9.

7. Leaf blade glabrous..8.

\+ Leaf blade covered above with crispate hairs. Ovaries 3, glabrous. ....4. **A. decipiens** Worosch. et Anfalov.

8. Leaf blade profile generally orbicular or orbicular-reniform, laciniated into 3 narrow, 0.5-1.5 cm broad segments in broad unaffected part; midsegment rhombic, pinnatipartite into linear-lanceolate lobes; lateral segments more broad, divided almost down

to base into 2-3 unequal lobes, each of them subdivided into 2nd or 3rd order lobes. Ovaries (4) 5, pubescent with hairs standing apart. Plant of Mongolian Altay..6. **A. gubanovii** Luferov et Worosch.

\+ Leaf blade profile generally pentagonal, laciniated into 5 broad rhomboid, 2-3.5 cm broad segments in broad unaffected part; each segment divided into 3 large lobes bearing large teeth. Ovaries 3, glabrous. Junggar plant..............................10. **A. nemorum** M. Pop.

9. Leaf blades laciniated into narrow-linear, 1.5-3 mm broad segments and lobes. Entire plant glabrous, rarely pubescent with fine crispate hairs in inflorescence. Flowers purple..7. **A. karakolicum** Rapaics.

\+ Leaf blades laciniated into broader (5-10 mm) segments and lobes. Entire plant more or less pubescent. Flowers light blue...15. **A. talassicum** M. Pop.

10. Leaf blades divided down to base. Their segments petiolulate...11.

\+ Leaf blades less laciniated...12.

11. Stem straight, not climbing, glabrous. Leaf petiole glabrous. Blade of nectary petal 2.5-4 mm broad..................9. **A. kusnezoffii** Reichb.

\+ Plant with stems climbing or sinuate in upper part, diffuse and compactly, patently pilose in inflorescence. Leaf petiole also patently pilose. Blade of nectary petal highly inflated up to 6-10 mm breadth..27. **A. volubile** Pall. ex Koelle.

12. Radical leaves present, cauline leaves 1 or few..............................13.

\+ Radical leaves absent and lower cauline leaves wilt even at anthesis...16.

13. Leaf blades laciniated almost down to base. Stem leafy upward. 3-12 flowers in inflorescence. Sinkiang plant...13. **A. sinchiangense** W.T. Wang.

\+ Leaf blades palmatisected but not for more than 1/3 of their length...14.

14. Leaves essentially radical and 1-2 each at stem base. 3-5 flowers in racemose inflorescence. Qinghai and Tibet plant...16. **A. tanguticus** (Maxim.) Stapf.

\+ Stems more or less leafy..15.

15. Lip of nectary petal divided into 2 subfiliform branches. Junggar plant...12. **A. rotundifolium** Kar. et Kir.

\+ Lip of nectary petal entire, with only 2 teeth at tip. South. Tibet plant...21. **A. naviculare** (Bruhl) Stapf.

16. Stem simple, erect. Middle and upper leaves short-petiolate or sessile, compact, palmatisected. Upper sepal broad-conical, carinate, falcate. Blade of nectary petal with short spur (Subgenus 2. *Anthora*)...17.

+ Stem simple or branched. Leaves with distinct petioles..............20.

17. Upper perianth segment (sepal) tall, broad-conical, with compactly pubescent beak, yellow................................18. **A. anthoroideum** DC.

+ Upper perianth segment (sepal) carinate or falcate, glabrous or faintly pubescent; perianth yellowish green, light blue or purple. ....18.

18. Upper perianth segment falcate; perianth light blue. South. Tibet. plant..20. **A. ludlowii** Exell.

+ Upper perianth segment carinate; perianth yellowish green (sometimes purple in *A. gilvum*)..19.

19. Inflorescence with pubescence of appressed hairs.......................... ....19. **A. gilvum** (Maxim.) Hand.-Mazz.

+ Inflorescence pubescent with hairs standing apart........................ ....22. **A. pendulum** N. Busch.

20. Rachis and pedicels glabrous..21.

+ Rachis and pedicels pubescent..22.

21. Blade of nectary petal about 1 mm broad.....1. **A. ambiguum** Reichb.

+ Blade of nectary petal more than 2 mm broad.. ....14. **A. soongaricum** Stapf.

22. Upper perianth segment conical. Perianth purple or pink................. ....8. **A. kongboense** Lauener.

+ Upper perianth segment carinate or falcate.......................................23.

23. Ovaries pubescent............................11. **A. richardsonianum** Lauener.

+ Ovaries glabrous...24.

24. Inflorescence compact...25.

+ Inflorescence lax.............................2. **A. baicalense** Turcz. ex Rapaics.

25. Rachis and pedicels patently pilose. Anther filaments glabrous. ....26.

+ Rachis and pedicels with short appressed pubescence or subglabrous; raceme many-flowered, compact, often branched; flowers bluish violet; anther filaments pilose. Leaves compact, greyish, laciniated into linear lobes up to 2.5 cm broad; stem stout, compactly leafy right up to inflorescence.. ....17. **A. turczaninowii** Worosch.

26. Flowers dingy violet with bright spots; hood very narrow, carinate, 4-5 mm tall; nectar blade 5 mm long. Follicles 5. Inflorescence many-flowered, compact. Stem stout, compactly covered with adjoining short-petiolate leaves; leaf blade laciniated into narrow (1.5-2.5 mm broad) linear lobes..................3. **A. chasmanthum** Stapf.

+ Flowers blue-violet, without spots; hood low, hemispherical, 8-9 mm tall; blade of nectary 7-10 mm long. Follicles 3. Stem thin with leaves standing apart; leaf blade laciniated into fairly broad segments with lanceolate lobes.............5. **A. glandulosum** Rapaics.

Subgenus 1. **Aconitum** (*Aconitum* subgen. *Tuberaconitum* Rapaics)

1. **A. ambiguum** Reichb. Ill. Spec. Acon. Gen. (1823) tab. 23; Fl. Sin. 27 (1979) 287; Voroshilov in Byull. Mosk. obshch. isp. prir., otd. biol. 93, 1 (1988) 90; Fl. Intramong. ed. 2, 2 (1990) 563; Luferov in Sosud. rast. Dal'n. Vost. 7 (1995) 52. —**Ic.**: Fl. Sin. 27, tab. 61, fig. 1-2; Fl. Intramong. ed. 2, 2, tab. 226, fig. 6-8.

Described from Siberia. Type, sketch (Reichb. l.c., tab. 23).

In birch groves, meadows, among bushes on slopes, pebble beds along rivers and brooks.

IA. Mongolia: ***East. Mong.*** (Mongolia chinensis, in reditu e China, 1841—Kirilow), ***Depr. Lakes*** (Ubsa lake region, Dzussylan mountains, July 12; Ulan-Natsyn river, July 16; Kharkira river valley, 2400 m, July 21, 1879—Pot.).

General distribution: East. Sib., Far East, Nor. Mong. (Fore Hubs., Hent., Hang., Mong.-Daur.), China (Dunbei).

2. **A. baicalense** Turcz. ex Rapaics in Növ. Közl. 6, 5 (1907) 14; Luferov in Sosud. rast. Dal'n. Vost. 7 (1995) 52; Gub. Konsp. Fl. Vneshn. Mong. [Conspectus of Flora of Outer Mongolia]) (1996) 47. —*A. czekanovskyi* Steinb. in Fl. URSS, 7 (1937) 733.

Described from Siberia. Type in Budapest (BP).

IA. Mongolia: "Vost. Mong. [East. Mong.]"—Gub. l.c.

General distribution: West. Sib., East. Sib., Nor. Mong., China (Dunbei).

3. **A. chasmanthum** Stapf in Holmes in Mus. Rep. Pharm. Soc. Brit. (1903) 2; id. in Ann. Bot. Gard. (Calcutta) 10 (1905) 142; Grub. Opred. rast. Mong. [Key to Plants of Mongolia] (1982) 110; Fl. Xizang. 2 (1985) 35; Fl. Sin. 27 (1979) 306; Gub. Konsp. Fl. Vneshn. Mong. [Conspectus of Flora of Outer Mongolia] (1996) 47. —**Ic.**: Stapf (1905) pl. 96; Grub. Plate 51, fig. 232.

Described from Himalayas. Type in Kew (K) or Vienna (WU).

On river banks, scrubs, meadows, meadowy and steppe slopes in mountain-steppe and alpine belts.

IA. Mongolia: ***Val. Lakes*** (20-22 km north of Dulan-Khairkhan somon, south-east. slope of Khan-Khushin-Ula mountains, 2500 m, July 25, 1952—Davazamc), ***Gobi-Alt.*** (Ikhe-Bogdo mountain range, Bityuten-amo creek valley, Aug. 12, 1927—A. and M. Simukov's; south. slope of Ikhe-Bogdo mountain range, upper course of left tributary of Ichetu-Gol, July 2, 1945—Yun.).

IIA. Junggar: ***Jung. Alat.*** (Chubaty pass, 2700-3000 m, July 1978—A. Reg.), ***Tien Shan*** (Dzhirgalan mountain, 1500-1800 m, July 26, 1878—A. Reg.; Yuldus, Sept. 1878—Fet.; Chunkur-daban pass [Irenkhabirga mountain range], between Shasta and Dzhungaka, 3000 m, June 13; Borborogusun river, 2700 m, June 15; Mongoto gorge, 3000-3400 m, July 4—1879, A. Reg.).

III. Tibet ("Cent. and South."—Fl. Sin. l.c.).

General distribution: West. Sib. (Tuva), Nor. Mong. (Hang.), Himalayas (Kashmir).

4. **A. decipiens** Worosch. et Anfalov in Bot. Journ. 28, 1 (1943) 27; Gub. Konsp. Fl. Vneshn. Mong. [Conspectus of Flora of Outer Mongolia] (1996) 47.

Described from Altay. Type in Moscow (MW).

IA. Mongolia ("Mong. Alt."—Gub. l.c.).
General distribution: Nor. Mong. (Hang.).

5. **A. glandulosum** Rapaics in Növ. Közl. 6, 5 (1907) 152; Voroshilov in Byull. Mosk. obshch. isp. prip., otd. biol. 93, 1 (1998) 90; Gub. Konsp. Fl. Vneshn. Mong. [Conspectus of Flora of Outer Mongolia] (1996) 47. —*A. altaicum* Steinb. in Fl. URSS, 7 (1937) 731, 222; Grub. Opred. rast. Mong. [Key to Plants of Mongolia] (1982) 110. —*A. smirnovii* Steinb. l.c. 731, 222; Claves pl. Xinjiang, 2 (1983) 281; Fl. Sin. 27 (1979) 300. —*A. napellus* β. *alpinum* Regel in Index Sem. Horti Petrop. (1861) 45, p.p.; Kryl. Fl. Zap. Sib. 5 (1951) 1149. —**Ic.:** Fl. SSSR, 7, Plate 14, fig. 1; Fl. Sin. 27, l.c. tab. 62, fig. 11 and tab. 77, fig. 4-6 (*A. smirnovii*).

Described from Siberia. Type in Budapest (BP). Isotypes in St.-Petersburg (LE).

On river banks, meadowy slopes, larch groves and dwarf birch thickets, forb meadows, alpine tundra in alpine and parts of forest belt.

IA. Mongolia: ***Khobd.*** (Kharkira mountain group, Borborogusun river, Saksaya tributary (Kobdo river system), July 2, 1903—Gr.-Grzh.; Turgen' mountain range, Turgen'-gol valley 7 km above estuary, July 17, 1971—Grub., Ulzij. and Dariima), ***Mong. Alt.*** (Mal. Kairta river valley, July 16, 1908—Sap.; Urtu-gol river valley, Aug. 17; Khara-Adzarga mountains, Sakhir-Sala river valley, Aug. 24; same site, around Khairkhan-Duru river, Aug. 26—1930, Pob.; south. bank of Indertin-gol river valley, July 24, 1947—Yun.; east. extremity of Achitnur lake basin, Ulyasutuin river valley, Aug. 7, 1947—Tarasov; upper Kharagaitu-gol, left bank tributary of Bulugun, Aug. 24, 1947—Yun.; Khasaktu-Khairkhan, nor. slope of Tsagan-Irmykul in upper Khunkerin-ama, 2700 m, Aug. 23, 1972—Grub. et al.; Uenchiin Gola basin, Armont'en-gola valley (right tributary of Khargaityn-gola) 3 km from estuary, Aug. 14, 1979—Grub., Muld. and Dariima; "Altay, Burchum"—Claves pl. Xinjiang), ***Gobi-Alt.*** (Ikhe-Bogdo mountain, south. macroslope, 3500 m, Aug. 4, 1973—Isach. and Rachk.).
IIA. Junggar ("Jung. Gobi"—Gub. l.c.).
General distribution: West. Sib. (Altay), East. Sib., Nor. Mong. (Fore Hubs., Hent., Hang., Mong.-Daur.).

Note. Following V.N. Voroshilov (l.c.), *A. altaicum* (described from Altay) and *A. smirnovii* (described from East. Siberia) have been treated as *A. glandulosum*.

6. **A. gubanovii** Luferov et Worosch. Byull. Mosk. obshch. isp. prir., otd. biol. 96, 4 (1991) 111; Gub. Konsp. Fl. Vneshn. Mong. [Conspectus of Flora of Outer Mongolia] (1996) 47.

Described from Mongolia (Mong. Altay). Type in Moscow (MW). Isotype in St.-Petersburg (LE).

In meadows, among shrubs, in forest and alpine belts of mountains.

IA. Mongolia: ***Mong. Alt.*** (Mong. Altay, Ikh-Dzhergalant river basin (right tributary of Bulgan river), 15 km north-west of Bulgan settlement in Bayan-Ulegeisk ajmaq [administrative territorial unit in Mongolia], No. 2225a, July 27, 1988, Gub. and Kam., isotypus !).

IIA. Junggar ("Jung. Gobi"—Gub. l.c.).

General distribution: endemic.

7. **A. karakolicum** Rapaics in Növ. Közl. 6, 5 (1907) 149; Shteinb. in Fl. SSSR, 7 (1937) 233; Fl. Kirgiz. 6 (1955) 64; Fl. Kazakhst. 4 (1961) 55; Pakhom. in Opred. rast. Sr. Azii [Key to Plants of Cent. Asia] 3 (1972) 186; Fl. Sin. 27 (1979) 310; Claves pl. Xinjiang, 2 (1983) 282. —*A. karakolicum* var. *karakolicum* et var. *patentiiplium* W.T. Wang in Fl. Sin. 27 (1979) 312, 610; Claves pl. Xinjiang, l.c. 282. —**Ic.:** Fl. Kazakhst. 4, Plate 7, fig. 7; Fl. Sin. 27, tab. 72, fig. 1-4.

Described from Nor. Tien Shan. Type in Budapest (BP). Isotype in St.-Petersburg (LE).

On exposed slopes, forest borders and gorges among forests.

IIA. Junggar: ***Tien Shan*** ("northern slopes of Tien Shan"—both varieties—Claves pl. Xinjiang; "north-western mountain regions"—var. *patentiiplium*—Claves pl. Xinjiang).

General distribution: Jung.-Tarb. (Jung. Ala Tau), Nor. Tien Shan.

8. **A. kongboense** Lauener in Notes Bot. Gard. Edinb. 25, 1 (1963) 17; Fl. Sin. 27 (1979) 288; Fl. Xizang. 2 (1985) 32. —**Ic.:** Lauener, fig. 1 N and 4 B; Fl. Sin. 27, tab. 62, fig. 14, tab. 65, fig. 1-2; Fl. Xizang. 2, tab. 8, fig. 1-2.

Described from Tibet. Type in London (BM). Isotype in Edinburgh (E).

In alpine meadows, among shrubs.

IIIB. Tibet: ***South.*** ("Lhasa, Gunbutszyanda"—Fl. Xizang. l.c.).

General distribution: China (South-West.).

9. **A. kusnezoffii** Reichb. Ill. Spec. Acon. Gen. (1823) tab. 21; Franch. Pl. David. 1 (1884) 22; Forbes and Hemsl. Index Fl. Sin. 1 (1886) 20; Maxim. Enum. pl. Mong. 1 (1889) 30; Palibin in Acta Horti Petrop. 14, 1 (1895) 107; Shteinb. in Fl. SSSR, 7 (1937) 217; Kitag. Lin. Fl. Mansh. (1939) 209; Fl. Intramong. 2 (1978) 221; Fl. Sin. 27 (1979) 269; Grub. Opred. rast. Mong. [Key to Plants of Mongolia] (1982) 110; Kadota, Rev. Acon. subgen. *Aconitum* East Asia (1987) 95, map II; Fl. Intramong. ed. 2, 2 (1990) 558; Gub. Konsp. Fl. Vneshn. Mong. [Conspectus of Flora of Outer Mongolia] (1996) 48. —**Ic.:** Reichb. tab. 21; Fl. Intramong. 2, tab. 116; Fl. Sin. 27, tab. 58, fig. 6-10; Kadota, fig. 30; Fl. Intramong. ed. 2, l.c. tab. 227.

Described from Nor. China. Type, sketch (Reichb. l.c. tab. 21).

In meadows, scrubs on mountain slopes and forest borders.

IA. Mongolia: ***East. Mong.*** (Dolon-Nor, Aug. 1/2, 1891—Putiata).

General distribution: East. Sib., Far East, Nor. Mong. (Cis-Hing.), China (Dunbei, Nor., Nor.-West., Cent., South-West.), Korean peninsula, Japan.

10. **A. nemorum** M. Pop. in Bull. Soc. natur. Moscou, 44 (1935) 131; Shteinb. in Fl. SSSR, 7 (1937) 236; Fl. Kirgiz. 6 (1955) 67; Fl. Kazakhst. 4 (1961) 56; Pakhom. in Opred. rast. Sr. Azii [Key to Plants of Mid. Asia] 3 (1972) 187; Fl. Sin. 27 (1979) 313; Claves pl. Xinjiang, 2 (1983) 233. —**Ic.:** Fl. Kazakhst. 4, Plate 7, fig. 4; Fl. Sin. 27, tab. 71, fig. 7-10.

Described from Nor. Tien Shan (vicinity of Alma-Ata). Type in Alma Ata (AA). Isotype in St.-Petersburg (LE).

In subalpine meadows, on borders of and in sparse spruce forests and juniper groves.

IIA. Junggar: ***Jung. Alat.*** (Chubaty pass, 2400-2700 m, Aug. 2, 1878—A. Reg.; in Toli district, No. 1272 and 1337, Aug. 6 and 7; 20 km south of Ven'tsyuan', No. 1507, Aug. 14; 15 km nor.-west of Ven'tsyuan' town, No. 1649, Aug. 29—1957, Kuan), ***Tien Shan*** (Talkibash mountain near Sairam, July 20 and 22—1877, A. Reg.; on Shary-su river, 1200-2400 m, June 20, 1878—A. Reg.; Sairam lake, July 18 and 20, 1878—Fet.; Bogdo mountain, 1800-2100 m, July 24, 1878—A. Reg.; Kokkamyr upland, 2700-3000 m, July 1878—A. Reg.; Westl. Arystyn Nebenbach, 2400-2700 m, July 8; Monguto, Irenkhabirga mountain range, Aug., 1878—A. Reg.; Montes Bogdo-Ola, in der Lehm und Kiessteppe, Aug. 2-3, 1908—Merzb.; Urumchi region, upper Tasenku river, Biangou locality, Sept. 25, 1929—Pop.; on Danyu river, No. 2163, July 21; Chzhaosu district, between Syata and Ven'tsyuan', No. 3428, Aug. 13; 10 km nor. of Chzhaosu, No. 3295, Aug. 14; mountains south of Sin'yuan' town, No. 1122, Aug. 22—1957, Kuan; 6-7 km south-east of Aksu settlement, Tekes river valley, No. 1523, Aug. 24, 1957—Yun. and Yuan'; on Urumchi-Kucha highway, No. 6049, July 21; Narat pass on Kunges river, No. 6573, Aug. 7-1958, Lee and Chu; upper Ili river, nor. foothills of Narat mountain range gently descending into Tsanma valley (left tributary of Kunges), 2250 m, Aug. 7, 1958—Yun. and Yuan'), ***Jung. Gobi*** (south.: Tsitai-Beidashan', No. 5208, Sept. 29, 1957 —Kuan), ***Dzhark.*** (Suidun town, July 16, 1877—A. Reg.; Pilyuchi gorge, June 21, 1878—Fet.).

General distribution: Jung.-Tarb., Nor. and Cent. Tien Shan; Mid. Asia (West. Tien Shan).

Note. The type specimen was found in a spruce (*Picea schrenkiana*) forest. The first description pointed to the affinity of *A. nemorum* with *A. volubile*, the former differing from the latter in broader leaf segments and non-climbing stem.

11. **A. richardsonianum** Lauener in Notes Bot. Gard. Edinb. 25, 1 (1963) 3; Fl. Sin. 27 (1979) 292; Fl. Xizang. 2 (1985) 32. —**Ic.:** Lauener, l.c. fig. 1A, 2A; Fl. Sin. 27, tab. 65, fig. 3-5; Fl. Xizang. 2, tab. 8, fig. 3-4.

Described from South. Tibet. Type in London (BM).

In alpine meadows.

IIIB. Tibet: ***South.*** ("Lhasa"—Fl. Xizang. l.c.).

General distribution: China (South-West.).

12. **A. rotundifolium** Kar. et Kir. in Bull. Soc. natur. Moscou, 15 (1842) 139; Stapf in Ann. Bot. Gard. (Calcutta) 10 (1905) 149; Shteinb. in Fl. SSSR,

7 (1937) 229; Fl. Kirgiz. 6 (1955) 59; Fl. Kazakhst. 4 (1961) 53; Ikonnik. Opred. rast. Pamira [Key to Plants of Pamir] (1963) 115; Pakhom. in Opred. rast. Sr. Azii [Key to Plants of Mid. Asia] 3 (1972) 184; Fl. Sin. 27 (1979) 188; Claves pl. Xinjiang, 2 (1983) 281. —**Ic.**: Fl. Kirgiz. 6, Plate 7, fig. 1; Fl. Kazakhst. 4, Plate 6, fig. 2; Ikonnik. Plate 12, fig. 5-7; Fl. Sin. 27, tab. 34, fig. 1-2.

Described from Jung. Ala Tau. Lectotype in St.-Petersburg (LE).

On rocky slopes in alpine and subalpine belts.

IIA. Junggar: ***Tarb.*** (south. slope of Saur mountain range, valley of Karagaitu river, Bain-Tsagan creek valley, June 23, 1957—Kuan), ***Jung. Alat.*** (on Kegen river, 1800 m, July 23, 1878—Fet.; upper Borotal, 3000 m, Aug.; Junggar Ala Tau, 2100-2400 m, Aug.—1878, A. Reg.; in Ven'tsyuan' town region, No. 2027, No. 4629, Aug. 25, 1957 —Kuan), ***Tien Shan*** (Sairam, Talki-bash, July 19; Dzhagastai, 1500-2100 m, Aug. 9—1877, A. Reg.; Kyzemchek mountains near Sairam lake, 2100-2700 m, July 31; Kokkamyr upland, 2400-2700 m, July 31; Chubaty pass, 2400-2700 m; on Urtaksary river, 1200-2100 m, Aug. 20—1878, A. Reg.; Shara baguchi, 1878—Fet.; on Ladyk river, 1878 —Larionov; Nilki gorge (Kash river), June 30; Irenkhabarga mountains, 3000-3400 m, July 4; same site, Monguto, 2700 m, Aug. 2; Kash river, 2700 m, July 15; Aryslyn, 2700 m, July 17; Kash, 2700 m, Aug. 17-18—1879, A. Reg.; Bol. Yuldus, 2400 m, Aug. 5, 1893--Rob.; Kaundu Tal, end of Aug. 1903; Tal Khaptn-su, Sept. 1-5, 1907; bei Zagan Ussun, 3500 m, Sept. 21, 1908—Merzb.; nor. slopes of Talkinsk mountain range, on south-east. bank of Sairam-nor lake, 2150 m, Aug. 18, 1953—Mois.).

General distribution: Jung.-Tarb.; Nor. and Cent. Tien Shan; Mid. Asia, Fore Asia (Nor.-West. Pakistan), Himalayas (Kashmir).

13. **A. sinchiangense** W.T. Wang in Fl. Sin. 27 (1979) 605, 182; Claves pl. Xinjiang, 2 (1983) 281. —**Ic.**: Fl. Sin. 27, tab. 34, fig. 7-8.

Described from Junggar. Type in Beijing (PE).

Along grassy slopes.

IIA. Junggar: ***Tien Shan*** ("Sinchiang: Hoshuo, 2620 m, No. 7080, Aug. 5, 1958, typus"—Wang, l.c.).

General distribution: endemic.

14. **A. soongaricum** Stapf in Ann. Bot. Gard. (Calcutta) 10 (1905) 141; Shteinb. in Fl. SSSR, 7 (1937) 232; Fl. Kirgiz. 6 (1955) 63; Fl. Kazakhst. 4 (1961) 54; Pakhom. in Opred. rast. Sr. Azii [Key to Plants of Mid. Asia] 3 (1972) 185; Fl. Intramong. 2 (1978) 224; Fl. Sin. 27 (1979) 308; Claves pl. Xinjiang, 2 (1983) 281. —*A. napellus* β. *alpinum* f. *soongaricum* Regel in Bull. Soc. natur. Moscou, 34 (1861) 105. —**Ic.**: Stapf, pl. 95; Fl. Kazakhst. 4, Plate 7, fig. 6; Fl. Intramong. 2, tab. 118, fig. 5-11; Fl. Sin. 27, tab. 72, fig. 5-6 and tab. 74, fig. 4.

Described from Pakistan (Gilgit) and Turkestan. Type in Kolkata (Calcutta) (CAL). Isotypes in St.-Petersburg (LE) and London (K).

On forest borders, subalpine forb meadows.

IIA. Junggar: *Cis-Alt.* ("Altay"—Claves pl. Xinjiang), ***Tien Shan*** (Modun-obo river, nor. slope of Dzhair mountain range, 1400 m, Aug. 3, 1953—Mois.; "Tien Shan, Barkul'"—Claves pl. Xinjiang).

General distribution: Jung.-Tarb., Nor. Tien Shan; Mid. Asia (Talas river valley), West. Sib., Himalayas (Kashmir).

15. **A. talassicum** M. Pop. in M. Pop. et Andross. Rastit. Zapov. Guralash. i Zaaminskogo lesnich. [Vegetation of Guralash Nature Reserve and Trans-Aminsk Forestry] (1936) 8; Shteinb. in Fl. SSSR, 7 (1937) 231; Fl. Kirgiz. 6 (1955) 60; Fl. Kazakhst. 4 (1961) 56; Pakhom. in Opred. rast. Sr. Azii [Key to Plants of Mid. Asia] 3 (1972) 186. —*A. talassicum* var. *villosulum* W.T. Wang in Fl. Sin. 27 (1979) 610, 312; Claves pl. Xinjiang, 2 (1983) 282. —**Ic.:** Fl. Kirgiz. 6, Plate 7, fig. 2; Fl. Kazakhst. 4, Plate 7, fig. 5; Fl. Sin. 27, tab. 72, fig. 8-9 (var. *villosulum*).

Described from Tien Shan (Talass Ala Tau). Type in Alma Ata (AA).

In moist meadows in subalpine zone, river valleys, juniper thickets.

IIA. Junggar: ***Dzhark.*** ("Ili region"—Claves pl. Xinjiang).

General distribution: Nor. Tien Shan; Mid. Asia (mountainous).

16. **A. tanguticum** (Maxim.) Stapf in Ann. Bot. Gard. (Calcutta) 10 (1905) 151, in adnot.; Fl. Sin. 27 (1979) 182; Fl. Xizang. 2 (1985) 22. —*A. rotundifolium* var. *tanguticum* Maxim. Fl. Tangut. 1 (1889) 26; Grum-Grzh. Opis. put. Zap. Kit. [Report of Expedition in West. China] 3 (1907) 476; Finet et Gagnep. in Bull. Soc. Bot. Fr. 51 (1904) 516. —*A. rotundifolium* var. *trichocarpum* Hand.-Mazz. in Acta Horti Gotoburg. 13 (1939) 91; Fl. Xizang. l.c.: 22. —**Ic.:** Fl. Sin. 27, tab. 33, fig. 1-4; Fl. Xizang. tab. 5, fig. 1-2.

Described from Qinghai (Nanshan) and Tibet (Weitzan). Syntypes in St.-Petersburg (LE). Plate III, fig. 3.

In alpine meadows, scrubs, rock talus.

IIIA. Qinghai: ***Nanshan*** (Regio alpina jugi a fl. Tetung S versus, No. 307, July 18/30, syntypus !; terra Tangutorum (prov. Kansu), in sylvis densis [Rangkhta-gol], July 27/ Aug. 8, syntypus !; pratis alpinis jugi a fl. Tetung boream versus siti, Aug. 6/18, syntypus !—1872; jugum a Nanshan ad Don-Kyru ad fl. Rakogol, 10,000-11,000 ft [3000-3400 m], pratis alpinis in humo frequens, No. 509, July 7/19, syntypus !; declivitas N jugi a fl. Tetung, reg. alp. 10,000-12,000 ft [3000-3700 m], in umbria rupium frequens, No. 638, July 19/31—1880, Przew., syntypus !; Babo-kho river, Aug. 9, 1890—Gr.-Grzh.; vicinity of Dan'gertin town, 2700 m, Aug. 26, 1901—Lad.; Mon'yuan', glacial moraine in Ganshig river sources, tributary of Peishikhe river, 3900-4300 m, Aug. 18, 1958—Dolgushin).

IIIB. Tibet: ***Weitzan*** (Regio alpina lateris S jugi Burchan-Budda, 13,000-15,500 ft m.s.l. [4000-4700 m] pratis alpinis, frequens, No. 391, Aug. 1/13, 1884—Przew., syntypus et isosyntypus !; Golubaya river basin, nor. descent of Chamudug-la pass, 4800 m, July 26, 1900; Burkhan-Budda mountain range, Khatu gorge, 4000 m, July 17, 1901—Lad.; "Bizhu"—Fl. Xizang. l.c.)

General distribution: West. Sib., China (Nor.-West., South-West.).

Note. Type material is a hybrid in the structure of inflorescence.

17. **A. turczaninowii** Worosch. in Byull. Glavn. bot. sada AN SSSR, 64 (1967) 36; Voroshilov in Byull. Mosk. obshch. isp. prir., otd. biol. 93, 1

(1988) 90; Gub. Konsp. Fl. Vneshn. Mong. [Conspectus of Flora of Outer Mongolia] (1996) 48. —*A. baicalense* (Regel) Turcz. ex Steinb. in Fl. URSS, 7 (1937) 223, 732, non Turcz. ex Rapaics, 1907.

Described from East. Siberia (Trans-Baikal). Type in St.-Petersburg (LE).

In forest belt and lower alpine sections.

IA. Mongolia: ***East. Mong.*** (Dariganga, Shiliin-Bogdo-Ula volcano, nor. slope, Aug. 11, 1970—Grub., Ulzij. et al.).

General distribution: East. Sib., Nor. Mong. (Hent., Hang., Mong.-Daur., Cis-Hing.), China (Nor.).

Subgenus 2. **Anthora** (DC.) Peterm.

18. **A. anthoroideum** DC. Syst. 1 (1817) 366; Fl. Kazakhst. 4 (1961) 53; Pakhom. in Opred. rast. Sr. Azii [Key to Plants of Mid. Asia] 3 (1972) 185; Fl. Sin. 27 (1979) 323; Claves pl. Xinjiang, 2 (1983) 283; Grub. Opred. sosud. rast. Mong. [Key to Vascular Plants of Mongolia] (1982) 110; Gub. Konsp. Fl. Vneshn. Mong. [Conspectus of Flora of Outer Mongolia] (1996) 47. —*A. anthora* L. β *anthoroideum* Regel, Pl. Radd. 1 (1861) 71, in adnot., Maxim. Fl. Tangut. 1 (1889) 25; Maxim. Enum. pl. Mong. 1 (1889) 29. —*A. anthora* auct. non L.: Finet et Gagnep. in Bull. Soc. Bot. Fr. 51 (1904) 513; Kryl. Fl. Zap. Sib. 5 (1931) 1147, p.p.; Shteinb. in Fl. SSSR, 7 (1937) 190; p.p. —**Ic.:** Fl. Kazakhst. 4, Plate 7, fig. 3; Fl. Sin. 27, tab. 74, fig. 10-11, tab. 76, fig. 1-3.

Described from Siberia. Type in Geneva (G).

In meadows and slopes, at 1400-2000 m alt.

IA. Mongolia: ***Khobd.*** (upper Oigura river, alpine tundra, July 9, 1905—Sap.), ***Mong. Alt.*** (south. Altay, Pevtsov; "Mong. Alt."—Gub. l.c.).

IIA. Junggar: ***Cis-Alt.*** (on Kongoity river, tributary of Chern. Irtysh, Sept. 18, 1876—Pot.; Altay, No. 3323, Aug. 6, 1956—Ching.), ***Tarb.*** (north of Chuguchak [= Dachen] town, No. 1517 and No. 2905, Aug. 12; Dachen, No. 1614, Aug. 13—1957, Kuan), ***Jung. Alat.*** (Chubaty pass (Sairam), 2400-2700 m, Aug. 2, 1878—A. Reg.), ***Tien Shan*** (Sairam, July 20; Talki gorge, 1800-2100 m, July 25—1877; Kokkamyr upland, 2100 m, July 29, 1878—A. Reg.), ***Dzhark.*** (Piluchi river, June 21, 1878—Fet.).

IIIA. Qinghai: ***Nanshan*** (South. Tetung mountain range, July 25, 1872; same site, July 30, 1880; North. Tetung mountain range, alpine belt, 3000-3700 m, July 31, 1880—Przew.; Yamatyn-umru-ula mountains, 3600-3900 m, June 13, 1894—Rob.), ***Amdo*** (Kadiger monastery, May 25, 1885—Pot.).

IIIB. Tibet: ***Weitzan*** (on Bychyu river, Yangtse tributary, June 25; July 20; south. bank of Russkoe lake, 4100 m, July 28—1884, Przew.; Bychyu river, 4300 m, July 11; Yangtse river basin, Nko-gun brook, Rkhombo-mtso lake, 4000 m, Aug. 6—1900, Lad.).

General distribution: Jung.-Tarb.; West. Sib., East. Sib., Caucasus, Europe (East.), Nor. Mongolia (Hangay).

19. **A. gilvum** (Maxim.) Hand.-Mazz. in Acta Horti Gotoburg. 13 (1939) 87. —*A. anthora* var. *gilvum* Maxim. Fl. Tangut. 1 (1889) 25; Borodina-Grabovskaya in Catalogue Type Specim. Centr. Asia (LE) (2000) 189.

—*A. flavum* Hand.-Mazz. l.c. (1939) 86; Fl. Intramong. 2 (1978) 226; Fl. Xizang. 2 (1985) 36; Fl. Sin. 27 (1979) 318. —**Ic.:** Fl. Intramong. 2, tab. 115, fig. 10-14; Fl. Sin. 27, tab. 74, fig. 9, tab. 75, fig. 1-5.

Described from Nanshan. Lectotype and syntypes in St.-Petersburg (LE).

In alpine meadows.

IIIA. Qinghai: *Nanshan* (silvis frondosis regionis inferioris alpinum N a fl. Tetung, Aug. 1872—Przew., lectotypus ! (Borodina-Grabovskaya, l.c.); decliv. N jugi S a fl. Tetung. reg. alpina, 10,000-12,000 ft [3000-3700 m], July 20/Aug. 2—syntypus !; jugum boreale finitimum, Aug. 8/20—1880, Przew., syntypus !; vicinity of Dan'-gertin town, 2700 m, Aug. 22, 1901—Lad.).

IIIB. Tibet: *Chang Tang* ("Ban'ge"—Fl. Xizang. l.c.), *Weitzan* Oppid. Dajigu et traject. jugi Nan-shan septentrionem versus Bychy (Goluboi and Bychyu river basins, 4300 m, July 11, 1900—Lad., "An'do"—Fl. Xizang. l.c.), *South.* ("Mochzhugunka"—Fl. Xizang. l.c.).

General distribution: China (South-West.).

20. **A. ludlowii** Exell. in J. Bot. (London) 64 (1926) 218; Fl. Sin. 27 (1979) 321; Fl. Xizang. 2 (1985) 37. —**Ic.:** Fl. Sin. 27, tab. 76, fig. 12-13, Fl. Xizang. 2, tab. 9, fig. 4 (fl.).

Described from South. Tibet (Gyantze). Type in London (BM). In alpine meadows.

IIIB. Tibet: *South.* ("Tszyantszy, Lankatsza, Naidun, Luntsza, Nan'mulin"—Fl. Xizang. l.c.).

General distribution: Himalayas.

21. **A. naviculare** (Bruhl) Stapf in Ann. Bot. Gard. (Calcutta) 10 (1905) 154; Tamura in Acta Phytotax. Geobot. (Kyoto) 15, 6 (1954) 192; Fl. Sin. 27 (1979) 186; Fl. Xizang. 2 (1985) 22. —*A. ferox* var. *naviculare* Bruhl in Ann. Bot. Gard. (Calcutta) 5 (1896) 11, pl. 111, fig. 2 and 14. —**Ic.:** Stapf. pl. 101; Fl. Sin. 27, tab. 34, fig. 3-4; Fl. Xizang. 2, fig. 3-4.

Described from Himalayas. Type? sketch in first description.

In alpine belt.

Tibet: *South.* ("Lhasa"—Tamura, l.c.; "Lhasa, Naidun, Mochzhugunka"—Fl. Xizang. l.c.).

General distribution: China (North.-West., South-West.), Himalayas (east.).

22. **A. pendulum** N. Busch in Bull. Jard. Bot. St. Pétersb. 5, 4 (1905) 135; Fl. Sin. 27 (1979) 319; Fl. Xizang. 2 (1985) 36. —**Ic.:** Fl. Sin. 27, tab. 74, fig. 7-8 and tab. 75, fig. 6-9; Fl. Xizang. 2, tab. 9, fig. 5-7.

Described from Tibet. Type in St.-Petersburg (LE).

In alpine meadows, among rocks.

IIIA. Qinghai: ("Qinghai"—Fl. Xizang. l.c.).

IIIB. Tibet: *Weitzan* (Semina cl. Ladygin lecta, typus et isotypus !; Bizhu, Sosyan', An'do"—Fl. Xizang. l.c.).

General distribution: China (Nor., Nor.-West., Cent., South-West.), Himalayas.

Note. Described from Tibet from V.F. Ladygin's collection without exact location of find.

Subgenus 3. **Gymnaconitum** (Stapf) Rapaics

23. **A. gymnandrum** Maxim. in Bull. Ac. Sci. St. Pétersb. 23 (1877) 308; Forbes and Hemsl. Index Fl. Sin. 1 (1886) 20; Maxim. Fl. Tangut. 1 (1889) 26; Finet et Gagnep. in Bull. Soc. bot. Fr. 51 (1904) 500; Stapf in Ann. Bot. Gard. (Calcutta) 10 (1905) 178; Gr.-Grzh. Opis. put. Zap. Kit. [Report of Expedition to West. China] 3 (1907) 476; Hao in Bot. Jahrb. 68 (1938) 595; Tamura in Acta Phytotax. Geobot. 15, 6 (1954) 192; Svenson in Brittonia, 8, 1 (1954); 55 Fl. Sin. 27 (1979) 324; Fl. Xizang. 2 (1985) 38; Borodina-Grabovskaya in Catalogue Type Specim. Centr. Asia (LE) (2000) 189. —**Ic.:** Fl. Tangut. 1, tab. 6, fig. 1-6; Fl. Sin. 27, tab. 77, fig. 1-3; Fl. Xizang. 2, tab. 10, fig. 1-2

Described from Qinghai (Nanshan). Lectotype and syntype in St.-Petersburg (LE). Plate III, fig. 1.

In alpine meadows, meadowy gorge slopes, alpine cereal grass-forb steppe.

IIIA. Qinghai: ***Nanshan*** (in agris circa templem Tchöibsen frequens, No. 233, July 3/15, 1872, Przew.—syntypus !; declivio australi jugi a fl. Tetung meridiem versus, July 14/26, 1872, Przew.—lectotypus ! (Borodina-Grabovskaya, l.c.); upland plain (3000 m) near Kuku Nor lake, June 19; between Nanshan and Donkyru mountains, on Rakogol river, 3000-3400 m, July 22; North. Tetung mountain range, 3000 m, Aug. 17—1880, Przew.; on Itel'-Gol river [Sanchuansk valley] [April] 1885—Sambarcha; south. bank of Kuku Nor, July 24, 1890—Gr.-Grzh.; south bank of Kuku Nor lake, 3000 m, Aug. 18, 1901—Lad.; Kukunor lake, Sasygyn area, Aug. 12, 1908—Czet.; "Kokonor: Sining, bei Schang-wu-chuang Hao"—Hao, l.c.; Mon'yuan', Ganshiga river valley, left tributary of Peishikha river, 3350-3720 m [Aug. 8] 1958; 24 km south of Xining, Aug. 4; 89 km west of Xining, 3450 m, Aug. 5—1959, Petr.).

IIIB. Tibet: ***Weitzan*** (Golubaya river basin, Khichu river, 4400 m, July 13, Kabzhikamba village on Khichu river, 40000 m; upper Ichyu river, 4000 m, Aug. 1—1901, Lad.; "Tszyanda, Gunbutszyanda"—Fl. Xizang. l.c.), ***South.*** ("around Lhasa, Da and Paal villages, Aug. 23; Khomo-sese town, Sept. 6"—Tamura, l.c., "between Darjeeling and Lhasa"—Svenson, l.c.; "Lyuchzhugunka"—Fl. Xizang. l.c.").

General distribution: China (South-West., Nor.-West.), Himalayas (east.).

Subgenus 4. **Lycoctonum** (DC.) Peterm.

24. **A. apetalum** (Huth) B. Fedtsch. in Fl. URSS, 7 (1937) 200; Claves pl. Xinjiang, 2 (1983) 279. —*Delphinium apetalum* Huth in Bot. Jahrb. 20 (1895) 398; Fl. Kazakhst. 4 (1961) 50; Fl. Sin. 27 (1979) 160. —*Aconitum monticola* Steinb. in Fl. URSS, 7 (1937) 730, 209; Fl. Kazakhst. 4 (1961) 52; Pakhom. in Opred. rast. Sr. Azii [Key to Plants of Central Asia] 3 (1972) 188; Fl. Sin. 27

(1979) 173; Claves pl. Xinjiang, 2 (1983) 280. —**Ic.:** Fl. SSSR, 7, Plate 13, fig. 5; Fl. Kazakhst. 4, Plate 6, fig. 1; Fl. Sin. 27, tab. 24, fig. 6-7 (sub *A. monticola*), tab. 26, fig. 1-3.

Described from Junggar (Kul'dzha). Type in St.-Petersburg (LE). Plate III, fig. 2. Map 1.

In undergrowth of spruce forest, on borders, meadowy slopes in alpine belt, 1700-1900 m.

IIA. Junggar: ***Tien Shan*** (at Chinese picket [Aug. 11] 1875—Larionov; Issykul'-Muzart foothills, 1600-1800 m [July] 1877, A. Reg.; in upper Dzhirgalan [tributary of lower Tekes], Sept. 1878—Fet.; Kumdaban, 2700-3000 m, May 30, 1879—A. Reg.; from Borgata to Khapchagai, from northern side of Kash, 1800-2100 m, July 6; Khapchagai-Aryslyn, 1800-2100 m, July 7; Monguto, Irenkhabirga mountains, 2700 m, July 9; Ausgang der Aryslyn-schlucht, 8000-9000 ft [2400-2700 m], July 10—holotypus !; Arystyn foothills, 1800-2400 m and 2400-2700 m, July 19; Arystandaban, Kunges, 2700 m, Aug. 22—1879, A. Reg.; Kapsalan mountains, Kensu spring, 1800 m, June 16, 1893—Rob.; Kul'dzha, Ili region, Gunlyu"—Claves pl. Xinjiang, l.c.).

General distribution: Jung.-Tarb.; West.Sib. (Altay).

25. **A. barbatum** Patr. ex Pers. Synops. Pl. 2 (1807) 83; Maxim. Enum. pl. Mong. 1 (1889) 30; Kryl. Fl. Zap. Sib. 5 (1931) 1153; Shteinb. in Fl. SSSR, 7 (1937) 204; Kitag. Lin. Fl. Mansh. (1939) 20; Fl. Kazakhst. 4 (1961) 52; Fl. Sin. 27 (1979) 176 (var. *hispidum*); Grub. Opred. rast. Mong. [Key to Plants of Mongolia] (1982) 110; Claves pl. Xinjiang, 2 (1983) 280 (var. *hispidum* (DC.) DC. et var. *puberulum* Ledeb.); Fl. Intramong. ed. 2, 2 (1990) 549 (var. *hispidum*); Gub. Konsp. Fl. Vneshn. Mong. [Conspectus of Flora of Outer Mongolia] (1996) 47. —*A. hispidum* DC. Syst. 1 (1817) 367. —*A. barbatum* var. *hispidum* (DC.) DC. Prodr. 1 (1824) 58. —*A. barbatum* var. *puberulum* Ledeb. Fl. Ross. 1, 1 (1841) 67. —**Ic.:** Fl. Kazakhst. 4, Plate 6, fig. 2; Fl. Intramong. 2, tab. 115, fig. 1-9; Fl. Sin. 27, tab. 31, fig. 1-3, 4-8 (var. *puberulum*); Grub. Plate 51, fig. 235; Fl. Intramong. ed. 2, l.c., tab. 222, fig. 1-4.

Described from Siberia. Type in London (K).

In meadowy steppes, forest borders and sparse birch and larch groves, rubbly and rocky slopes.

IIIA. Mongolia: ***Khobd.*** (Turgen' mountain rnage, Turgen'-Gola valley, 5 and 7 km from estuary, July 16 and 17, 1971—Grub., Ulzij. et al.; 37 km north-west of Ulangom, east. macroslope of Mong. Altay, July 5, 1977—Karam., Sanczir et al.), ***Mong. Alt.*** (Taishiri-Ula mountain range, nor. slope, July 15, 1877—Pot.; 8 km south-east of Tszazaktu-khana site, Aug. 9; Khara-Tszarga mountains, Sakhir-Sala river valley, Aug. 24; same site, Khairkhan-Duru river, Aug. 26—1930, Pob.; Taishiri-Ula mountain range, nor. slope, July 11 and 12, 1945—Yun.), ***Cent. Khalkha*** (Kharukhe river sources (Ara-Dzhargalanta river), Uste town, Aug. 12, 1925—Krasch. and Zam.), ***East. Mong.*** (Mong. Chin., 1841—Kirilow), ***Depr. Lakes*** (Ubsa lake region, Ulan-Natsyn river valley, July 15, 1879—Pot.; 2-3 km west of Numurlyk somon, July 20, 1945—Yun.), ***Val. Lakes*** (Laman-Gegen monastery, July 17, 1926—Tug.), ***Gobi-Alt.*** (Dundu-Saikhan mountains, nor. slope, upper belt, July 4, 1909—Czet.; Dundu-Saikhan mountain range, meadow on nor. slope, July 30, 1956—Davazamc and A. Yakshina; Dundu-Saikhan mountain range, south. slope, under main peak at 2825 m, July 23, 1970—Grub. et al.).

IIA. Junggar: ***Tien Shan*** ("Balishen' (Barkul')"—(var. *hispidum*), "nor. Sinkiang"—(var. *puberulum*)—Claves pl. Xinjiang, l.c.).

General distribution: Europe (East.), West. Sib., East. Sib., Far East, Nor. Mong. (Fore Hubs., Hent., Hang., Mong.-Daur.), China (Dunbei, Nor.-West., Cent.), Japan.

26. **A. leucostomum** Worosch. in Byull. Glavn. bot. sada AN SSSR, 11 (1952) 62; Fl. Kazakhst. 4 (1961) 49; Fl. Kirgiz. dop. 1 (1967) 69; Pakhom. in Opred. rast. Sr. Azii [Key to Plants of Mid. Asia] 3 (1972) 187; Fl. Sin. 27 (1979) 171; Claves pl. Xinjiang, 2 (1983) 279; Gub. Konsp. Fl. Vneshn. Mong. [Conspectus of Flora of Outer Mongolia] (1996) 48. —*A. septentrionale* auct. non Koelle; Grub. Opred. rast. Mong. [Key to Plants of Mongolia] (1982) 110; Gub. l.c. 48 —*A. lycoctonum* β. *septentrionale* auct. non Ser. in DC.: Maxim. Enum. pl. Mong. 1 (1889) 29. —*A. lycoctonum* γ. *vulgaria* auct. non Regel: Franch. Pl. David. 1 (1884) 23; Maxim. Fl. Tangut. 1 (1889) 26; id. Enum. pl. Mong. 1 (1889) 29. —*A. lycoctonum* auct. non L.: Finet et Gagnep. in Bull. Soc. Bot. Fr. 51 (1904) 501. —*A. excelsum* auct. non Reichb. —**Ic.:** Byull. Glavn. bot. sada AN SSSR, 11, fig. 1 a, b, c; Fl. Sin. 27, tab. 29, fig. 1-3.

Described from Jung. Ala Tau. Type in Moscow (MW).

On mountains, in forest and subalpine meadows, forest borders, 1400-2500 m alt.

IA. Mongolia: ***Khobd.*** (20 km west-south-west of Ulangom, Mukhur-Ulyastu-Gol gorge, July 6, 1977—Karam., Sanczir et al.), ***Mong. Alt.*** (25-30 km north of Koktogoi, right bank of Kairta river, Kuidyn valley, July 15, 1959—Yun. and Yuan'; Sanginin-Gol river basin (upper Chern. Irtysh), 25 km south-east of Dayan-Nur post, 1800-2100 m, July 16-17, 1988—Kam., Bud. et al.), ***Cent. Khalkha*** (Grub. l.c.; Gub. l.c.), ***Depr. Lakes*** (Ubsa lake region, Dzusylan gorge, July 13, 1879—Pot.; 32 km south-east of Barun-Tukhun, nor. macroslope of Khan-Khukhei mountain range, 2200 m, July 23, 1973—Banzragch, Karam. et al.).

IIA. Junggar: ***Cis-Alt.*** (Kandagatai, Sept. 15, 1876—Pot.; Koktogoi, No. 2012, Aug. 17 and No. 2069, Aug. 18, 1956, Ching; 25-30 km north of Koktogoi, right bank of Koirta river, Kuidin valley, No. 1420, July 15, 1959—Yun. and Yuan'; in Shinta region, from Altay mountains, No. 10730, July 21, 1959—Kuan.), ***Tarb.*** (Dachen, No. 1630, Aug. 13, 1957—Kuan), ***Jung. Alat.*** (in Toli district, Albakzin mountains, No. 2543, Aug. 6 and No. 2699, Aug. 7; mountains south of Ven'tsyuan' town, No. 1130, Aug. 22; 20 km north of Ulastai, No. 3844, Aug. 28; nor. bank of Sairam lake, No. 4708, Aug. 28; 10 km north of Ulastai, No. 3933, Aug. 29; 15 km north-west of Ven'tsyuan' [= Arasan], No. 1618, Aug. 29—1957, Kuan), ***Tien Shan*** (upper Kunges river, 1200 m, June 24—1877, Przew.; Talki gorge, 1800 m, July 18; south. bank of Sairam lake, 1200 m, July 20, 1877—A. Reg.; Khonakhai mountain, 1500-2100 m, June 16; Chapchal pass, June 28; Sairam-Nor mountain, July 14; Sairam lake, July 23; Bogdo mountain, 2100-2400 m, July 24; Kokkamyr upland, 2700 m [s.d. (without date)] July—1878; Bach Nilki, near Kash, 2100 m, June 8; south. tributary of Karagol, 2700 m, June 16; Borgaty, 1500-1800 m, July 4; between Dzhirgalan and Aryslyn, 2400 m, July 8—1879, A. Reg.; Tekche river, Borokhoro mountain range, July 10, 1899—Korol'kov), ***Depr. Lakes*** (Sairam-Nur, 2 km north-east of Sairam pass, No. 1678, Aug. 19, 1957—Yun. et al.; south. branch of Barkul' lake, mountains, No. 2219, Sept. 27, 1957—Kuan; upper Ili river, Tsanma valley—left tributary of Kunges, No. 397, July 7; same site from Tsanma to Dagdyn-Gol area, No. 489, July 7; Narat mountain in Kunges, No. 6563 and 6566, Aug. 7—1958, Yun and Yuan'), ***Jung. Gobi*** (nor.: Dzhandlyk river valley, Chingil'

[=Tsinkhe] river system, July 28, 1906—Sap.), *Zaisan* (Burchum, No. 3274, Sept. 6, 1956—Ching), *Dzhark.* (Suidun, July 16, 1877—A. Reg.; Urtas-Aksu, Kul'dzha region, June 17; Pilyuchi gorge near Kul'dzha, June 21, 1848—Fet.).

IIIA. Qinghai: ***Nanshan*** (South. Tetung mountain range, July 9, 1872; Yusup-Khatyma river, 2700-3000 m, July 23, 1880—Przew.; Choibsenkhit temple, Aug. [30] 1901—Lad.).

General distribution: Jung.-Tarb., Nor. and Cent. Tien Shan; West. Sib., East. Sib., Mid. Asia, Nor. Mong. (Fore Hubs., Hent., Hang., Mong.-Daur), China (Dunbei, Nor., Nor.-West., Cent., South-West.), Japan.

27. **A. volubile** Pall. ex Koelle, Spic. observ. Aconit. [1787] 21; Finet et Gagnep. in Bull. Soc. Bot. Fr. 53 (1905) 505, p.p.; Kryl. Fl. Zap. Sib. 5 (1931) 1150; Shteinb. in Fl. SSSR, 7 (1937) 213; Kitag. Lin. Fl. Mansh. (1939) 212; Fl. Kazakhst. 4 (1961) 58; Fl. Sin. 27 (1979) 258; Grub. Opred. rast. Mong. [Key to Plants of Mongolia] (1982) 110; Y.Z. Zhao, Acta Sci. Nat. Univ. Intermong. 14 (1983) 221; Kadota, Rev. Acon. Subgen. *Aconitum* East Asia (1987) 91; Fl. Intramong. ed. 2, 2 (1990) 565; Gub. Konsp. Fl. Vneshn. Mong. [Conspectus of Flora of Outer Mongolia] (1996) 48. —*A. volubile* var. *latisectum* Regel, Pl. Radd. 1 (1861) 92, in adnot.; Maxim. Fl. Tangut. 1 (1889) 26. —**Ic.:** Fl. Kazakhst. 4, Plate 7, fig. 1; Fl. Sin. 27, tab. 55; fig. 3-5; Kadota, l.c. fig. 29; Fl. Intramong. ed. 2, l.c. tab. 229, fig. 5-7.

Described from Siberia ("Siberia"). Type in London (BM).

On slopes in forests, among coastal shrubs.

IA. Mongolia: ***Mong. Alt.*** (mountain nor. of Yamant-Ula, July 20, 1977—Karam., Sanczir et al.; "Khara-Adzarga"—Grubov, 1982).

IIIA. Qinghai: ***Nanshan*** (North. Tetung mountain range, nor. slope, Sept. 1, 1872; same site, forest belt, 2400-2700 m, Aug. 14, 1880; South. Tetungsk mountain range, Aug. 7, 1880—Przew.).

General distribution: Europe (East.), West. Sib. (Altay), East. Sib., Far East, Nor. Mong., China (Dunbei), Korean peninsula, Japan.

* * *

Apart from the above-listed species, the following have also been cited in the literature for South. Tibet:

A. dolichostachyum W.T. Wang in Fl. Sin. 27 (1979) 606, 287 ("Gunbutszyanda").

A. polyschistum Hand.-Mazz. in Acta Horti Gotoburg. 13 (1939) 100 ("near Lhasa, mt. Dela et Sharla").

A. pseudokongboense W.T. Wang et L.Q. Li in Acta Bot. Yunnan, 8, 3 (1986) 260 ("Gongbogyamda, 93°30′ E, 30° N").

7. **Delphinium** L.

Sp. pl. (1753) 530

1. Perennial grasses with bulbs; perianth segments and staminodes pale or bright yellow..54. **D. biternatum** Huth.

+ Rhizomatous grasses; perianth segments and staminodes of different colour, from light blue to dark violet, sometimes blackish purple..2.
2. Nectary petals and staminodes dark cinnamon, shorter than perianth segments, dark blue, violet, rarely dark cinnamon in colour..3.
+ Nectary petals and staminodes of same colour as perianth segments (light blue, dark blue, dark purple, etc.)...........................41.
3. Perianth segments dark cinnamon with purple shade........................ ....36. **D. triste** Fisch. ex DC.
+ Perianth segments dark blue or violet...4.
4. Perianth segments and staminodes hollow, poorly differentiated into blade and spur (Section 1. *Aconitoides*)...................................5.
+ Perianth segments and staminodes yellow- or white-haired inside; differentiated into blade and spur..6.
5. Stems and leaves covered (on margin) with yellow glandular hairs.................................1. **D. leiophyllum** (W.T. Wang) W.T. Wang.
+ Stems without yellow glandular hairs....2. **D. trichophorum** Franch.
6. Inflorescence corymbose, usually small or 1-flowered.....................7.
+ Inflorescence racemose...19.
7. Follicles 4 or 5..8.
+ Follicles 3...12.
8. Leaves trifid rather shallowly...9.
+ Leaves trifid for 2/3 or more of their length...................................10.
9. Stems 6-10 cm tall; leaf blade up to 4 cm broad.. ....21. **D. lacostei** Danguy.
+ Stems 40-50 cm tall; leaf blade 6-9.5 cm broad.. ....38. **D. yechengense** C.Y. Yang et B. Wang.
10. Flowers single, terminal; ovaries 4.. ....6. **D. chumulangmaense** W.T. Wang.
+ Flowers 2-6 each in corymbose inflorescence; ovaries 5..................11.
11. Segments of radical leaves petiolate.. ....25. **D. pseudoglaciale** W.T. Wang.
+ Segments of radical leaves sessile. Spur of perianth segment subulate, 2.1-2.4 cm, about 3 mm in diam. at base................................ ....28. **D. pylzowii** Maxim.
12. Leaves trisected rather shallowly...13.
+ Leaves trisected for 2/3 or more of their length................................16.
13. Spur of perianth segments conical or tubular-conical, 7-12 mm in diam. at base.............................5. **D. chrysotrichum** Finet et Gagnep.
+ Spur of perianth segments terete or subulate, 2-4 (4.5) mm in diam. at base...14.
14. Bracts contacting or subcontacting with flower.................................... ....3. **D. albocoeruleum** Maxim.

+	Bracts at some distance from flower	15.
15.	Sepals 2-2.5 cm ... 26. **D. pseudopulcherrimum** W.T. Wang.	
+	Sepals 1.5-1.7 cm ... 31. **D. siwanense** Franch.	
16.	Bracteoles leaflike, 3-lobate ... 7. **D. chungbaense** W.T. Wang.	
+	Bracteoles entire, linear or lanceolate	17.
17.	Flowers single	18.
+	Flowers in 2-7-flowered corymbose inflorescence. Lower sepals obovoid, orbicular at tip ... 26. **D. pseudopulcherrimum** W.T. Wang.	
18.	Spur terete-subulate or subulate, 1.7-2.6 cm ... 4. **D. candelabrum** Ostenf.	
+	Spur terete, 1-1.3 cm ... 32. **D. tangkulaense** W.T. Wang.	
19(6)	Raceme compact, compact-flowered, short-terete, terete or globose	20.
+	Raceme more or less lax	22.
20.	Leaves trifid almost down to base, with segments also deeply incised ... 27. **D. pulanense** W.T. Wang.	
+	Leaves trisected, with lobes separated into segments	21.
21.	Raceme with 20-40 flowers, short-terete ... 10. **D. densiflorum** Duthie ex Huth.	
+	Raceme with 10-20 flowers, terete ... 16. **D. inconspicuum** Serg.	
22.	Spur sacciform or conical, 5-7 mm, roughly 1/4 of sepal ... 34. **D. tetragynum** W.T. Wang.	
+	Spur subulate or terete-subulate, as long as or somewhat longer than sepal, rarely slightly shorter	23.
23.	Leaves radical or 1 or 2 cauline leaves present, gathered at stem base	24.
+	Leaves radical as well as cauline; stem uniformly foliate	28.
24.	Bracts entire	25.
+	Bracts laciniated ... 11. **D. dissectum** Huth.	
25.	Bracteoles with 1.5-2.2 mm long hairs along margin ... 15. **D. iliense** Huth.	
+	Bracteoles with about 0.8 mm long hairs along margin	26.
26.	Raceme about 4 cm long, with 2 or 3 flowers; rachis with yellow glandular hairs; bracteoles away from flower; spur of sepal terete, subulate, about 1.3 cm ... 18. **D. kaschgaricum** C.Y. Yang et B. Wang.	
+	Raceme more elongated and many-flowered	27.
27.	Flowers not appressed to rachis. Inflorescence glabrous. Lobes of leaf blade sharp-tipped ... 8. **D. crassifolium** Schrad. ex Ledeb.	
+	Flowers appressed to rachis. Inflorescence with pubescence of glandular hairs. Lobes of leaf blade blunt-tipped ... 17. **D. kansuense** W.T. Wang.	

28. Bracteoles carinate-ovoid....................23. **D. naviculare** W.T. Wang.
+ Bracteoles flat, not carinate....................29.
29. Pedicels glabrous....................30.
+ Pedicels pubescent....................32.
30. Bracts right at base of flower...........19. **D. korshinskyanum** Nevski.
+ Bracts away from flower....................31.
31. Bracts narrow-lanceolate or linear-oblanceolate, long-ciliate along margin....................15. **D. iliense** Huth.
+ Bracts narrow-linear, glabrous....................30. **D. sinoelatum** C.Y. Yang et B. Wang.
32. Bracts 2-3 mm broad....................33.
+ Bracts 1 mm broad....................36.
33. Stems covered compactly with yellow glandular hairs....................14. **D. hillcoatiae** Munz.
+ Stems without glandular hairs....................34.
34. Bracts away from flower. Stem branched; peduncles up to 2.7 cm long; spur of sepal somewhat longer than sepal....................13. **D. ellipticoovatum** W.T. Wang.
+ Bracts right at base of flower....................35.
35. Spur 1.4-1.5 cm long, 3-4 mm thick. Pedicels 2.5-6 cm long....................9. **D. dasyanthum** Kar. et Kir.
+ Spur 1.4-2 cm long, about 2.5 mm thick. Pedicels 1-2.5 cm long....................22. **D. mirabile** Serg.
36. Sepals without glandular hairs on outer surface....................37.
+ Sepals with glandular hairs on outer surface....................38.
37. Ovaries, pedicels and sepals compact-bristly on outer surface....................35. **D. tianschanicum** W.T. Wang.
+ Ovaries glabrous; pedicels and sepals short-pubescent on outer surface....................12. **D. eglandulosum** C.Y. Yang et B. Wang.
38. Leaf blade with long-cuspidate tips of lobes....................29. **D. shawurense** W.T. Wang.
+ Leaf blade with short-cuspidate tips of lobes....................39.
39. Bracts right at base of flower. Ovaries compact-pubescent....................33. **D. tarbagataicum** C.Y. Yang et B. Wang.
+ Bracts away from flower....................40.
40. Spur as long as or slightly shorter than sepal....................37. **D. wuqiaense** W.T. Wang.
+ Spur longer than sepal....................20. **D. kunlunshanicum** C.Y. Yang et B. Wang.
41(2). Leaf blade tripartite or sometimes trisected .. ..42.
+ Leaf blade 3-palmatisected down to base....................50.
42. Leaf lobes overlapping; sepals with distinct nervation; spur about 1/4 of sepal, sacciform or conical, 6-7 mm in diam. at base....................

....24. **D. nordhagenii** Wendelbo.
+ Leaf lobes not overlapping; sepals without distinct nervation; spur terete or subulate, longer or slightly shorter than sepal, less than 4 mm in diam. at base..43.
43. Ovaries 4 or 5 (6)................................47. **D. kingianum** Bruhl ex Huth.
+ Ovaries 3..44.
44. Spur shorter than or nearly as long as sepal.......................................45.
+ Spur longer than sepal, rarely nearly as long.......................................46.
45. Lower part of stem more or less covered with patent and downturned white setaceous hairs; stem glabrous above and compact-pubescent with short glandular hairs only under inflorescence. Bracteoles with compact glandular pubescence. Leaf blades glabrous..39. **D. aemulans** Nevski.
+ Stem and bracteoles glabrous. Leaf blades with glandular pubescence..............................40. **D. barlykense** Lomonosova et Khan.
46. Bracts right, or almost right, at base of flower......................................
..............................50 **D. pseudocyananthum** C.Y. Yang et B. Wang.
+ Bracts away from flower..47.
47. Leaf blade tripartite to 5 mm depth from base.......................................48.
+ Leaf blade trisected for more than 5 mm above base.......................
....44. **D. gyalanum** Marq. et Airy Shaw.
48. Pedicels pubescent..49.
+ Pedicels glabrous..............................42. **D. cheilanthum** Fisch. ex DC.
49. Rachis glabrous..............................49. **D. obcordatilimbum** W.T. Wang.
+ Rachis pubescent..................45. **D. hsinganense** S.F. Li et Z.F. Fang.
50(41). Primary leaf segments lobed; segments of higher order ovoid or narrow-ovoid..51.
+ Primary leaf segments partite or laciniated; segments of higher order lanceolate or linear..54.
51. Inflorescence corymbose or 1-flowered.....4. **D. candelabrum** Ostenf.
+ Inflorescence racemose..52.
52. Nectary blade entire........................47. **D. kingianum** Bruhl ex Huth.
+ Nectary blade bilobed to bipartite..53.
53. Bracts at very base of flower..
........................51. **D. sinoamabile** N.I. Malyutin ex Borod.-Grabovsk.
+ Bracts away from flower. Pedicel glabrous..
....53. **D. sparsiflorum** Maxim.
54. Inflorescence corymbose; somewhat umbellate or 1-flowered at least in upper part..55.
+ Inflorescence racemose..57.
55. Ovaries 5...41. **D. coeruleum** Jacq. ex Cambess.
+ Ovaries 3..56.
56. Peduncles with diffuse pubescence...............43. **D. grandiflorum** L.

\+ Peduncles with compact pubescence. Entire plant with white pubescence. Sepals 1.6-1.7 cm long....48. **D. mollipilum** W.T. Wang.

57\. Leaves glabrous; flowers more or less concealed; blade of nectary petal without hairs..52. **D. souliei** Franch.

\+ Leaves pubescent; flowers more or less open; blade of nectary petal with yellow hairs..58.

58\. Bracts right, or almost right, at base of flower...39. **D. aemulans** Nevski.

\+ Bracts away from flower..59.

59\. Rachis of racemose inflorescence glabrous...49. **D. obcordatilimbum** W.T. Wang.

\+ Rachis of racemose inflorescence with diffuse pubescence...46. **D. kamaonense** Huth.

Subgenus **Delphinastrum** (DC.) Peterm.

Section 1. **Aconitoides** W.T. Wang

1\. **D. leiophyllum** (W.T. Wang) W.T. Wang in Bull. Bot. North-East Forest. Inst. 8 (1980) 23; Fl. Xizang. 2 (1985) 43. —*D. forrestii* Diels var. *leiophyllum* W.T. Wang in Fl. Sin. 27 (1979) 611, 352. —**Ic.:** Fl. Xizang. 2, tab. 11, fig. 1-3.

Described from South. Tibet (Linchou). Type in Shaanxi (NWFC).

On rubbly slopes, among shrubs, 4500-4800 m alt.

IIIB. Tibet: ***Weitzan*** ("Tszyachi (Dachu)"—Fl. Xizang. l.c.), ***South.*** ("Tibet, Linchou, 4300-4500 m, No. 1910, Oct. 3, 1972"—W.T. Wang, 1980, l.c.).

General distribution: endemic.

2\. **D. trichophorum** Franch. in Bull. Soc. Philom. Paris, sér. 8, 5 (1893) 166; Huth in Bot. Jahrb. 20 (1895) 410; Finet et Gagnep. in Bull. Soc. Bot. Fr. 51 (1904) 494; Munz in J. Arn. Arb. 48 (1967) 492; Fl. Sin. 27 (1979) 354; Fl. Xizang. 2 (1985) 43. —**Ic.:** J. Arn. Arb. 48, fig. 7 D, E; Fl. Xizang. 2, tab. 10, fig. 3-4; Fl. Sin. 27, tab. 80, fig. 1-6.

Described from South-West. China. Type in Paris (P).

In alpine meadows, among grasses and shrubs on river banks, 3500-5400 m alt.

IIIA. Qinghai: "East. and South-East. Qinghai"—Fl. Xizang. l.c.".

IIIB. Tibet: ***Weitzan*** (Dzhagyn-Gol river, 4400 m, July [7]; Yantszytszyan (Golubaya) river basin, Nko-gun' river, 4000 m; same site, on Nko-gun', Golubaya and Khichyu rivers, vicinity of Chokerku monastery, 3600-4400 m, Aug. 6, 1900—Lad.; "Bizhu"—Fl. Xizang. l.c.).

General distribution: China (Nor.-West., South-West).

Section 2. **Elatopsis** Huth

3. **D. albocoeruleum** Maxim. in Bull. Ac. Sci. St. Pétersb. 23 (1877) 307; Forbes and Hemsl. Index Fl. Sin. 1 (1886) 19; Maxim. Fl. Tangut. 1 (1889) 23; Huth in Bot. Jahrb. 20 (1895) 409; Finet et Gagnep. in Bull. Soc. Bot. Fr. 51 (1904) 494; Gr.-Grzh. Opis. put. Zap. Kit. [Report of Expedition in West. China] 3 (1907) 476; Hao in Bot. Jahrb. 68 (1938) 597; Munz in J. Arn. Arb. 48 (1967) 503; Fl. Sin. 27 (1979) 380; Y.Z. Zhao in Acta Sci. Natur. Univ. Intramong. 18, 2 (1987) 285; Y.Z. Zhao, l.c. 19, 4 (1988) 675; Fl. Intramong. ed. 2, 2 (1990) 539. —*D. przewalskii* Huth, l.c. 407; J. Arn. Arb. 48, 529; Y.Z. Zhao, l.c. (1988) 675. —*D. albocoeruleum* var. *pumilum* Huth, l.c. 409. —*D. albocoeruleum* var. *przewalskii* (Huth) W.T. Wang in Fl. Sin. 27 (1979) 381; Pl. Vasc. Helanshan [Vascular Plants of Helanshan] (1986) 103; Y.Z. Zhao, l.c. (1987) 685; Fl. Intramong. ed. 2, l.c. 539. —*D. albocoeruleum* var. *latilobum* Y.Z. Zhao, l.c. 676; Fl. Intramong. ed. 2, l.c. 539. —*D. pylzowii* auct. non Maxim.: J. Arn. Arb. 48, 532, p.p. —**Ic.**: Maxim. 1889, pl. 4; J. Arn. Arb. 48, fig. 8 B: Fl. Sin. 27, tab. 84, fig. 15-16, tab. 87, fig. 6-8; Pl. Vasc. Helanshan [Vascular Plants of Helanshan] tab. 14 (*D. albocoeruleum* var. *przewalskii*); Fl. Intramong. ed. 2, 2, tab. 218, fig. 1-2 (*D. albocoeruleum*), 3-4 (*D. albocoerileum* var. *przewalskii*), 5 (*D. albocoeruleum* var. *latilobum*).

Described from Qinghai. Type in St.-Petersburg (LE). Isotypes in London (K) and Paris (P).

In forest and alpine belts, on rocks and banks of mountain brooks, 3000-5000 m alt.

IA. Mongolia: ***Alash. Gobi*** (mont. Alaschan, in vallibus solo humoso argilloso, rarum, No. 405, Sept. 26-Oct. 8, 1871, Przew., syntypus *D. przewalskii* Huth!; mont. Alaschan, pars media declivit. occidentalis, in pratis alpinis, rarum, No. 206, July 5/17, 1873, Przew., syntypus *D. przewalskii* Huth!; "Alashan mountain range"—Y.Z. Zhao, 1987).

IB. Qaidam: ***Mount.*** (Dulan-khit temple, 3300 m, under spruce and juniper, Aug. 8, 1901—Lad.).

IIIA. Qinghai: ***Nanshan*** (Jugum N a fl. Tetung, regio alpina, in sylvis frequens, No. 403, Aug. 12/24, 1872—Przew., holotypus et isotypus!; on rocks in alpine belt of North. Tetung mountain range, 3000 m, Aug. 17, 1880—Przew.; South. Kukunor mountain range, rocks of nor. sloope, around Tszagastyn-kotol' pass, 3600 m, Aug. 17; Chorten-tan temple, Gan'chan-gomba and Sin'chen, in forests up to 3000 m, Aug. 30, 1901, Lad.; "Babakho river valley, 3000 m, Aug. 9, 1905"—Gr.-Grzh. l.c.; "Kokonor, Jahe-mari gebirge, 4000 m"—Hao, l.c.).

IIIB. Tibet: ***Weitzan*** (Dzhaotchi river, July 12; 4000 m, July 26—1884, Przew., Yamatyn-umru mountains, Yamatyn-umru area, 3600-4000 m, June 13, 1894—Rob.; Golubaya basin, on top of Chamudugla pass, 4800 m, July 27, 1900; nor. slope of Burkhan-Budda mountain range, Khatu gorge, 4600 m, July 17, 1901—Lad.).

General distribution: China (Nor.-West., South-West.).

4. **D. candelabrum** Ostenf. in Hedin, S. Tibet, 6, 3 (1922) 80; Munz in J. Arn. Arb. 48 (1967) 509; Fl. Sin. 27 (1979) 377; Fl. Xizang. 2 (1985) 52. —**Ic.**: J. Arn. Arb. 48, fig. 8, 1; Hedin, l.c., pl. 2, fig. 2-3; Fl. Xizang. 2, tab. 11, fig. 6.

Described from Tibet. Type in Stockholm (S).

On sandy-rubbly slopes, in alpine belt, 5100-5300 m alt.

IIIA. Qinghai ("Qinghai"—Fl. Xizang. l.c.).

IIIB. Tibet: ***Chang Tang*** ("Ban'ge"—Fl. Xizang. l.c.), ***Weitzan*** ("33°32' N, 88°52' E, 5127 m"—Hedin, S. Tibet, l.c.; "Bizhu"—Fl. Xizang. l.c.), ***South.*** ("Nan'mulin"—Fl. Xizang. l.c.).

General distribution: China (South-West.), Himalayas (east.).

5. **D. chrysotrichum** Finet et Gagnep. in Bull. Soc. Bot. Fr. 51 (1904) 488; Fl. Xizang. 2 (1985) 48. —*D. tsarongense* Hand.-Mazz. in Anz. Akad. Wiss. Wien, 59 (1922) 245, p.p.; Munz in J. Arn. Arb. 48 (1967) 493. —*D. chrysotrichum* var. *tsarongense* (Hand.-Mazz.) W.T. Wang in Fl. Sin. 27 (1979) 367.

Described from South-West. China. Type in Paris (P).

On fixed talus, 4800-5000 m alt.

IIIB. Tibet: ***Weitzan*** ("Bizhu"—Fl. Xizang. l.c.).

General distribution: China (South-West.).

6. **D. chumulangmaense** W.T. Wang in Acta Phytotax. Sin. 12, add. 1 (1965) 98; Fl. Sin. 27 (1979) 369; Fl. Xizang. 2 (1985) 50. —**Ic.:** Fl. Sin. 27, tab. 78, fig. 7-8.

Described from Himalayas (Dzhomolungma mountain). Type in Beijing (PE).

On northern rubbly slopes, 5000 m alt.

IIIB. Tibet: ***South.*** ("Dinzhi"—Fl. Xizang. l.c.).

General distribution: Himalayas.

7. **D. chungbaense** W.T. Wang in Fl. Sin. 27 (1979) 612, 371; Fl. Xizang. 2 (1985) 50. —**Ic.:** Fl. Sin. 27, ic. 20; Fl. Xizang. 2, tab. 14, fig. 3-4.

Described from South. Tibet. Type in Beijing (PE).

On slopes, 5600 m alt.

IIIB. Tibet: ***South.*** ("Dzhunba"—Fl. Xizang. l.c.; "Tibet: Chungba, No. 6677, Aug. 16, 1975—typus"—Fl. Sin. l.c.).

General distribution: endemic.

8. **D. crassifolium** Schrad. ex Ledeb. Fl. Ross. 1, 1 (1841) 62; Maxim. Enum. pl. Mong. 1 (1889) 28; Huth in Bot. Jahrb. 20 (1895) 401; Finet et Gagnep. in Bull. Soc. Bot. Fr. 51 (1904) 489; Nevsk. in Fl. SSSR, 7 (1937) 154; Kitag. Lin. Fl. Mansh. (1939) 219; Munz in J. Arn. Arb. 49 (1968) 135; Grub. Opred. rast. Mong. [Key to Plants of Mongolia] (1982) 109 (cum auct. Schrad.); Y.Z. Zhao in Acta Sci. Natur. Univ. Intramong. 19, 4 (1988) 674; Fl. Intramong. ed. 2, 2 (1990) 535. —*D.crassifolium* Schrad. ex Spreng. Gesch. Bot. 2 (1818) 201, nom. nud.; Gub. Konsp. Fl. Vneshn. Mong. [Conspectus

of Flora of Outer Mongolia] (1996) 49. —**Ic.:** Fl. SSSR, 7, Plate 10, fig. 4; J. Arn. Arb. 49, fig. 16 K; Grub. Plate 50, fig. 230; Fl. Intramong. ed. 2, 2, tab. 216, fig. 5.

Described from East. Siberia. Type in St.-Petersburg (LE).

In larch forests and their borders, meadows, among shrubs.

IA. Mongolia: ***Khobd.*** (Turgen' mountain range, 7 km above Turgen'-Gola estuary, grassy larch grove, July 17, 1971—Grub. et al.; east. macroslope of Mong. Altay, 37 km nor.-west of Ulangom, cereal grass-forb meadow clearance, July 5, 1977—Karam. et al.), ***Mong. Alt.*** (8 km south-east of Tszasakty-Khana, larch forest, Aug. 9, 1930—Pob.; Taishiri-Ula mountain range, nor. slope, larch forest, 5-8 km away from Yusun-Bulak, July 12, 1945—Yun.).

General distribution: East. Sib., Far East, Nor. Mong. (Fore Hubs., Hang., Fore Hing.), China (Dunbei).

9. **D. dasyanthum** Kar. et Kir. in Bull. Soc. natur. Moscou, 15 (1842) 138; Huth in Bot. Jahrb. 20 (1895) 396; Finet et Gagnep. in Bull. Soc. Bot. Fr. 51 (1904) 494; Nevsk. in Fl. SSSR, 7 (1937) 132; Fl. Kazakhst. 4 (1961) 38; Munz in J. Arn. Arb. 48 (1967) 482; Pakhom. in Opred. rast. Sr. Azii [Key to Plants of Mid Asia] 3 (1972) 170. —*D. winklerianum* Huth in Bot. Jahrb. l.c. 419, syn. nov.; Fl. Sin. 27 (1979) 387; Claves pl. Xinjiang, 2 (1983) 286. —**Ic.:** Fl. SSSR, 7, Plate 9, fig. 3; Fl. Kazakhst. 4, Plate 5, fig. 5; J. Arn. Arb. 48, fig. 6 D.

Described from East. Kazakhstan (Jung. Ala Tau). Lectotype in St.- Petersburg (LE).

On rocky slopes of mountains.

IIA. Junggar: ***Jung. Alat.*** ("Dsungaria orient., 1826—Turcz."—Huth, l.c., Kazan pass, 2700-3300 m, Aug. 10-11, 1878; Borotala, June 20 [1881]—A. Reg.; Borotala river basin, below Koketau pass, Aug. 21, 1909—Lipsk.), ***Tien Shan*** (south. bank of Borkul' lake, in ravine, No. 4951, Sept. 27, 1957—Kuan).

General distribution: Jung.-Tarb., Nor. Tien Shan; Mid. Asia.

10. **D. densiflorum** Duthie ex Huth in Bot. Jahrb. 20 (1895) 393; Rehd. in J. Arn. Arb. 14 (1993) 11; Munz in J. Arn. Arb. 48 (1967) 483; Fl. Sin. 27 (1979) 357. —*D. brunonianum* Royle var. *densa* Maxim. Fl. Tangut. 1 (1889) 24; Borodina-Grabovskaya in Catalogue type specim. Centr. Asia (LE) (2000) 195. —**Ic.:** Huth, tab. 6, fig. 4; J. Arn. Arb. 48, fig. 6 E; Fl. Sin. 27, tab. 84, fig. 1-4.

Described from Himalayas and Qinghai (Nanshan). Syntype in St.- Petersburg (LE). Plate IV, fig. 2.

On rocks in alpine belt.

IIIA. Qinghai: ***Nanshan*** (jugum a fl. Tetung boream versus, in altissimis pratis alpinis, sub rupibus, frequens, No. 396 [on rocks, alpine zone, south of Tetung river], Aug. 3/15, 1872—syntypus et isosyntypus!; Decliv. N jugis a fl. Tetung ad latera fragmentis rupium lecta, frequens [nor. slopes of South. Tetung mountain range, on rocks], July 21-Aug. 2, 1880—syntypus *D. brunonianum* var. *densa* Maxim.—Przew.), *Amdo* ("between Labran and Yellow river, Rok"—Rehder, l.c.).

General distribution: Himalayas (Kashmir).

Note. We selected syntype specimens of this species (No. 396, Aug. 3/5, 1872) as lectotype (isolectotype is also available) of *D. brunonianum* var. *densa* Maxim. (Borodina-Grabovskaya, l.c.).

11. **D. dissectum** Huth in Bot. Jahrb. 20 (1895) 403; Munz in J. Arn. Arb. 48 (1967) 516; Grub. Opred. rast. Mong. [Key to Plants of Mongolia] (1982) 109; Gub. Konsp. Fl. Vneshn. Mong. [Conspectus of Flora of Outer Mongolia] (1996) 49. —**Ic.:** J. Arn. Arb. 48, fig. 9 D; Grub. Plate 50, fig. 231.

Described from Nor. China. Type in St.-Petersburg (LE).

In steppes, meadows, forest borders, among shrubs.

IA. Mongolia: ***Cent. Khalkha*** (Sharkhai-khuduk river valley, on river bed, July 24, 1909—Czet.; Dzhargalante river basin: water divide between Ara-Dzhargalante and Uber-Dzhargalante rivers, rubbly sand, Aug. 10; sources of Kharukhe river (Ara-Dzhargalante river), northern slopes of Uste mountain, subalpine zone, Aug. 12; Uber-Dzhargalante river, rubbly southern slopes of Agit mountain, Aug. 26—1925, Krasch. and Zam.; vicinity of Ikhe-tukhum-Nor lake, Murzuk-Michigun mountains, No. 518; Nomgani-Ama gorge, No. 606; Minchun, Nomoga mountains, No. 623—1926, Zam.; Buridu-Ula mountains, July 16, 1952—Davazamc).

General distribution: Nor. Mong. (Fore Hubs., Hang., Mong.-Daur.), China (Nor.).

12. **D. eglandulosum** C.Y. Yang et B. Wang in Acta Phytotax. Sin. 30, 1 (1992) 85. —**Ic.:** Acta Phytotax. Sin. 30, 1, tab. 2, fig. 1-6.

Described from Nor. Sinkiang. Type in Urumchi (XJA).

On forest borders, 1300 m above sea level.

IIA. Junggar: ***Cis-Alt.*** ("Burqin, Biliudi"—C.Y. Yang et B. Wang, l.c.).

General distribution: endemic.

13. **D. ellipticoovatum** W.T. Wang in Acta Bot. Sin. 10, 3 (1962) 83; Munz in J. Arn. Arb. 49 (1968) 141; Fl. Sin. 27 (1979) 387; Claves pl. Xinjiang. 2 (1983) 286. —**Ic.:** Fl. Sin. 27 tab. 89, fig. 13-15.

Described from North-West. Sinkiang. Type in Beijing (PE).

On mountain slopes.

IIA. Junggar: ***Jung. Alat.*** ("Ven'tsyuan' district"—Claves pl. Xinjiang, l.c.).

General distribution: endemic.

14. **D. hillcoatiae** Munz in J. Arn. Arb. 49 (1968) 144; Fl. Sin. 27 (1979) 360; Fl. Xizang. 2 (1985) 46. —**Ic.:** J. Arn. Arb. 49, l.c. fig. 17, G; Fl. Sin. 27, tab. 82, fig. 10-13; Fl. Xizang. 2, tab. 14, fig. 4-6.

Described from East. Himalayas. Type in London (BM).

On grassy slopes near villages, 3400-3700 m alt.

IIIB. Tibet: ***South.*** ("Lhasa"—Fl. Xizang. l.c.).
General distribution: Himalayas.

15. **D. iliense** Huth in Bot. Jahrb. 20 (1895) 402; Nevsk. in Fl. SSSR, 7 (1937) 155; Fl. Kirgiz. 6 (1955) 47; Fl. Kazakhst. 4 (1961) 41; Pakhom. in Opred. rast. Sr. Azii [Key to Plants of Mid. Asia] 3 (1972) 173; Fl. Sin. 27 (1979) 382; Claves pl. Xinjiang, 2 (1983) 285; Gub. Konsp. Fl. Vneshn. Mong. [Conspectus of Flora of Outer Mongolia] (1996) 49; Borodina-Grabovskaya in Catalogue type specim. Centr. Asia (LE) (2000) 196. —*D. turkestanicum* Huth, 1895, l.c. 420; Pakhom. 1972, l.c. 173. —*D. wentsaii* Y.Z. Zhao in Acta Sci. Natur. Univ. Intramong. 21, 1 (1990) 134. —*D. corymbosum* auct. non Regel: Huth, l.c. (1895) 409; Nevsk. in Fl. SSSR, 7 (1937) 155; Munz in J. Arn. Arb. 49 (1968) 133; Fl. Kazakhst. 4 (1961) 42. —**Ic.:** Fl. SSSR, 7, Plate 9, fig. 6; Fl. Kirgiz. 6, Plate 5; Fl. Kazakhst. 4, Plate 5, fig. 1; J. Arn. Arb. 49, fig. 16 M, I (sub nom. *D. corymbosum*); Fl. Sin. 27, tab. 89, fig. 1-4; Y.Z. Zhao, l.c. fig. 1 (1-7) sub nom. *D. wentsaii.*

Described from Junggar (Tien Shan). Lectotype and syntype in St.- Petersburg (LE).

In forest meadows, rubbly slopes, coastal scrubs.

IIA. Junggar: ***Jung. Alat.*** (Junggar Ala Tau, 2100-2400 m, 1878—A. Reg.), ***Tien Shan*** (northern slopes of Tien Shan, Sept. 12, 1875)—Pias.; Kunges river, 1000 m, June 30, 1877—Przew.; Talki gorge, July 18; Dzhagastai pass, 1500-2100 m, Aug. 9—1877; Tianschan. fl. Agias, 5000-6000 ft, June 25, 1878, lectotypus! (Borodina-Grabovskaya, l.c.) —A. Reg.; Urtak-Sary, west of Sairam lake, July 19; Sairam lake, July 23; west of Sairam lake, July 28—1878, Fet.; Nilki, 1500 m, June [30]; Borgaty, 5000-6000 ft, July 4, syntypus!—1879, A. Reg.; Ulastai, 1879—Lar.; Tekes river, 3300 m, June 2; Karkary area, 3000 m, meadow, June 23; Koksai river east of Sary-tyur pass, 3000 m, in meadow, June 24—1893, Rob.; Bogdo-Ula mountains, on nor. slope of pass toward lake, Aug. 30, 1898—Klem.; mittleres Dschanart Tal, June 14-17; Sarydschass zwischen Lager und Gletscherende, Aug. 2-8—1903; Syrt von Karabulak und Karadshon, July 30; mittleres und unteres Agias Tal, Aug. 11-20; oestliches K[c]haptn-su Tal, Sept. 1-5 —1907; Montes Bogdo-ola et opp. Urumtschi, am wege von Fucan zur Bogdo-ola, in der Lehm- und Kiessteppe, Aug. 2-3, 1908—Merzb.; upper Tasenku river, Biangou site, spruce forest, Sept. 25, 1929—Pop.; Alabash-Kul'dzha, Aug. 4, 1956—Sorokin), ***Jung. Gobi*** (Gub. l.c.).
General distribution: Jung.-Tarb. (Jung. Alat.), Nor. and Cent. Tien Shan; Mid. Asia (West. Tien Shan).

16. **D. inconspicuum** Serg. Not. Syst. (Tomsk) 4 (1930) 6; Kryl. Fl. Zap. Sib. 5 (1931) 1144; Nevsk. in Fl. SSSR, 7 (1937) 145; Fl. Kazakhst. 4 (1961) 40; Munz in J. Arn. Arb. 49 (1968) 146; Grub. Opred. rast. Mong. [Key to Plants of Mongolia] (1982) 109; Gub., Kam. in Byull. Mosk. obshch. isp. prir., otd. biol. 95, 1 (1990) 122; Gub. Konsp. Fl. Vneshn. Mong. [Conspectus of Flora of Outer Mongolia] (1996) 49. —*D. inconspicuum* f. *longistachys* Serg. l.c. 7. —**Ic.:** Serg. l.c.

Described from Kazakhstan and Mongolia (Altay). Lectotype and syntype in Tomsk (TK).

In alpine and subalpine meadows.

IA. Mongolia: ***Khobd.*** ("Turgen-Ula"—Grub. l.c., Gub. l.c.), ***Mong. Alt.*** (Karatyr river (Chern. Kobdo), forest near lake, Aug. 1, 1908—Sap.; waterdivide of Buyantu and Bulugun, 2 km south of Akhuntyn-daba pass on road to Kudzhurtu, 2840 m, upper Ulagchin-gola, boulder pile-up along river valley within sedge-*Cobresia* wasteland, July 2; Kobdo river 2 km beyond Zhilandy river estuary, right bank, Gurtin-ama creek valley, 2200-2300 m, larch forest on slope exposed northward, July 11, 1971, Grub. et al.; mountains along south. bank of Dayan-nur lake, dwarf birch thicket, on nor. slope, July 26, 1977—Karam. et al.; 15 km west of Khoto-Nur, Aug. 17, 1979—Gub. (MW); Tsengel-Khairkhan mountain range, Achagardag-daba pass, 25 km south-west of Sagsai settlement, 2600 m, July 14, 1988—Bud., Gub. et al.).

IIA. Junggar: ***Cis-Alt.*** (Sanginin-Gol river basin, upper Chern. Irtysh, 25 km south-east of Dayan-Nur post, 1800-2100 m, July 16-17, 1988—Kam., Bud., Gub. et al.), ***Jung. Gobi*** (east.: nor. macroslope of Baitag-Bogdo mountain range, Budun-Khargaityn-Gol gorge, 15 km east of Baitag-Bogdo post, Aug. 1, 1988—Gub., Kam. et al.).

General distribution: West. Sib. (Altay), East. Sib., Nor. Mong. (Hang.).

Note. Cited for the first time for Chinese flora.

17. **D. kansuense** W.T. Wang in Acta Phytotax. Sin. 12, add. 1 (1965) 100; Fl. Sin. 27 (1979) 364. —*D. crassifolium* Schrad. var. *tanguticum* Maxim. Fl. Tangut. 1 (1889) 22; Franch. in Bull. Soc. Philom. Paris, sér. 8, 5 (1893) 163. —**Ic.:** Fl. Sin. 27, tab. 82, fig. 5-9.

Described from Nor.-West. China (Gansu). Type in Lanzhou (NWTC).

IIIA. Qinghai: ***Nanshan*** (North. Tetung mountain range, Aug. 16; South. Tetung mountain range, in lower forest belt, in wet meadows, Sept. 1; same site, Sept. 11 —1872; North. Tetung mountain range, in alpine meadows along brooks, Aug. 16, 1880 —Przew.; vicinity of Dan'gertin' town, along mountain descents, in swamps, ploughed farms, 2700 m, Aug. 22, 1901—Lad.).

General distribution: China (Nor.-West.).

18. **D. kaschgaricum** C.Y. Yang et B. Wang in Bull. Bot. Res. (Harbin) 9, 1 (1989) 21. —**Ic.:** Bull Bot. Res. (Harbin) 9, 1, tab. 1, fig. 1-7.

Described from Sinkiang. Type in Urumchi (XJA).

On slopes, 3300 m.

IB. Kashgar: ***West.*** ("Artux Xian"—C.Y. Yang et B. Wang, l.c.).

General distribution: endemic.

19. **D. korshinskyanum** Nevski in Fl. URSS, 7 (1937) 724, 153; Fl. Sin. 27 (1929) 388; Y.Z. Zhao in Acta Sci. Natur. Univ. Intramong. 19, 4 (1988) 675; Fl. Intramong. ed. 2, 2 (1990) 535. —*D. elatum* var. *sericeum* W.T. Wang in Acta Bot. Sin. 10, 3 (1962) 81; id. in Fl. Sin. 27 (1979) 383, 613. —*D. elatum* auct. non L. Finet et Gagnep. in Bull. Soc. Bot. Fr. 51 (1904) 480; Danguy in Bull. Mus. Hist. Nat. (Paris) 19, 7 (1913) 43; Pakhom. in Opred. rast. Sr. Azii [Key to Plants of Mid. Asia] 3 (1972) 174, p.p.; Fl. Sin. 27 (1979) 383; Grub. opred. rast. Mong. [Key to Plants of Mongolia] (1982) 109; Claves pl. Xinjiang, 2 (1983) 285; Gub. Konsp. Fl. Vneshn. Mong. [Conspectus of Flora

of Outer Mongolia] (1996) 49. —**Ic.:** Fl. SSSR, 7, Plate 9, fig. 3; Fl. Kazakhst. 4, Plate 5, fig. 5; J. Arn. Arb. 49 (1968) fig. 16 O; 17 A, B; Fl. Sin. 27, tab. 89, fig. 1-4; Fl. Intramong. 2, tab. 113; Fl. Intramong. ed. 2, 2 tab. 216, fig. 1-4.

Described from Far East. Type in St.-Petersburg (LE).

In larch and mixed forests, their borders and glades, forest meadows, coastal scrubs; reaches up to upper boundary of forest belt.

IA. Mongolia: ***Khobd.*** (Dzusylan, July 13; on Kharkira river, July 23—1879, Pot.), ***Mong. Alt.*** (20 km west-south-west of Ulangom, Mukhur-Ulyasty-Gol gorge, gorge floor, in scrubs, July 6 and 9, 1977—Karam.; upper Dzhelty river, Sanginin-Gol valley, 1700 m, Aug. 14, 1979—Gub. (MW); Sanginin-Gol river basin (upper Chern. Irtysh), Aug. 16-17, 1988—Kam., Gub. et al. (MW).

General distribution: Jung.-Tarb., Cent. Tien Shan; West. Sib., East. Sib., Far East, Nor. Mong. (Fore Hubs., Hent., Hang., Fore-Hing.), China (Dunbei).

20. **D. kunlunshanicum** C.Y. Yang et B. Wang in Bull. Bot. Res. (Harbin) 9, 1 (1989) 23. —**Ic.:** Bull. Bot. Res. (Harbin) 9, 1, tab. 2, fig. 1-8.

Described from Sinkiang. Type in Urumchi (XJA).

IB. Kashgar: ***West.*** (Yecheng (Kargalik)—C.Y. Yang et B. Wang, l.c.).

General distribution: endemic.

21. **D. lacostei** Danguy, J. Bot. (Paris) sér. 2, I, 3 (1908) 50; Pamp. Fl. Carac. (1930) 109; Nevsk. in Fl. SSSR, 7 (1937) 124; Ikonnik. Opred. rast. Pamira [Key to Plants of Pamir] (1963) 113; Munz in J. Arn. Arb. 48 (1967) 486; Pakhom. in Opred. rast. Sr. Azii [Key to Plants of Mid. Asia] 3 (1972) 169; Fl. Tadzh. 4 (1975) 33. —**Ic.:** J. Arn. Arb. 48, fig. 6 H; Fl. SSSR, 7, Plate 9, fig. 2; Ikonnik. Plate 12, fig. 4-6; Fl. Tadzh. 4, Plate 6, fig. 2.

Described from Pakistan (Sasserla mountains). Type in Paris (P).

On rocky slopes, 4000-5500 m alt.

IIIC. Pamir (Kishk-arek river, 4500 m, July 23; Shor-luk river gorge, 4000-5500 m, July 28—1942, Serp.).

General distribution: East. Pam.; Fora Asia (Pakistan, Afghanistan), Himalayas (Kashmir).

Note. Cited for the first time for Chinese flora.

22. **D. mirabile** Serg. Not. Syst. (Tomsk) 4 (1930) 5; Munz in J. Arn. Arb. 49 (1968) 152; Gub., Kam. in Byull. Mosk. obshch. isp. prir., otd. biol. 96, 6 (1991) 114; Gub. Konsp. Fl. Vneshn. Mong. [Conspectus of Flora of Outer Mongolia] (1996) 49. —*D. laxiflorum* DC. var. β. *alpinum* Bge. Suppl. Fl. alt. (1835) 33. —**Ic.:** Serg. l.c.; J. Arn. Arb. 49, fig. 17 N.

Described from West. Siberia (Altay). Type in Tomsk (TK).

In larch forests.

IA. Mongolia: ***Khobd.*** (Turgen' mountain range, Turgen'-Gol valley, 7 km above estuary, July 17, 1971 Grub. et al.), ***Mong. Alt.*** (Tsengel-Khairkhan mountain range, Achagardag-Daba pass, 25 km south-west of Sagsai settlement, July 14; Sanginin-Gol river basin (upper Chern. Irtysh), 25 km south-east of Dayan-Nur post, July 17; Elt-Gol river basin (upper Chern. Irtysh) 30 km south of Altay settlement, July 22—1988. Kam., Gub. et al. (MW).

General distribution: West. Sib. (Altay), East. Sib.

Note. Species close to *D. inconspicuum* Serg. but differing in pubescent follicles and ovaries.

23. **D. naviculare** W.T. Wang in Acta Bot. Sin. 10, 3 (1962) 82; Munz in J. Arn. Arb. 49 (1968) 153; Fl. Sin. 27 (1979) 385; Claves pl. Xinjiang, 2, (1983) 286. —*D. naviculare* var. *lasiocarpum* W.T. Wang in Acta Bot. Sin. 10, 3, l.c. 82; Fl. Sin. 27, l.c. 387; Claves pl. Xinjiang, 2, l.c. 286. —**Ic.:** Fl. Sin. 27, tab. 89, fig. 5-8.

Described from Tibet. Type in Beijing (PE).

In undergrowth of spruce forests and among shrubs, banks of rivers and grassy mountain slopes, 1600-1700 m.

IB. Kashgar: ***Nor.***

IIA. Junggar: ***Jung. Alat., Tien Shan, Jung. Gobi.***

IIIB. Tibet.

General distribution: endemic.

24. **D. nordhagenii** Wendelbo in Nytt Mag. Bot. 3 (1954) 227; Munz in J. Arn. Arb. 48 (1967) 489; Fl. Sin. 27 (1979) 357; Fl. Xizang. 2 (1985) 45. —*D. brunonianum* var. *nordhagenii* (Wendelbo) R.A. Qureshi et M.N. Chaudhri, Pakistan Syst. 4, 1-2 (1988) 79. —**Ic.:** J. Arn. Arb. 48, fig. 7 A; Fl. Sin. 27, tab. 81, fig. 8-10; Fl. Xizang. 2, tab. 12, fig. 5-7.

Described from Nor.-West. Pakistan. Type in Oslo (O).

On loose talus, rocky slopes, pebble banks of brooks in high mountains, 4300-5400 m alt.

IIIB. Tibet: ***Chang Tang*** ("Gaidi, Zhitu"—Fl. Xizang. l.c.), ***South.*** ("Pulan"—Fl. Xizang. l.c.).

General distribution: Fore Asia (Karakorum), Himalayas.

25. **D. pseudoglaciale** W.T. Wang in Fl. Sin. 27 (1979) 612, 368; Fl. Xizang. 2 (1985) 49. —**Ic.:** Fl. Sin. 27, ic. 19; Fl. Xizang. 2, tab. 14, fig. 5-7.

Described from South. Tibet. Type in Beijing (PE).

On talus, 4900-5000 m alt.

IIIB. Tibet: ***South.*** ("Tibet: Nanmulin, 4900-5000 m, No. 6052, Sept. 2, 1975—typus"; "Nanmulin"—Fl. Xizang. l.c.).

General distribution: endemic.

26. **D. pseudopulcherrimum** W.T. Wang in Acta Phytotax. Sin. 12, add. 1 (1965) 100; id. in Fl. Sin. 27 (1979) 374; Fl. Xizang. 2 (1985) 51. —**Ic.:** Fl. Sin. 27, tab. 86, fig. 1-2; Fl. Xizang. 2, tab. 15 fig. 1-2.

Described from South. Tibet. Type in Beijing (IMD).

On slopes, 4000-5100 m alt.

IIIB. Tibet: ***South.*** ("Lhasa, Chzhunba, 4000 m, P.K. Hsiao et K.C. Hsia, 1884, typus"—W.T. Wang, 1965, l.c.; "Lhasa"—Fl. Xizang. l.c.).

General distribution: endemic.

27. **D. pulanense** W.T. Wang in Fl. Sin. 27 (1979) 611, 360; Fl. Xizang. 2 (1985) 46. —**Ic.:** Fl. Sin. 27, tab. 81, fig. 1-4; Fl. Xizang. 2, tab. 12, fig. 1-4.

Described from South. Tibet. Type in Beijing (PE).

In alpine belt, 5000 m alt.

IIIB. Tibet: ***South.*** ("Tibet, Pulan, No. 8289, July 24, 1976, typus"—W.T. Wang, 1979, l.c.).

General distribution: endemic.

28. **D. pylzowii** Maxim. in Gartenflora, 25 (1876) 289; id. in Bull. Ac. Sci. St. Pétersb. 23 (1877) 307; Forbes and Hemsl. Index Fl. Sin. 1 (1886) 19; Maxim. Fl. Tangut. 1 (1889) 21; Huth in Bot. Jahrb. 20 (1895) 408; Hemsl. Fl. Tibet (1902) 164; Finet et Gagnep. in Bull. Soc. Bot. Fr. 51 (1904) 495; Gr.-Grzh. Opis. put. Zap. Kit. [Report of Expedition in West. China] 3 (1907) 476; Munz in J. Arn. Arb. 48 (1967) 532; Fl. Sin. 27 (1979) 376. —*D. pylzovii* var. *trigynum* W.T. Wang in Acta Bot. Sin. 10, 3 (1962) 78; Fl. Xizang. 2 (1985) 52. —**Ic.:** Maxim. 1889, tab. 3; Curtis' Bot. Mag. tab. 8813; J. Arn. Arb. 48, fig. 10 J, K; Fl. Xizang. 2, tab. 15, fig. 3-4 (var. *trigynum*).

Described from Qinghai. Type and isotypes in St.-Petersburg (LE). Isotype in Paris (P). Plate IV, fig. 1.

In forests, among shrubs, on meadow slopes in alpine belt, 2000-3700 m alt.

IIIA. Qinghai: ***Nanshan*** (Jugum S a fl. Tetung. regio alpina, in pratis, nec alibi, frequens, No. 211, July 1/13, 1872—Przew., holotypus et 2 isotypi!; [Chertynton temple], No. 485, Aug. 29, 1872—Przew.; upland around Kuku-nor lake, 3000 m, July 19; between Nanshan and Donkyr towns, on Rakogol river, 3000-3700 m July 21; nor. slope of South. Tetung mountain range, 3000-3700 m, July 31; on Yusup-Khatyma river, 2700 m, July 23; North. Tetung mountain range [Yarlyn-Gol river] 3000 m, Aug. 16—1880, Przew.; Choibsen temple, 2100-2400 m, Aug. 30; Chorten-tan temple, 2100-2400 m, Sept. 8—1901, Lad.; "Khyi-kho river valley, 3600-4000 m, Aug. 20, 1905"—Gr.-Grzh. l.c.; "south. and south-east. Qinghai"—Fl. Sin. l.c.").

IIIB. Tibet: ***Chang Tang*** ("Goring valley, 90°25', 30°12'"—Hemsl. l.c.); ***Weitzan*** ("Dangchu valley, 92°12', 32°12'"—Hemsl. l.c.; "Sosyan'"—Fl. Xizang. l.c.).

General distribution: China (South-West.).

29. **D. shawurense** W.T. Wang in Fl. Sin. 27 (1979) 613; 383; Claves pl. Xinjiang, 2 (1983) 285. —*D. shawurense* var. *albiflorum* C.Y. Yang et B. Wang, 1992, Acta Phytotax. Sin. 30, 1 (1992) 91. —**Ic.:** Fl. Sin. 27, tab. 89, fig. 9-12.

Described from Sinkiang. Type in Beijing (PE).

In hill forests, 1800-2200 m alt.

IIA. Junggar: ***Tarb.*** ("Saur mountain range, Khobuk-sair"—Fl. Sin. l.c., Claves pl. Xinjiang, l.c.).

General distribution: endemic.

30. **D. sinoelatum** C.Y. Yang et B. Wang in Acta Phytotax. Sin. 30, 1 (1992) 88. —**Ic.:** Acta Phytotax. Sin. 30, 1, tab. 3, fig. 1-7.

Described from Nor. Sinkiang. Type in Urumchi (XJA).

In steppes, steppified slopes along forest fringes, at 1940 m above sea level.

IIA. Junggar: ***Tarb.*** ("Hoboksar Xian, Saur mountains"—C.Y. Yang et B. Wang, l.c.).
General distribution: endemic.

31. **D. siwanense** Franch. in Bull. Soc. Philom. Paris, sér. 8, 5 (1893) 162; Huth in Bot. Jahrb. 20 (1895) 410; Finet et Gagnep. in Bull. Soc. Bot. Fr. 51 (1904) 483; Munz in J. Arn. Arb. 48 (1967) 491; Fl. Sin. 27 (1979) 381; Y.Z. Zhao in Acta Sci. Natur. Univ. Intramong. 19, 4 (1988) 675; Fl. Intramong. ed. 2, 2 (1990) 537. —*D. leptopogon* Hand.-Mazz. in Acta Horti Gotoburg. 13 (1939) 58; Munz in J. Arn. Arb. 49 (1968) 149. —*D. siwanense* var. *leptopogon* (Hand.-Mazz.) W.T. Wang in Fl. Sin. 27 (1979) 381; Y.Z. Zhao in Acta Sci. Natur. Univ. Intramong. 19, 4 (1988) 675; Fl. Intramong. ed. 2, 2 (1990) 537. —*D. cheilanthum* auct. non Fisch. ex. DC.: Franch. Pl. David. 1 (1884) 22. —**Ic.:** J. Arn. Arb. 48, fig. 7 C; l.c. 49, fig. 17 J; Fl. Intramong. ed. 2, 2, tab. 217, fig. 1-3, 4 (var. *leptopogon*).

Described from Nor. China. Type in Paris (P).

In mountain forests and floodplain meadows.

IA. Mongolia: ***East. Mong.*** (Datsin'shan' mountain range, vicinity of Khukh-Khoto, Baotu"—Fl. Intramong. ed. 2, l.c.), ***East. Gobi*** ("Ulantsab"—Fl. Intramong. ed. 2, l.c.), ***Alash. Gobi*** (Y.Z. Zhao, l.c.).
General distribution: China (Nor., Nor.-West.).

32. **D. tangkulaense** W.T. Wang in Acta Phytotax. Sin., 12, add. 1 (1965) 99; Fl. Sin. 27 (1979) 371; Fl. Xizang. 2 (1985) 50. —*D. chrysotrichum* var. *pygmaeum* Ostenf. in Hedin, S. Tibet, 6, 3 (1922) 79. —**Ic.:** Hedin, pl. 1, fig. 5 and pl. 2, fig. 1; Fl. Sin. 27, tab. 85, fig. 1-2; Fl. Xizang. 2, tab. 14, fig. 1-2.

Described from Tibet. Type in Beijing (PE).

In mountain meadows, on talus and in floodplains of brooks, 4700-5400 m.

IIIA. Qinghai ("south-west. Qinghai"—Fl. Xizang. l.c.).
IIIB. Tibet: ***Chang Tang*** (Tankulashan', No. 18, Y.T. Kan-typus; Tumyn'kela, 4950 m, No. 2390, C.H. Yang), ***Weitzan*** ("An'do"—Fl. Xizang. l.c.; "38°03′ N, 90°47′ E, Aug. 24, 1900—Hedin"—Hedin, l.c.), ***South.*** ("Chzhunba, Shuankhu"—Fl. Xizang. l.c.).
General distribution: endemic.

33. **D. tarbagataicum** C.Y. Yang et B. Wang in Bull. Bot. Res. (Harbin) 9, 1 (1989) 24. —**Ic.:** Bull. Bot. Res. (Harbin) 9, 1, tab. 3, fig. 1-6.

Described from Nor. Sinkiang. Type in Urumchi (XJA).

On slopes, 1800 m alt.

IIA. **Junggar:** ***Tarb.*** ("Tacheng Xian, Qoqiek"—C.Y. Yang et B. Wang).
General distribution: endemic.

34. **D. tetragynum** W.T. Wang in Bull. Bot. Res. (Harbin) 3, 1 (1983) 26.
Described from South-West. Sinkiang. Type in Beijing (PE).
On rubble slopes, 4500 m alt.

IIIC. **Pamir:** "Tazkorgan Xian"—W.T. Wang, l.c.
General distribution: endemic.

35. **D. tianschanicum** W.T. Wang in Acta Bot. Sin. 10, 3 (1962) 85; Munz in J. Arn. Arb. 49 (1968) 163; Fl. Sin. 27 (1979) 388; Claves pl. Xinjiang, 2 (1983) 287. —**Ic.:** J. Arn. Arb. 49, fig. 18 L; Fl. Sin. 27, tab. 88, fig. 10-13.
Described from Sinkiang (Bogdoshan' mountain range). Type in Beijing (PE).
On forest borders, gorges and meadow slopes, 1700-2700 m alt.

IIA. **Junggar:** ***Tien Shan*** ("Tien Shan"—Claves pl. Xinjiang, l.c.; "south. and nor. slopes of Tien Shan"—Fl. Sin. l.c.).
General distribution: ? Tien Shan.

36. **D. triste** Fisch. ex DC. Syst. 1 (1817) 362; Finet et Gagnep. in Bull. soc. Bot. Fr. 51 (1904) 492; Nevsk. in Fl. SSSR, 7 (1937) 157; Munz in J. Arn. Arb. 48 (1967) 539; Grub. Opred. rast. Mong. [Key to Plants of Mongolia] (1982) 109. —**Ic.:** Fl. SSSR, 7, Plate 10, fig. 6; J. Arn. Arb. 48, fig. 11 F.
Described from East. Siberia (Dauria). Type in St.-Petersburg (LE). Isotype in Paris (P).
In steppe meadows.

IA. **Mongolia:** ***Cent. Khalkha*** (Tsinkir-Gol river valley around Tsinkir-Dulang, hill steppe, along gentle trails of creek valley, July 23, 1949—Yun.), ***Gobi-Alt.*** (Dundu Saikhan mountains, south. slope, midbelt, July 6, 1909—Czet.).
General distribution: East. Sib., Nor. Mong. (Fore Hubs., Hent., Hang., Mong.-Daur.).

Note. Buds and just-blossomed flowers dark- to light-blue; all parts of fully blossomed flower (except spur) darken to a deep brown colour.

37. **D. wuqiaense** W.T. Wang in Bull. Bot. Res. (Harbin) 3, 1 (1983) 32.
Described from South-West. Sinkiang. Type in Beijing (PE).
On slopes, 3300 m alt.

IB. **Kashgar:** ***West.*** ("Wuqia Xian"—W.T. Wang. l.c.).
General distribution: endemic.

38. **D. yechengense** C.Y. Yang et B. Wang in Acta Phytotax. Sin. 30, 1 (1992) 82. —**Ic.:** Acta Phytotax. Sin. 30, 1, tab. 1, fig. 1-5.
Described from West. Kashgar. Type in Urumchi (XJA).
On mountain slopes, 3800-4300 m alt.

IB. Kashgar: *West.* ("Yecheng"—C.Y. Yang et B. Wang, l.c.).
General distribution: endemic.

Section 3. **Delphinastrum**
(*Delphinium* sect. *Delphinastrum* DC.)

39. **D. aemulans** Nevski in Fl. URSS, 7 (1937) 725, 161; Munz in J. Arn. Arb. 48 (1967) 502; Pakhom. in Opred. rast. Sr. Azii [Key to Plants of Mid. Asia] 3 (1972) 175; Fl. Sin. 27 (1979) 422; Claves pl. Xinjiang, 2 (1983) 287. —**Ic.:** J. Arn. Arb. 48, fig. 8 A.

Described from East. Kazakhstan (Jung. Alat.). Type in St.-Petersburg (LE).

On mountain slopes, 1400 m alt.

IIA. Junggar: "north-western mountainous regions of Sinkiang province"—Claves pl. Xinjiang, l.c.; in Toli district, No. 1357 and 2700, Aug. 7, 1957—Kuan.
General distribution: Jung.-Tarb.

40. **D. barlykense** Lomonosova et Khan in Bot. Journ. 70, 1 (1985) 111, Revushkin, Vysokogorn. fl. Alt. [Alpine Flora of Altay] (1988) 66; Gub., Kam. et al., Byull. Mosk. obshch. isp. prir., otd. biol. 95, 1 (1990) 119; Gub. Konsp. Fl. Vneshn. Mong. [Conspectus of Flora of Outer Mongolia] (1996) 49.

Described from West. Siberia (Tuva). Type in Novosibirsk.

Isotype in St.-Petersburg (LE).

IA. Mongolia: *Mong. Alt.* (Gub. 1996, l.c.).
General distribution: West. Sib. (Tuva).

41. **D. coeruleum** Jacq. ex Cambess. in Jacq. Voy. bot. 4 (1844) 7; Huth in Bot. Jahrb. 20 (1895) 463; Hemsl. Fl. Tibet (1902) 163; Finet et Gagnep. in Bull. Soc. Bot. Fr. 51 (1904) 494; Ostenf. et Pauls. in Hedin, S. Tibet, 6, 3 (1922) 79; Svenson in Brittonia, 8, 1 (1954) 55; Munz in J. Arn. Arb. 48 (1967) 508; Fl. Sin. 27 (1979) 453; Fl. Xizang. 2 (1985) 58. —*D. grandiflorum* auct. non L.: Tamura in Acta Phytotax. Geobot. (Kyoto) 15, 6 (1954) 194. —**Ic.:** Jacq. ex camb. l.c. tab. 6; J. Arn. Arb. 48, tab. 8, fig. 4; Fl. Sin. 27, tab. 106, fig. 1-4; Fl. Xizang. 2, tab. 17, fig 1-4.

Described from Himalayas. Type, sketch (Jacq. ex Camb., l.c. tab. 6).

On rocky and rubbly slopes and talus, among fields, 2000-5000 m alt.

IIIA. Qinghai: *Nanshan* (Mon'yuan', Ganshiga river valley, left tributary of Peishikhe river, 3350-3720 m—1858, Dolgushin; "Qinghai"—Fl. Xizang. l.c.).

IIIB. Tibet: *Weitzan* ("Bizhu"—Fl. Xizang. l.c.; "35°30′ N, 92°11′ E, Sept. 6-18, 1896; 31°40′ N, 88°22′ E, Naktsong-tso, Sept. 11, 1901—Hedin"—Hedin, l.c.), *South.* ("between Darjeeling and Lhasa"—Svenson, Jacquem, l.c., "Lhasa, Naidun, Nan'mulin, Linzhou, Pulan', Chzhada"—Fl. Xizang. l.c.; "Rakas Tal"—Hemsl. l.c.; "vicinity of Shigatze and Lhasa"—Tamura, l.c.).
General distribution: China (Nor.-West., South-West.), Himalayas (east.).

42. **D. cheilanthum** Fisch. ex DC. Syst. 1 (1817) 352; DC. Prodr. 1 (1824) 53; Forbes and Hemsl. Index Fl. Sin. 1 (1886) 19; Huth in Bot. Jahrb. 20 (1895) 466; Finet et Gagnep. in Bull. Soc. Bot. Fr. 51 (1904) 483; Nevsk. in Fl. SSSR, 7 (1937) 162; Fl. Kazakhst. 4 (1961) 44; Munz in J. Arn. Arb. 48 (1967) 511; Pakhom. in Opred. rast. Sr. Azii [Key to Plants of Mongolia] (1982) 110; Y.Z. Zhao in Acta Sci. Natur. Univ. Intramong. 19, 4 (1988) 676; Gub. Konsp. Fl. Vneshn. Mong. [Conspectus of Flora of Outer Mongolia] (1996) 49. —*D. cheilanthum* var. *pubescens* Y.Z. Zhao in Fl. Intramong. ed. 2, 2 (1990) 712, 541. —*D. cheilanthum* var. *middendorfii* auct. non Trautv.: Maxim. Enum. pl. Mong. 1 (1889) 27. —**Ic.:** Fl. SSSR, 7, Plate 10, fig. 8; Fl. Kazakhst. 4, Plate 5, fig. 2; J. Arn. Arb. 48, fig. 8 K; Fl. Sin. 27, tab. 90, fig. 7-11; Grub. l.c., Plate 50, fig. 229.

Described from East. Siberia (Dauria). Type in St.-Petersburg (LE).

On rock, placers, rocks meadow slopes and moraine, in upper mountain belt, meadows, forest borders and scrubs in river valleys in larch forest belt, 2500-2600 m alt.

IA. Mongolia: ***Khobd.*** (20 km west-south-west of Ulangom, Mukhur-Ulyasty-Gol gorge, July 9, 1977—Karam. et al.), ***Mong. Alt.*** (valley between main mountain range [Mong. Altay] and Adzhi-Bogdo, July 2; around Dolon-Nor lake, July 8; Tsitsirlik- Gol, in fields, July 9—1877, Pot.; in steppe between Olon-Nor and Taishir-Ol mountains, July 17, 1894—Klem.; nor. trail of Taishirin-Ula, 7-8 km east-south-east of Yusun-Bulak, July 11, 1945; Bulugun river basin, upper Ketsu-Sairin-Gol river, July 26, 1947 —Yun.; in upper Ulyasutain-Gol, June 28, 1973—Golubk.; Uenchiin-Gol basin, valley of Arshantyn-gol (right tributary of Khargaityn-Gol) 3 km from estuary, 2500-2600 m, Aug. 14, 1979—Grub., Dariima et al.; Munkh-Khairkhan mountain range, 75 km south-west of Kobdo, near Ulan-Daba pass around Olgoi-Nur lake, 3000 m, Aug. 11, 1982 —Gub. (MW); Bulgan-Gol river basin, upper gorge of Ulyastyin-Gol river, July 9, 1984 —Kam. and Dariima; basin of Ikh-Dzhergalant river (right tributary of Bulgan river), 15 km north-west of Bulgan settlement, July 27, 1988—Gub. and Kam.), ***Cent. Khalkha*** (Grub. l.c., Gub. l.c.), ***East. Mong.*** ("Mongolia chinensis, 1842, Kirilow), ***Gobi-Alt.*** (Dundu Saikhan mountains, nor. slope, upper belt, on wet sites, July 5 and 13, 1909—Czet.; same site, Aug. 19, 1931—Ik.-Gal.; Gurbun Saikhan, 2100 m, 1925—Chaney; Ikhe-Bogdo mountain range, Birtyuten-ama, alpine zone, Aug. 12, 1927—M. Smirnov; Bain Tsagan mountains, Aug. 9; Dzun Saikhan town, Aug. 25—1931. Ik-Gal.; Bain Tsagan mountains, Aug. 9; Dzun Saikhan town, 25—1931, Ik.-Gal.; Bain Tsagan mountains, July-Aug. 1933—Khurlat et al.; nor. part of Dzun-Saikhan mountain range, June 19; nor. slope of Ikhe-Bogdo mountain range, upper Shishkhid creek valley, mountain steppe belt, June 30; same site, upper Seveul creek valley, rocky steppe, July 2—1945, Yun.; Ikhe-Bogdo mountain range, Narin-Khurili creek valley, nor. slope, between rocks, June 28, 1945—Yun.; Ikhe-Bogdo-ula mountain range, south-east. slope, Narin-Khurilet gorge, on rocks, July 28, 1948—Grub.; Dzun-Saikhan mountain range, nor. slope along Dalan-Dzadagad—Ëlo-ama creek valley road near pass, 2400 m, forb mountain steppe, July 21; same site, Ëlyn-ama gorge, July 22; Dundu-Saikhan mountain range, south. slope of main peak, 2825 m, July 23—1970, Grub., Ulzij. et al.; 35 km east-nor.-east of Bayan-Dalai somon, Dund-Saikhan-Ula mountain range, in rocky gorge, July 23, 1972—Gur. and Rachk.; south macroslope of Ikhe-Bogdo town, 3800 m, Aug. 4, 1973—Isach. and Rachk.; Baga-Bogdo mountain range, Tsagan-Burlas river valley, 2500 m, July 24, 1978—Ogureeva (MW).

IIA. Junggar: *Cist-Alt.* (15-20 km nor.-west of Shara-Sume (on Kran river), shrubby meadow steppe, No. 1173, July 7, 1959—Yun. and Yuan').

General distribution: Jung.-Tarb.; East. Sib., Far East, Nor. Mong. (Fore-Hubs., Hent., Hang., Mong.-Daur.), China (Dunbei).

43. **D. grandiflorum** L. Sp. pl. (1753) 531; Forbes and Hemsl. Index Fl. Sin. (1886) 19; Maxim. Fl. Tangut. 1 (1889) 21; id. Enum. pl. Mong. 1 (1889) 27; Hemsl. in J. Linn. Soc. (London) Bot. 31(1894) 108; Huth in Bot. Jahrb. 20 (1895) 461; Hemsl. Fl. Tibet (1902) 163; Finet et Gagnep. in Bull. Soc. Bot. Fr. 51 (1904) 582; Gr.-Grzh. Opis. put. Zap. Kit. [Report of Expedition in West. China] 3 (1907) 476; Danguy in Bull. Mus. Hist. Nat. (Paris) 19, 7 (1913) 5; Kitag. Lin. Fl. Mansh. (1939) 219; Munz in J. Arn. Arb. 49 (1968) 517; Fl. Intramong. 2 (1978) 217; Fl. Sin. 27 (1979) 445; Grub. Opred. rast. Mong. [Key to Plants of Mongolia] (1982) 109; Fl. desert Sin. 1 (1985) 511; Pl. Vasc. Helanshan (1986) 103; Y.Z. Zhao in Acta Sci. Natur. Univ. Intramong. 19, 4 (1988) 676; Fl. Intramong. ed. 2, 2 (1990) 543; Gub. Konsp. Fl. Vneshn. Mong. [Conspectus of Flora of Outer Mongolia] (1996) 49. —*D. grandiflorum* var. *chinensis* Fisch. ex DC. Syst. 1 (1817) 351. —*D. chinense* (Fisch. ex DC.) Fisch. ex DC. Prodr. 1 (1824) 53; Kryl. Fl. Zap. Sib. 5 (1931) 1137; Fl. Kazakhst. 4 (1961) 44. —*D. grandiflorum* var. *glandulosum* W.T. Wang in Acta Bot. Sin. 10, 3 (1962) 273; Fl. desert Sin. l.c. 513. —**Ic.:** Curtis' Bot. Mag. tab. 1686; J. Arn. Arb. 49, fig. 9 E; Fl. Sin. 27, tab. 104, fig. 1-5; Grub. Opred. rast. Mong. [Key to Plants of Mongolia] Plate 50, fig. 228; Fl. desert Sin. 1, tab. 188, fig. 1-6, 7 (var. *glandulosum*); Fl. Intramong. ed. 2, 2, tab. 221, fig. 1-4, 5 (var. *pilosum*).

Described from East. Siberia. Type in London (Linn.).

In steppes, meadows, forest fringes, scrubs.

IA. Mongolia: ***East. Mong.*** ("steppe de la vallee de Kailar, No. 1554, June 26, 1896—Chaffanjon"—Danguy: Dariganga, nor. slope of Shiliin-Bogdo-Ula, Aug. 11, 1970—Grub. et al.), ***Gobi-Alt.*** ("Dzun-Saikhan"—Grub. l.c.), ***Alash. Gobi*** (Pl. Vasc. Helanshan, l.c.).

IC. Qaidam: ***Mountain.*** ("Tsaidam, side of slope at 14,800 ft. Thorold"—Hamsl. 1902, l.c.: Dulankhit temple, near Tsagan-Nor lake, 3200 m, Aug. 12, 1901—Lad.).

IIIA. Qaidam: *Amdo* (on hillocky steppe between border with nor. mountain range and mountain chain parallel to it, in moist sites, July 4, 1872—Przew.; Dzhakhar mountains, Mudzhik river, July 6, 1890—Gr.-Grzh.), ***Nanshan*** (North. Tetung mountain range, middle forest belt, 2400-2700 m, Aug. 13; South. Tetung mountain range, lower forest belt, 2300 m, Aug. 18 - 1880, Przew.; on Itel'-Gol river [San'chuanskaya valley], June-July 1885—Pot.; "North. Tetungs mountain range, 2700-4000 m, Aug. 8, 1890—Przew."—Gr.-Grzh. l.c.; Kuku-nor lake, vicinity of Donkir town, Aug. 11, 1908—Czet.; 37 km south of Kulon, eastern extremity of Nanshan, Kulon-Lanzhou road crossing, mountain steppe, Aug. 12, 1958; pass 86 km west of Xinin, cereal grass-forb steppe with alpine elements, Aug. 5, 1959—Petr.).

IIIB. Tibet: ***South.*** ("Lhasa, Nan'mulin, Lankatsza"—Fl. Xizang. l.c.).

General distribution: Jung.-Tarb. (Tarbagatai); West. Sib. (Altay), East. Sib., Far East, Nor. Mong., China (Dunbei, Nor., Nor.-West., East., Cent., South-West., South.), Korean peninsula.

44. **D. gyalanum** Marq. et Airy Shaw in J. Linn. Soc. (London) Bot. 48 (1929) 156; Munz in J. Arn. Arb. 49 (1968) 97; Fl. Sin. 27 (1979) 394; Fl. Xizang. 2 (1985) 53. —*D. kawaguchii* Tamura in Acta Phytotax. Geobot. (Kyoto) 15, 6 (1954) 194; Munz in J. Arn. Arb. 49, l.c. 147. —*D. lasiocarpum* Tamura in Acta Phytotax. Geobot. (Kyoto) l.c. 196; Munz in J. Arn. Arb. 49, l.c. 102. —**Ic.:** J. Arn. Arb. 49, fig. 14 C; Fl. Sin. 27, ic. 21; Acta Phytotax. Geobot. (Kyoto), fig. 2 (sub nom. *D. kawaguchii*); Acta Phytotax. Geobot. (Kyoto), fig. 3 (sub nom. *D. lasiocarpum*).

Described from South-East. Tibet. Type in Kew (K). Isotype in Edinburgh (E).

In alpine meadows, rubbly slopes, among shrubs, field hedges, 3000-4500 m alt.

IIIB. Tibet: ***South.*** ("Lhasa, Lankhatsza, Nan'mulin, Shigatsze"—Acta Phytotax. Geobot. (Kyoto) l.c. sub nom. *D. kawaguchii* et *D. lasiocarpum*).

General distribution: China (South-West.), Himalayas (east.).

45. **D. hsinganense** S.H. Li et Z.F. Fang in Fl. Pl. Herb. Chin. Bor.-Or. 3 (1975) 114; Fl. Sin. 27 (1979) 395; Y.Z. Zhao in Acta Sci. Natur. Univ. Intramong. 19, 4 (1988) 676; Fl. Intramong. ed. 2, 2 (1990) 545. —**Ic.:** Fl. Intramong. ed. 2, 2 tab. 219, fig. 1-4.

Described from China (Dunbei). Type in Shenyang (IFP).

On meadow slopes in gorges, 2100-2400 m alt.

IA. Mongolia: ***East. Mong.*** ("Kukunbuir, Kheilashan'"—Fl. Intramong. l.c.), ***Alash. Gobi*** ("Alashan"—Fl. Intramong. l.c.).

General distribution: China (Dunbei, Nor.-West.).

46. **D. kamaonense** Huth in Bull. Herb. Boiss. 1 (1893) 333; id. in Bot. Jahrb. 20 (1895) 460; Munz in J. Arn. Arb. 48 (1967) 521; Fl. Sin. 27 (1979) 448; Fl. Xizang. 2 (1985) 56. —*D. grandiflorum* var. *kamaonense* (Huth) Huth ex Bruhl et King in Ann. Bot. Gard. (Calcutta) 5 (1896) 98. —**Ic.:** J. Arn. Arb. 49, fig. 9 I, J.

Described from Himalayas. Type in ? Geneva (G). Isotypes in Edinburgh and Kew (E, K).

On meadow slopes, 2800-4100 m alt.

IIIB. Tibet: ***South.*** ("Pulan'"—Fl. Xizang. l.c.).

General distribution: Himalayas.

47. **D. kingianum** Bruhl ex Huth in Bot. Jahrb. 20 (1895) 469; Munz in J. Arn. Arb. 49 (1968) 100; Fl. Sin. 27 (1979) 391; Fl. Xizang. 2 (1985) 52. —**Ic.:** J. Arn. Arb. 49, l.c. fig. 14 E; Fl. Sin. 27, tab. 90, fig. 1-2.

Described from South. Tibet. Type, sketch (Bruhl, 1894, Ann. Roy. Bot. Gard. (Calcutta) 5, tab. 121; Huth, 1895, l.c. tab. VII, fig. 13).

On slopes, 4600-4800 m.

IIIB. Tibet: ***Chang Tang*** ("Gaidze"—Fl. Xizang. l.c.), ***South.*** ("Lhasa, Linzhou, Lankatsza"—Fl. Xizang. l.c.).

General distribution: Himalayas (east.).

48. **D. mollipilum** W.T. Wang in Acta Bot. Sin. 10, 3 (1962) 268; Fl. Sin. 27 (1979) 457; Munz in J. Arn. Arb. 48 (1967) 525; Y.Z. Zhao in Acta Sci. Natur. Univ. Intramong. 18, 2 (1987) 285; id. l.c. 19, 4 (1988) 678; Fl. Intramong. ed. 2, 2 (1990) 545. —**Ic.:** Fl. Intramong.ed. 2, 2, tab. 221, fig. 6-7.

Described from Nor.-West. China. Type in Beijing (PE).

On meadow slopes, in gorges, 2100-2400 m alt.

IA. Mongolia: ***Alash. Gobi*** ("Alashan mountain range"—Y.Z. Zhao, 1987; "Alashan, Kheilashan"—Fl. Intramong. ed. 2, l.c.).

General distribution: China (Nor.-West.).

49. **D. obcordatilimbum** W.T. Wang in Fl. Sin. 27 (1979) 616, 444; Fl. Xizang. 2 (1985) 56. —**Ic.:** Fl. Sin. 27, ic. 23; Fl. Xizang. 2, tab. 16, fig. 1-3.

Described from South-East. Tibet (Sikan). Type in Beijing (PE). In alpine meadows, about 3800 m.

IIIB. Tibet: ***Weitzan*** ("Leinyaotszi"—Fl. Xizang. l.c.).

General distribution: endemic.

Note. Reported at the boundary of Central Asia and South-East. Tibet.

50. **D. pseudocyananthum** C.Y. Yang et B. Wang in Acta Phytotax. Sin. 30, 1 (1992) 83. —**Ic.:** Acta Phytotax. Sin. 30, 1, tab. 1, fig. 6-12.

Described from Cis-Altay Junggar (Habahe). Type in Urumchi (XJA).

In mountains, coastal meadows, at about 1000 m alt.

IIA. Junggar: ***Zaisan*** ("Habane Xian"—C.Y. Yang et B. Wang, l.c.).

General distribution: endemic.

51. **D. sinoamabile** N.I. Malyutin ex Borod.-Grabovsk. nom. nov. —*D. amabile* C.Y. Yang et B. Wang in Acta Phytotax. Sin. 30, 1 (1992) 84, non Tidestr. 1925. —**Ic.:** Acta Phytotax. Sin. 30, 1, tab. 3, fig. 8-14.

Described from Sinkiang (Tarbagatai). Type in Urumchi (XJA).

On steppified slopes, among shrubs in forest fringes.

IIA. Junggar: ***Tarb.*** ("Hoboksar Xian, Saur mountains"—C.Y. Yang et B. Wang, l.c.).

General distribution: endemic ?

52. **D. souliei** Franch. in Bull. Soc. Philom. Paris, sér. 8, 5 (1893) 172; Huth in Bot. Jahrb. 20 (1895) 463; Munz in J. Arn. Arb. 49 (1968) 80; Fl. Sin. 27 (1979) 439. —**Ic.:** J. Arn. Arb. 49, fig. 12 G, H; Fl. Sin. 27, tab. 102, fig. 5-6.

Described from South-West. China. Type in Paris (P).

In meadowy slopes, 3500-4400 m alt.

IIIA. Qinghai: *Amdo* (grasslands between Labrang and Yellow River, near summit of Wanchen nira, 4000 m, July 29, 1926—Rock).

General distribution: China (South-West., Nor.-West.).

Note. Reported at the border of Central Asia.

53. **D. sparsiflorum** Maxim. in Bull. Ac. Sci. St. Pétersb. 23 (1877) 307; id. Fl. Tangut. 1 (1889) 23; Huth in Bot. Jahrb. 20 (1895) 421; Munz in J. Arn. Arb. 49 (1968) 116; Fl. Sin. 27 (1979) 420. —**Ic.:** Maxim. 1889, pl. 5; J. Arn. Arb. 49, fig. 15 J; Fl. Sin. 27, tab. 97, fig. 8-10.

Described from Qinghai. Type in St.-Petersburg (LE). Plate IV, fig. 3. Map. 3.

In spruce forests, 2400-2600 m alt.

IIIA. Qinghai: *Nanshan* (China occidentalis. Terra Tangutorum (prov. Kansu). Jugum S a fl. Tetung, in fomibus sylvarum secus rivulos, rarum [South. Tetung mountain range, in forest ravines on banks of brooks, rare], No. 182, June 27/July 9, 1872—Przew., holotypus et isotypus !, middle forest belt, on slopes south of Tetung river, 2600 m, Aug. 3; same site, 2400 m, Aug. 5—1880, Przew.).

General distribution: China (Nor.-West.).

Section 4. **Oligophyllon** Dimitrova

54. **D. biternatum** Huth in Bot. Jahrb. 20 (1895) 422; Nevsk. in Fl. SSSR, 7 (1937) 182; Fl. Kirgiz. 6 (1955) 52; Fl. Kazakhst. 4 (1961) 47; Munz in J. Arn. Arb. 48 (1967) 271; Pakhom. in Opred. rast. Sr. Azii [Key to Plants of Mid. Asia] 3 (1972) 179; Fl. Sin. 27 (1979) 459; Claves pl. Xinjiang, 2 (1983) 287. —**Ic.:** Fl. SSSR, 7, Plate 10, fig. 2; Fl. Kirgiz. 6, Plate 4, fig. 2a and 2b; Fl. Kazakhst. 4, Plate 4, fig. 4; J. Arn. Arb. 48, fig. 3 D.

Described from Mid. Asia. Isotype in St.-Petersburg (LE).

In finely hummocky sand in barren lands, submontane plains and trails of mountains, along ravine slopes and rubbly slopes of hillocks, rock placers and rocks in scrubs.

IIA. Junggar: *Tien Shan* (barren land in lower Kunges river, July 3, 1877—Przew.; east. Turkestan [without exact date] Aug. 1889—Grombch.; Tekes river, on rocks, 3300 m, June 2, 1893—Rob.; Nordrand der Tekes-steppe, zwischen Sumbe Chonachai und Piket Sumbe, June 24-25, 1907—Merzb.; on Ili-Dzhagastai road, No. 3151, Aug. 7, 1957 —Kuan), *Dzhark.* (in Ili region"—Claves pl. Xinjiang, l.c.).

General distribution: Jung.-Tarb. (Jung. Alat.), Nor. and Cent. Tien Shan; Mid. Asia, Fore Asia (Khofasan).

* * *

The following species have also been reported in Central Asian territory: from section **Elatopsis:**

D. glaciale Hook. f. et Thoms., Fl. Ind. 1 (1855) 53.

IIIB. Tibet: *Chang Tang* (Sendzha),

D. mollifolium W.T. Wang in Bull. Bot. Res. (Harbin) 3, 1 (1983) 31.

IIA. Junggar: *Tien Shan,*

D. nangziense W.T. Wang in Acta Phytotax. Sin. 29, 5 (1991) 458.

IIIB. Tibet: ***South.*** (Lhasa),

D. qinghaiense W.T. Wang in Acta Phytotax. Sin. 29, 5 (1991) 458. (South-West. ***Qinghai***),

D. taxkorganense W.T. Wang in Acta Bot. Univ. Yunnan, 15 (1993) 349. **IIIC. Pamir** (Tashkurgan),

from Section **Delphinastrum**:

D. gubanovii Friesen in Byull. Mosk. obshch. isp. prir., otd. biol. 95, 5 (1990) 130.

IA. Mongolia: ***Mong. Alt.,***

D. nangchienense W.T. Wang in Fl. Sin. 27 (1979) 616, 445.

IIIB. Tibet: ***Weitzan*** (Nankhen'),

D. pseudoaemulens C.Y. Yang et B. Wang in Acta Phytotax. Sin. 30, 1 (1992) 86.

IIA. Junggar: ***Tarb.,***

D. yuchuanii Y.Z. Zhao in Acta Sci. Nat. Univ. Intromongol. 20 (1989) 248. (SW Nei Mongol).

Subfamily 3. RANUNCULOIDEAE Hutch.

Tribe 1. Adonideae Langlet

8. Callianthemum C.A. Mey.

in Ledeb. Fl. alt. 2 (1830) 269, 336

1. Petals narrow, sublinear. Diminutive, 3-6 cm tall, plants; radical leaves more or less as long as stem, ovoid-oblong in form..2. **C. cachemirianum** Cambess.

\+ Petals broad, ovoid or oval. More large, (4) 6-20 cm tall, plants; radical leaves shorter than stem, ovoid, elliptical or oblong in form..2.

2. Leaves compact, thickish, subcoriaceous, with 3-6 pairs of pinnae bearing teeth of rounded terminal lobes convoluted along margins. Sepals green or reddish....................................1. **C. alatavicum** Freyn.

\+ Leaves very thin, with 2-4 pairs of pinnae, oblong or narrow-elliptical, with flat margins of narrow and long terminal teeth. Sepals pinkish or white..........................3. **C. sajanense** (Regel) Witas.

1. **C. alatavicum** Freyn in Bull. Herb. Boiss. 6 (1898) 882; Witas. in Verh. Zool.-Bot. Gesell. Wien, 49 (1899) 345; Schipcz. in Fl. SSSR, 7 (1937) 56; Persson in Bot. Not. (Lund) (1938) 283; Fl. Kirgiz. 6 (1955) 23; Fl. Kazakhst. 4 (1961) 18; Pakhom. in Opred. rast. Sr. Azii [Key to Plants of Middle Asia] 3 (1972) 189; Fl. Tadzh. 4 (1975) 15; Fl. Sin. 28 (1980) 242; Claves pl. Xinjiang,

2 (1983) 305. —*C. rutaefolium* auct. non C.A. Mey.: Maxim. Enum. pl. Mong. 1 (1889) 13. —**Ic.:** Fl. Kirgiz. 6, Plate 8, fig. 1; Fl. Kazakhst. 4, Plate 2, fig. 5; Fl. Tadzh. 4, Plate 2, fig. 5-8; Fl. Sin. 28, ic. 35.

Described from Jung. Ala Tau. Type in Vienna (W).

On talus, placers, rocks and alpine meadows.

IB. Kashgar: ***West.*** (Sarykol' mountain range, west of Kashgar, Bostan-Terek locality, alpine belt, pebble beds, July 11, 1929—Pop. between Upal and Koshkulak, No. 265, June 10, 1959—Lee et al.).

IIA. Junggar: ***Tarb.*** (around Mukurtai pass, 2700-3000 m, along placers, June 19, 1893—Rob.; Saur mountain range, south. slope, Karagaitu river valley, Bain-Tsagan creek valley, subalpine meadow, June 23, 1957—Yun. et al.), ***Tien Shan*** (Mal. Yuldus, 2600-2700 m, on humid slope, May 29, 1877—Przew.; Koshety-daban, in forest belt along both slopes, June 11, 1877—Pot.; Sairan lake, Kyzemchek, 3000 m, July—1878, A. Reg.; Yuldus, Sept. 1878—Fet.; Koshetydaban, in alpine belt, 1879—Przew.; south. tributary of Karagol, pass toward Nilka, 2700 m, June 16; Turgun-tsagan, July—1879, A. Reg.; Borokhoro mountain range, Chichkan, around Mukhurdai pass, 2700-3000 m, along placers, 1899—Korol'kov; Vorderes Dschanart Tal, June 14-17; Sary dschass, Aug. 2-8—1903; Passe zwischen Kin-su und Kurdai, July 3, 1907; zwischen Kungess Tal und Klein Juldus, also Tai-aschu Pass und Zanma Tal, May 5-10; Oberes Gross Juldus Tal am Wege zum Karagaitasch Pass, May 22-26; südliches Kiukonik Tal, Aufstieg zum Pass, Südseite, June 15—1908, Merzb.; Uch-Turfan, Bedel' pass, June 30, 1908—Divn.; 6 and 15 km south of Danyu, No. 363, 456, 509; along banks of Danyu river, on slope, No. 2115, 2135, 2148, July 21; Danyu-Daban, on waterdivide, No. 554, July 23—1957, Kuan; Bogdoshan alpine meadow, 3150 m, No. 5690, June 17, 1958—Lee and Chu; Mal. Yuldus, 2300 m, on valley alluvium, No. 6312, Aug. 2, 1958—Lee and Chu).

IIIB. Tibet: ***Chang Tang*** (Keriya (=Russky) mountain range, alp. belt, Kyuk-Egil' river valley, 3800-4000 m, on rocky placers, July 11, 1885—Przew.).

IIIC. Pamir (Billuli pass, on Chumbus river side, June 13, 1909—Divn.; "Pamir, Chichilgumbar, about 3250 m, July 21, 1935"—Persson, l.c.; Kokat pass, at sources of Yazach and Balung rivers, 3200-4000 m, July 1—1942, Serp.; nor. slope of King-Tau mountain range, upper portion of forest belt, short-grass meadows among spruce-juniper groves, 3400 m, June 10, 1959—Yun. and Yuan').

General distribution: Jung.-Tarb., Nor. and Cent. Tien Shan, East. Pam.; Europe (mount.), Mid. Asia (mount.), West. Sib. (Altay).

2. **C. cachemirianum** Cambess. in D. Dietr. Rep. Bot. 2 (1841-1843) 71; id. in Jacq. Voy. bot. 4 (1844) 5; Forbes and Hemsl. Index Fl. Sin. 1 (1886) 12; Maxim. Fl. Tangut. 1 (1889) 11; Hemsl. Fl. Tibet (1902) 162; Finet et Gagnep. in Bull. Soc. Bot. Fr. 55 (1904) 135. —*C. pimpinelloides* (D. Don) Hook. f. et. Thoms. Fl. Ind. 1 (1855) 26; Pamp. Fl. Carac. (1930) 108, cum auct. comb. Royle, Ill. Bot. Himal. Mount. 1 (1839) 53; Fl. Sin. 28 (1980) 245; Fl. Xizang. 2 (1985) 93. —*Ranunculus pimpinelloides* D. Don in Royle, l.c. 45. —*Callianthemum tibeticum* Witas. in Verh. Zool.-Bot. Gesell. Wien, 49 (1899) 330; Borodina-Grabovskaya in Catalogue type specim. Centr. Asia (LE) (2000) 193. —*C. imbricatum* Hand.-Mazz. in Acta Horti Gotoburg. 13 (1939) 132. —**Ic.:** Cambess. 1844, tab. 3; Fl. Sin. 28, tab. 82, fig. 4-7; Fl. Xizang. 2, tab. 26, fig. 8-10.

Described from Kashmir. Type in Montpellier (MPU). Plate 1, fig. 2.
On silty moist sand, clay cliffs.

IIIB. Tibet: ***Chang Tang*** ("80°30' E, 35°20' N"—Hemsl. l.c.; " Ban'che"—Fl. Xizang. l.c.), ***Weitzan*** (Tibet borealis, in arena limosa humida [vallis fl. Alak-nor-gol], No. 25, May 27, syntypus et isosyntypus *C. tibeticum* Witas. !; Tibet borealis, Montes ad fl. Botschu (affl. Jang tze), [Bychyu river], June 6/18, 1884, Przew., lectotypus *C. tibeticum* Witas. ! (Borodina-Grabovskaya, l.c.); south. slope of Amne-Kor mountain range, 4100 m, Yantszytszyan' (Goluboi river) basin, Gazunda area, on Denchyu river, 3400 m, April 22; Dzhagyn-Gol river, April 30; Yantszytszyan (Goluboi river) basin, descents of Nali-Sun pass, along Rabichyu brook entering Yaluntszyan, 4600 m, April 30; Yantszytszyan river basin (Goluboi river), Lamlunla pass and Lamlun area, 4300-4500 m, May 15—1901, Lad.; "Sosyan'"—Fl. Xizang. l.c.), ***South.*** (Dinzhi, Shuankhu, Lankadze, Chzhunba"—Fl. Xizang. l.c.).

General distribution: Fore Asia (Afghanistan), China (South-West.), Himalayas.

3. **C. sajanense** (Regel) Witas. in Verh. Zool.-Bot. Gesell. Wien, 49 (1899) 338; Gub. Konsp. Fl. Vneshn. Mong. [Conspectus of Flora of Outer Mongolia] (1996) 49. —*C. rutaefolium* C.A. Mey. var. *sajanense* Regel, Bull. Soc. natur. Moscou, 34 (1861) 52. —*C. angustifolium* Witas. l.c. (1899) 336; Kryl. Fl. Zap. Sib. 5 (1931) 1124; Schipcz. in Fl. SSSR, 7 (1937) 55; Fl. Kazakhst. 4 (1961) 18; Pakhom. Opred. rast. Sr. Azii [Key to Plants of Mid. Asia] 3 (1972) 189; Fl. Sin. 28 (1980) 243; Grub. Opred. rast. Mong. [Key to Plants of Mongolia] (1982) 108; Claves pl. Xinjiang, 2 (1983) 306.

Described from East. Siberia. Type in St.-Petersburg (LE).

In alpine meadows, on rocks and along rocky slopes, 1500-3800 m above sea level.

IA. Mongolia: ***Mong. Alt.*** (Belaya Kobdo river sources (Aksu), alp. tundra, June 29, 1906—Sap.; pass from Dayan-Nur lake basin to Irtysh basin, 15-20 km south of Dayan-Nur settlement, alpine meadow and dwarf birch thickets, 2400-2500 m, July 15, 1988—Kam., Bud., Gub. (MW).

IIA. Junggar: ***Cis-Alt.*** (Kran river, June 23, 1903—Gr.-Grzh.; Elt-Gol river basin (upper Chern. Irtysh), 30 km south of Altay settlement, July 22—1988, Kam., Bud., Gub. et al.), ***Tien Shan*** (Danu pass (3600 m), between Ulan-Usu valley and Danu-Gol, fine-rubble talus, July 19; Danu-Gol river alley, fine-rubble talus in alpine belt, July 21; upper Danu river, on ascent to Se-Daban pass, alpine belt, *Cobresia* meadow, July 22—1957, Yun. et al.).

General distribution: Jung.-Tarb.; West. Sib., East. Sib., Nor. Mong. (Fore Hubs., Hang.).

9. **Adonis** L.

Sp. pl. (1753) 547

1. Annual plants; flowers bright red.......................... 1. **A. aestivalis** L.
\+ Perennial plants; flowers of different colour.......................................2.
2. Flowers light blue, rarely whitish or pinkish; stems numerous in mature plants, form grove.................................3. **A. coerulea** Maxim.

+ Flowers yellow or white; stems one or more....................................3.
3. Flowers 15-25 mm in diam., yellow; sepals light green, sometimes with violet shade. Alpine plants..............2. **A. bobroviana** Simonov.
+ Flowers 30-50 mm in diam., yellow or white; sepals green or violet. Plants of plains or low mountains..4.
4. Flowers bright yellow with green sepals. Plants of forest-meadows, 20-40 cm tall with solitary or few flowers. Stems and pedicels glabrous..5. **A. sibirica** Patr. ex Ledeb.
+ Flowers white with violet sepals. Plants of steppes, 10-20 cm tall with many flowers. Stem and pedicels pubescent with sparse fine capitate hairs..4. **A. mongolica** Simonov.

1. **A. aestivalis** L. Sp. pl. ed. 2 (1762) 771; Bobr. in Fl. SSSR, 7 (1937) 537, p.p.; Fl. Kazakhst. 4 (1961) 132; Riedl. in Ann. Naturh. Mus. Wien, 66 (1963) 55; Kovalevsk. in Opred. rast. Sr. Azii [Key to Plants of Mid. Asia] 3 (1972) 230; Fl. Tadzh. 4 (1975) 150; Fl. Sin. 28 (1980) 253. —*A. aestivalis* var. *parviflora* M.B. Fl. Taur.-Cauc. 3 (1819) 378; Kryl. Fl. Zap. Sib. 5 (1931) 1220; Fl. Sin. l.c. 255. —*A. parviflora* (M.B.) Fisch. ex DC. Prodr. 1 (1824) 24; Fl. Kirgiz. 6 (1955) 113; Fl. Kazakhst. 4 (1961) 130; Kovalevsk. l.c. 230; Claves pl. Xinjiang, 2 (1983) 307. —**Ic.:** Fl. Kazakhst. 4, Plate 17, fig. 7 (*A. parviflorus*); Fl. Sin. 28, tab. 85, fig. 1-4 (*A. aestivalis* var. *parviflora*).

Described from South. Europe. Type in London (Linn.).

In meadows, under shade of rocks and shrubs in lower belt of mountains and as weed.

IIA. Junggar: ***Tien Shan*** (Kul'dzha, April 20, 1877; Sarybulak around Kul'dzha, 1200 m, April 24; Khorgos, 600-900 m, May; Cent. Khorgos, 1200-1500 m, May 15—1878; Piluchi river gorge, 1000 m, April 24, 1879—A. Reg.; high (about 200 m) right-side terrace of Manas river, near Chen-do-kho-ze river estuary, May 28, 1954—Mois.; Manas, Khunshan'-tszui, on slope, No. 632, June 6, 1957—Kuan).

General distribution: Aralo-Casp., Fore Balkh., Jung.-Tarb. (Jung. Ala Tau), Nor. and Cent. Tien Shan; Europe, Mediterr., Balk.-Asia Minor, Fore Asia, Caucasus, Mid. Asia, West. Sib., Nor. Amer.

2. **A. bobroviana** Simonov. in Novit. Syst. Pl. Vasc. 1968 (1968) 127; Fl. Sin. 28 (1980) 252; Fl. Intramong. ed. 2, 2 (1990) 480. —*Adonanthe bobroviana* (Simonov.) Chrtek et Slavikova in Preslia, 50 (1978) 24. —*Adonis apennina* auct. non L.: Maxim. Fl. Tangut. 1 (1889) 11. —**Ic.:** Fl. Sin. 28, tab. 85, fig. 5-8; Fl. Intramong. ed. 2, 2, tab. 194, fig. 1-2.

Described from Nor.-West. China. Type in St. -Petersburg (LE). Plate V, fig. 5.

In arid steppes, along river banks, on clayey and loamy soil.

IA. Mongolia: ***Ordos*** ("Paiqui to'ing the, E. Licent (K)"—Simonovicz, l.c. 128).

IIIA. Qinghai: ***Nanshan*** (Godyavoputan river valley, No. 468, July 17; Pabatson river valley, July 24; Binou village, July 25—1908, Czet.; on road from Talachi to

Sinulipu village, April 8, 1909—Demidenko), *Amdo* (upper course of Huang He river, 3000-3400 m, gorge, May 9, 1880—Przew.; valley of Urun'vu river, May 12, 1885 —Pot.).

General distribution: China (Nor.-West.).

3. **A. coerulea** Maxim. in Bull. Ac. Sci. St. Pétersb. 23 (1877) 306; Forbes and Hemsl. Index Fl. Sin. 1 (1886) 12; Maxim. Fl. Tangut. 1 (1889) 10; Hemsl. in J. Linn. Soc. (London) Bot. 30 (1894) 107; Hemsl. Fl. Tibet (1902) 162; Gr.-Grzh. Opis. put. Zap. Kit. [Report of Expedition in West. China] 3 (1907) 455; Fl. Sin. 28 (1980) 249; Fl. Xizang. 2 (1985) 95. —*A. coerulea* Maxim. f. *integra* W.T. Wang in Acta Phytotax. Sin. 6 (1957) 371; Fl. Sin. l.c. 249 and f. *puberula* W.T. Wang in Fl. Sin. 28 (1980) 249, 352. —**Ic.:** Maxim. 1889, tab. 1, fig. 20-30; Fl. Sin. 28, tab. 83, fig. 7-10; Fl. Xizang. 2, tab. 26, fig. 11-13.

Described from Qinghai. Type in St.-Petersburg (LE).

On subalpine meadows, in association with shrubs, 4300-5200 m alt.

IB. Qaidam: ***Mount.*** ("Wide valleys, 5000 m, Thorold"—Hemsl. 1894").

IIIA. Qinghai: ***Nanshan*** (jugum alpinum S a fl. Tetung, regio media, silvis frondosis, rara, No. 36, May 13/25, 1873—Przew., holotypus !; near Donkyr monastery, Chadzhi gorge, May 12 and 14, 1890—Gr.-Grzh.; Suleikhe river, 3400 m, June 2; Yamatyn-umru mountains, 3700-4000 m, Aug. 13—1894, Rob.; Chadzhi area, 2700 m, May 1905—Gr.-Grzh.), ***Amdo*** (in gorges of Syan'-si-bei mountain range, 3400-3700 m, May 29, 1880—Przew.).

IIIB. Tibet: ***Weitzan*** (along nor. slope of Burkhan-Budda mountain range, in Nomokhun-Gol gorge, June 8; near summit of waterdividing mountain range between Huang He river and Goluboi river, June 11; along descents of left bank of Golubaya river, among shrubs, July 11—1884, Przew.; Dzhagyn-Gol river, May 19-20; Burkhan-Budda mountain range, nor. slope of Nomokhun gorge, 4300 m, May 24; left bank of Dzhagyn-Gol river, clay mounds, 4100 m, July 3; Yantszy-tszyan river basin (Golubaya river), Chokabu-vrun area, 4300 m, July 10; sources of I-chyu river, Rkhombo-mtso lake, coastal mounds, 4000 m, Aug. 5—1900; Denchyu river, 3400 m, April 5, bank of Varmun-chyu and Shurchyu rivers, May 19-20; Makhmukh-chyu river, 4100 m, May 21—1901, Lad.; "Sosyan'"—Fl. Xizang. l.c.), ***South.*** ("Shuankhu"—Fl. Xizang. l.c.).

General distribution: China (Nor.-West., South-West.).

Note. Along with light blue or light lilac flowers commonly found and characteristic of this species, some collectors mention white (Przewalsky, June 8, 1884; Roborowsky, Aug. 13, 1894; Ladygin, Aug. 5, 1900 and April 5 and May 19-20, 1901) or pink (Przewalsky, June 11, 1884) forms; sometimes, they are found together with typical forms.

4. **A. mongolica** Simonov. in Novit. Syst. Pl. Vasc. (Leningrad) 1968 (1968) 125. —*Adonanthe mongolica* (Simonov.) Chrtek et Slavikova in Preslia, 50 (1978) 24. —**Ic.:** Grub. 1982, Plate 56, fig. 256.

Described from Nor. Mongolia (Hangay). Type in St.-Petersburg (LE). Plate V, fig. 7.

In forb steppes on mountains.

IA. Mongolia: ***Cent. Khalkha*** (Dzhargalanta river basin, between Botoga and Agit mountains, Sept. 13, 1925—Krasch. and Zam.; Khaltyn-Dava, 53 km south-east of Ulan-Bator, mountain steppe, May 27; 60 km from Ulan-Bator on road to Sain-Shandu, forb steppe, May 29—1941, Yun.).

General distribution: Nor. Mong. (Fore Hubs., Hang., Mong.-Daur.).

Note. *A. mongolica* is close to *A. sibirica* with which, while occupying other ecological niches (steppe plant, unlike forest-meadow *A. sibirica*), shares overlapping distribution range. Notwithstanding significant geographic disjunction, *A. mongolica* possibly exhibits some characteristics of *A. bobroviana* (habit, significant number of stems and flowers, pubescence of stem and pedicel with fine capitate hairs, violet-coloured sepals).

5. **A. sibirica** Patr. ex Ledeb. Del. Sem. Horti Dorp., Suppl. 2 (1824) 1; Kryl. Fl. Zap. Sib. 5 (1931) 1223; Bobr. in Fl. SSSR, 7 (1937) 530, p.p.; Fl. Kazakhst. 4 (1961) 128; Fl. Sin. 28 (1980) 252; Grub. Opred. rast. Mong. [Key to Plants of Mongolia] (1982) 116; Claves pl. Xinjiang, 2 (1983) 307; Fl. Intramong. ed. 2, 2 (1990) 480; Gub. Konsp. Fl. Vneshn. Mong. [Conspectus of Flora of Outer Mongolia] (1996) 48. —*Adonanthe sibirica* (Patr. ex Ledeb.) Spach, Hist. Natur. vég. 8 (1839) 229. —*Adonis apennina* auct non. L. —**Ic.:** Fl. Kazakhst. 4, Plate 17, fig. 1; Fl. Sin. 28, tab. 84, fig. 6-7; Fl. Intramong. ed. 2, 2, tab. 194, fig. 3-4.

Described from Altay. Type in St.- Petersburg (LE). Plate V, fig. 6.

On slopes, forest borders.

IA. Mongolia: ***Khobd.*** (Gub. l.c.), ***Cent. Khalkha*** ("nor.-west.: Ugomer"—Grub. and Gub. l.c.).

IIA. Junggar: ***Tien Shan*** (in Toli district, Albakzin mountains, on slope, No. 2587, Aug. 7, 1957—Kuan).

General distribution: Jung.-Tarb.; East. Europe, West. and East. Sib., Nor. Mong. (Fore Hubs., Hent., Hang., Mong.-Daur.: west.), China (Dunbei).

Note. See also note under *A. mongolica*.

Tribe 2. **Anemoneae** DC.

Subtribe 1. **Anemoninae** Spach

10. **Anemone** L.

Sp. pl. (1753) 538

1. Radical leaves absent..2.
+ Radical leaves formed, one or more in rosette..................................3.
2. Leaves opposite, set in pairs in nodes of dichotomously branching stem (spathe and indusium formed of 2 leaves). Perianth segments (4) 5...5. **A. dichotoma** L.

+ Leaves gathered in 3's in the form of spathe surrounding peduncle. Perianth segments 8-12....................2. **A. altaica** Fisch. ex. C.A. Mey.

3. Plants with bulb-shaped rhizome..4.

+ Plants without underground bulb..5.

4. Radical leaf segments on fairly long (2-15 mm) petiolules................ ....10. **A. petiolulosa** Juz.

+ Radical leaf segments sessile or subsessile.. ....7. **A. gortschakowii** Kar. et Kir.

5. Peduncles single; spathe comprising 3 petiolate or sessile leaves. ....6.

+ Peduncles (1) 2-5, gathered in umbel, surrounded by spathe of 4 sessile leaves or shoot even branched above spathe of 2 opposite leaves..10.

6. Leaves of spathe on visible petiolules and differ little in form from radical leaves..7.

+ Leaves of spathe sessile and differ noticeably from radical leaves. ....8.

7. Plant 20-40 cm tall. Stem compact, erect. Flower large, 25-45 mm in diam. Nuts covered with long soft crispate hairs......................... ....14. **A. sylvestris** L.

+ Plant 5-24 cm tall. Stem tender, arcuately curved downward. Flower 10 mm in diam. Nuts with short silky pubescence............................... ....6. **A. exigua** Maxim.

8. Radical leaves nearly entire, divided only upward into 3 (5) obtuse lobes, elliptical, orbicular-spatulate or spatulate in profile, longer than broad. Nuts on short pedicels.. ....15. **A. trullifolia** Hook. f. et Thoms.

+ Radical leaves laciniated into lobes down to base..............................9.

9. Radical leaves pinnately 5-lobed. Root neck thin. Ovaries and fruits glabrous. Flowers reddish brown or dark lilac, nearby black, rarely yellowish white......................................8. **A. imbricata** Maxim.

+ Radical leaves 3-lobed. Root neck thick due to petiole remnants of dead leaves. Ovary with fine rigidulous white or reddish-white or reddish hairs. Perianth segments white with light blue spot at base...9. **A. obtusiloba** D. Don.

10. Shoot branched above a pair of leafy bracts. Nuts without wings and not complanate...11.

+ Peduncles gathered in umbel. Nuts invariably totally glabrous, complanate and surrounded by broad thin wing, with short curved beak at tip..12.

11. Large, 60-100 cm tall plant; leaves large and tripartite, with tomentose pubescence downward. Flowers pink to purple, large,

25-30 mm in diam. Nuts small, 1-2 mm long, compactly pubescent, with very short curved beak....16. **A. vitifolia** Buch.-Ham. ex DC.

\+ Plant 30-50 cm tall with 3-lobed radical leaves, pubescent with appressed rigid hairs. Flowers white, 7-13 mm in diam. Nuts 7-10 mm long, glabrous, with long curved beak at tip.. ....12. **A. rivularis** Buch.-Ham. ex DC.

12\. Leaf blade profile orbicular or orbicular-reniform..........................13.

\+ Leaf blade profile elongated..15.

13\. Leaf blade quinate, all lobes sessile or on developed short petioles; latter, in turn, laciniated into long, oblong or sublinear cuspidate lobes. Leaves of spathe large, resembling radical leaves in the form of cleavage and teeth. Perianth segments glabrous outside. Stems and leaf stalks pubescent with long distant hairs..................... ....3. **A. crinita** Juz.

\+ Leaf blades ternate. Leaves of spathe small.......................................14.

14\. Leaf lobes incised rather shallowly into broad overlapping segments rounded at tip. Petiole with large white membranous sheath. Umbel with 2-4 flowers or flowers single. Perianth segments invariably pubescent outside with appressed hairs. Stems and leaf stalks pubescent with distant hairs........................ ....1. **A. alaschanica** (Schipcz.) Borod.-Grabovsk.

\+ Leaf lobes incised almost down to base into oblong segments, subobtuse at tip. Sheath imperceptible. Umbel with 3-6 flowers. Perianth segments usually glabrous. Stems and petioles of leaves pubescent with downturned hairs.................13. **A. schrenkiana** Juz.

15\. Leaf blades ternate, midlobe on petiole up to 2 cm long, lateral lobes on short petioles or sessile. Stems and petioles of leaves pubescent with appressed hairs...............11. **A. protracta** (Ulbr.) Juz.

\+ Leaf blades ternate; all lobes on identically short (0.3-0.5 mm) petioles. Stems and petioles of leaves pubescent with distant hairs. ....4. **A. demissa** Hook. f. et Thoms.

1\. **A. alaschanica** (Schipcz.) Borod.-Grabovsk. comb. et stat. nov. —*A. narcissiflora* var. *alaschanica* Schipcz. in Acta Horti Bot. Univ. Jurjev. 13, 2 (1912) 100. —*A. narcissiflora* auct. non L.: Maxim. Enum. pl. Mong. 1 (1889) 9, p.p.; Walker in Contribs. U.S. Nat. Herb. 28, 4 (1941) 615.

Described from Alashan mountain range. Lectotype and syntypes in St.-Petersburg (LE). Plate V, fig. 1. Map 2.

On shaded rock slopes, in gorges, in lower, middle and upper mountain belts.

IA. Mongolia: ***Alash. Gobi*** (mont. Alaschan, pars media, declivit. occidentalis, in rupibus regionis alpinae, rara, No. 130, June 23/July 5, 1873—Przew., syntypus et isosyntypus !; Alashan mountain range, Tsuburgan-Gol gorge, No. 68, April 30, 1908—

Czet., lectotypus ! (Borodina-Grabovskaya, selected here, LE); same site, Yamata gorge, nor. slope, middle and lower belts, No. 86, May 4—syntypus !; same site, Tszosto gorge, May 12, 1908—Czet., syntypus !; "Ha La Hu Kou, No. 57, 58, 1923, Ching"—Walker, l.c.).

General distribution: endemic.

Note. *A. alaschanica* plants have 3-lobed leaf blades, orbicular-reniform in profile, with rather shallowly incised, nearly entire lobes with broad segments rounded at tip; large white membranous sheaths at petiole base; compared to radical leaves, phyllary short; 2-4 flowers in umbel or flowers single; perianth segments intensely pubescent outside. K. Maximowicz (l.c.) pointed to the original characteristics of Alashan plants.

2. **A. altaica** Fisch. ex C.A. Mey. in Ledeb. Fl. alt. 2 (1830) 362; Kryl. Fl. Zap. Sib. 5 (1931) 1159; Juz. in Fl. SSSR, 7 (1937) 248; Fl. Kazakhst. 4 (1961) 60; Fl. Sin. 28 (1980) 12; Claves pl. Xinjiang, 2 (1983) 298. —**Ic.:** Fl. SSSR, 7, Plate 15, fig. 3 a, b; Fl. Kazakhst. 4, Plate 8, fig. 8; Fl. Sin. 28, tab. 2, fig. 4-6.

Described from Altay. Type in St.-Petersburg (LE).

In forests, forest borders and grasslands.

IIA. Junggar: ***Cis-Alt.*** ("Altay mountains"—Claves pl. Xinjiang, l.c.).

General distribution: Arctic (Yenisey river estuary), Europe (East.), West. Sib. (Altay), East. Sib.

3. **A. crinita** Juz. in Fl. URSS, 7 (1937) 739, 274; Fl. Kazakhst. 4 (1961) 65; Grub. Opred. rast. Mong. [Key to Plants of Mongolia] (1982) 111; Claves pl. Xinjiang, 2 (1983) 299; Fl. Intramong. ed. 2, 2 (1990) 468; Gub. Konsp. Fl. Vneshn. Mong. [Conspectus of Flora of Outer Mongolia] (1996) 48. —*A. narcissiflora* subsp. *crinita* (Juz.) Tamura in Acta Phytotax. Geobot. (Kyoto) 17 (1958) 115; Fl. Intramong. 2 (1978) 230; Fl. Sin. 28 (1980) 48. —*A. narcissiflora* auct. non L.: Kryl. Fl. Zap. Sib. 5 (1931) 1162, p.p. —**Ic.:** Fl. Kazakhst. 4, Plate 8, fig. 7; Fl. Intramong. 2, tab. 121, fig. 1-5; Fl. Sin. 28, tab. 13, fig. 8; Fl. Intramong. ed. 2, 2, tab. 189, fig. 6-8.

Described from Altay. Type in St.-Petersburg (LE). Plate V, fig. 2.

In forest borders, banks of brooks, grassy slopes in subalpine belt.

IA. Mongolia: ***Khobd.*** (on west. slope of Ulan-Daban, in forest borders, in sparse birch grove, June 5, 1895—Klem.), ***Mong. Alt.*** (in mountains between Mogoi and Bidzhiin rivers, Aug. 6, 1896—Klem.).

General distribution: Jung.-Tarb. (Tarbagatai mountain range); West. Sib., East. Sib., Far East, Nor. Mong., China (Dunbei, Nor.), Korean peninsula, Nor. Amer.

4. **A. demissa** Hook. f. et Thoms. Fl. Ind. 1 (1855) 23; id. Fl. Brit. Ind. 1 (1872) 9; Bruhl in Ann. Bot. Gard. (Calcutta) 5 (1896) 81; Ulbr. in. Bot. Jahrb. 37 (1906) 267; Fl. Sin. 28 (1980) 48; Fl. Xizang. 2 (1985) 82; Pl. Vasc. Helanshan (1986) 103. —*A. narcissiflora* var. *demissa* (Hook. f. et. Thoms.) Finet et Gagnep. in Bull. Soc. Bot. Fr. 51 (1904) 73; Fl. Intramong. ed. 2, 2

(1990) 468. —**Ic.:** Ann. Bot. Gard. (Calcutta) 5, tab. 107; Fl. Sin. 28, tab. 14, fig. 4-6; Fl. Xizang. 2, tab. 22, fig. 1; Fl. Intramong. ed. 2, 2, tab. 189, fig. 1-3.

Described from East. Himalayas (Sikkim). Type in London (K).

On exposed rock placers, in subalpine and alpine meadows.

IA. Mongolia: ***Alash. Gobi*** (Pl. Vasc. Helanshan. l.c.).

IIIB. Tibet: ***South.*** ("Lhasa, Nan'mulin, Linzhou"—Fl. Xizang. l.c.).

General distribution: China (South-West.), Himalayas (east.), Indo-Mal.

5. **A. dichotoma** L. Sp. pl. (1753) 540; Maxim. Enum. pl. Mong. 1 (1889) 10; Ulbr. in Bot. Jahrb. 37 (1906) 261; Danguy in Bull. Mus. Hist. Nat. (Paris) 19, 7 (1913) 2; Kryl. Fl. Zap. Sib. 5 (1931) 1161; Juz. in Fl. SSSR, 7 (1937) 268; Kitag. Lin. Fl. Mansh. (1939) 213; Fl. Intramong. 2 (1978) 228; Fl. Sin. 28 (1980) 56; Grub. Opred. rast. Mong. [Key to Plants of Mongolia] (1982) 111; Fl. Intramong. ed. 2, 2 (1990) 463; Gub. Konsp. Fl. Vneshn. Mong. [Conspectus of Flora of Outer Mongolia] (1996) 48. —**Ic.:** Fl. Sin. 28, tab. 16, fig. 1-4; Fl. Intramong. ed. 2, 2, tab. 187, fig. 1-2.

Described from Siberia. Type in London (Linn.).

In meadows.

IA. Mongolia: ***East. Mong.*** (Khailar town, Nunlin'tun' village, No. 640, June 10, 1951—Wang; "environs de Kailar, 750 m, sables, June 23, 1896—Chaffanjon"—Danguy, l.c.; Khalkhin-gol river near Khercam, June 10, 1925—V. Kazakevich).

General distribution: Europe (south-east of East. Europe), West. Sib., East. Sib., Far East, Nor. Mong. (Hent., Hang., Mong.-Daur., Cis-Hing.), China (Dunbei), Korean peninsula, Japan.

6. **A. exigua** Maxim. in Bull. Ac. Sci. St. Pétersb. 23 (1877) 306; Forbes and Hemsl. Index Fl. Sin. 1 (1886) 11; Maxim. Fl. Tangut. 1 (1889) 9; Ulbr. in Bot. Jahrb. 37 (1906) 233; Fl. Sin. 28 (1980) 11. —**Ic.:** Maxim. Fl. Tangut. tab. 2, fig. 7-15; Fl. Sin. 28, tab. 1, fig. 5-8.

Described from Qinghai (Nanshan). Type in St.-Petersburg (LE).

In forests, midbelt of mountains.

IIIA. Qinghai: ***Nanshan*** (Alpes S a fl. Tetung, silvis frondosis, regionis undique frequens, No. 26, May 11/23, 1873—Przew., holotypus !; around Runvyr-Ger monastery, in Runvyr-Lunva river valley, in coniferous forest, May 23, 1885—Pot.).

General distribution: China (Nor.-West., Cent., South-West.).

7. **A. gortschakowii** Kar. et Kir. in Bull. Soc. natur. Moscou, 15 (1842) 131; Juz. in Fl. SSSR, 7 (1937) 260; Fl. Kirgiz. 6 (1955) 72; Fl. Kazakhst. 4 (1961) 61; Ikonnik. Opred. rast. Pamira [Key to Plants of Pamir] (1963) 115; Kovalevsk. in Opred. rast. Sr. Azii [Key to Plants of Mid. Asia] 3 (1972) 192; Fl. Sin. 28 (1980) 34; Claves pl. Xinjiang, 2 (1983) 297. —**Ic.:** Fl. SSSR, 7, Plate 16, fig. 3; Fl. Kazakhst. 4, Plate 8, fig. 3; Ikonnik. Plate 13, fig. 4; Fl. Sin. 28, tab. 8, fig. 4-6.

Described from Jung. Ala Tau. Type in St.-Petersburg (LE).

On mountain-steppe slopes and among shrubs in foothills zone.

IIIC. Pamir ("Pamir"—Claves pl. Xinjiang, l.c.).

General distribution: Fore Balkh. (Betpakdala), Jung.-Tarb. (Jung. Ala Tau), Nor. Tien Shan; Mid. Asia (West. Tien Shan, Pamiro-Alay).

8. **A. imbricata** Maxim. Fl. Tangut. 1 (1889) 8; Hemsl. Fl. Tibet (1902) 162; Ulbr. in Bot. Jahrb. 37 (1906) 243; Rehder in J. Arn. Arb. 14 (1933) 12; Fl. Sin. 28 (1980) 55; Fl. Xizang. 2 (1985) 83; Borodina-Grabovskaya in Catalogue type specim. Centr. Asia (LE) (2000) 191. —*A. obtusiloba* subsp. *imbricata* (Maxim.) Bruhl in Ann. Bot. Gard. (Calcutta) 5, (1896) 79. —**Ic.:** Maxim. tab. 22, fig. 1-6.

Described from Tibet. Lectotype in St.-Petersburg (LE).

In alpine meadows, along mountain descents, under rocks.

IIIA. Qinghai: ***Nanshan*** (Yamatynumru mountains, moist alpine meadow, 4000-4300 m, June 27, 1894—Rob.).

IIIB. Tibet: ***Chang Tang*** (south. slope of Amnen-Kor mountain range, in moss and along moist hummocky mountain descent, 4100-4300 m, June 8, 1900—Lad.; "foothills of Dangla mountains, north-western extremity of range, 90°35′ E, 33°40′ N, 5000 m, June 1927—Rockhill"—Hemsl. l.c., "Ban'ge"—Fl. Xizang. l.c.), ***Weitzan*** (regio super. alpina montium secus fl. Jang-tze ad latera prope campos lapidosis frequens, No. 94, June 8/20—lectotypus ! (Borodina-Grabovskaya, l.c.); ripa sinistra fl. Jang-tze in rupibus frequens, No. 46, June 11/23, syntypus !—1884, Przew.; nor. slope of Burkhan-Buda mountain range, Nomokhun gorge, 4300 m, May 21; Russkoe lake, 4100 m, on clayey descents of coastal mounds, June 19; Yan-Tszy-Tszyana basin, Chokabu-Vrun area, 4300 m, in shade under rocks, July 10; Dzhagyn-Gol river, July 10—1990, Lad.; at the emergence of Yellow river from Russkoe lake, along descents of soft boulders, 4100 m; mounds along east. bank of Russkoe lake, 4100 m—June 2, 1901, Lad.; alpine region between Radja and Jupar range; among rocks in alpine meadows of Wajola, June 1926—Rock; "Eastern Tibet: Radja and Yellow river gorges; alpine region between Radja and Jupar ranges; Jupar range; Amnyi Machen range—Rock"—Rehder l.c.; "Sosyan', Bizhu"—Fl. Xizang. l.c.), ***South.*** ("Goring valley, 90°25′ E, 30°12′ N, 5000 m—Littledale" —Hemsl. l.c., "Lhasa"—Fl. Xizang. l.c.).

General distribution: endemic.

9. **A. obtusiloba** D. Don, Prodr. Fl. Nepal (1825) 194; Forbes and Hemsl. Index Fl. Sin. 1 (1886) 11; Maxim. Fl. Tangut. 1 (1889) 7; Ulbr. in Bot. Jahrb. 37 (1906) 241; Gr.-Grzh. Opis. put. Zap. Kit. [Report of Expedition in West. China] 3 (1907) 455; Juz. in Fl. SSSR, 7 (1937) 256; Walker in Contribs. U.S. Nat. Herb. 28, 4 (1941) 615; Fl. Kirgiz. 6 (1955) 71; Fl. Kazakhst. 4 (1961) 61; Kovalevsk. Opred. rast. Sr. Azii [Key to Plants of Mid. Asia] 3 (1972) 190; Fl. Sin. 28 (1980) 36; Fl. Xizang. 2 (1985) 80. —**Ic.:** Fl. SSSR, 7, Plate 16, fig. 2; Fl. Sin. 28, tab. 8, fig. 11; Fl. Xizang. 2, tab. 23, fig. 3.

Described from Himalayas. Type in London (BM). Plate I, fig. 4.

In moist alpine and coastal meadows, moist riverine pebble beds.

IIA. Junggar: ***Tien Shan*** (Manas river basin, left bank, Ulan-Usu river valley, at confluence with Dzhartas, upper forest boundary, subalpine meadow on nor. slope, July 18, 1957—Yun. et al.; 36 km south-east of Nyutsyuan'tszy, in subalpine meadows, No. 165, July 19, 1957—Kuan).

IIIA. Qinghai: ***Nanshan*** (pass through South.-Tetung mountain range, alpine belt, July 2, 1872; Tetung river, deciduous forest in midbelt, May 14, 1873; same site, 1880 —Przew.; "Shin-Chen town, Chzhusa pass, Xining Alps, 2600-3400 m, May-June 1905" —Gr.-Grzh., l.c.; "La Chiung Kou, 1923, Ching"—Walker, l.c.; 70 km south-east of Chzhan'e, Matisy temple, July 12, 1958—Petr.), ***Amdo*** (South Kukunor mountain range, Balekun area, 3000-3200 m, alpine meadow, June 6-7; Mudzhik river, 2900 m, moist meadow near sources, June 16, 1880—Przew.; Nuryn-Dzhamba river, May 11; same site, Urun'vu river valley, May 12; same site, Lanchzha-Lunva river valley, May 14; same site, Ngvarsi river valley, May 16—1885, Pot.).

IIIB. Tibet: ***Chang Tang*** (Keria mountain range, Kyukegil' river, 3800-4000 m, alpine meadow, July 1, 1885—Przew.), ***Weitzan*** (upper Huang He river, 3000 m, near sources [around Khyrmy Barunzasak], May 9, 1884—Przew.; mountains along Bychyu river, on rocks, June 15; mountains along Talachyu river, on pebble bed, June 19; left bank of Yangtze river, June 11 and 13—1884, Przew.; Yangtze river basin, Nyamtso area, Dynechyu river, 4000 m, July 15; same site, Donra area, July 17; Dzhagyn-Gol river, July 17—1900; Yangtze river basin, Entokgomba temple on Yaluntszyana river bank, 3500 m, April 26; Russkoe lake and Sergchyu river bank (right tributary of Huang He river), 4100 m, May 27—1901, Lad.; "Radja and Yellow river gorges: grassy slopes south of Radja; opposite bank of river, 3200 m, May 25; western mt., west of river, opposite Radja, 4000 m, June—1926, Rock"; "Sosyan"—Fl. Xizang. l.c.), ***South.*** ("Lhasa, Pulan', Chzhumba"—Fl. Xizang. l.c.).

General distribution: Nor. and Cent. Tien Shan; Nor. Mong. (Hang.), China (Nor.-West., South-West.), Himalayas, Indo-Mal., Nor. -West. India.

10. **A. petiolulosa** Juz. in Fl. URSS, 7 (1937) 836, 259; Fl. Kirgiz. 6 (1955) 75; Fl. Kazakhst. 4 (1961) 61; Kovalevsk. in Opred. rast. Sr. Azii [Key to Plants of Mid. Asia] 3 (1972) 191; Claves pl. Xinjiang, 2 (1983) 298. —**Ic.:** Fl. SSSR, 7, Plate 16, fig. 4; Fl. Kazakhst. 4, Plate 8, fig. 2.

Described from Mid. Asia (West. Tien Shan). Type in St.-Petersburg (LE).

In spruce forests, borders and meadows.

IIA. Junggar: ***Tien Shan*** ("Tien Shan—Ili region"—Claves pl. Xinjiang, l.c.).

General distribution: Fore Balkh. (Betpakdala), Nor. Tien Shan (Karatau); Mid. Asia.

11. **A. protracta** (Ulbr.) Juz. in URSS, 7 (1937) 273; Fl. Kirgiz. 6 (1955) 71; Fl. Kazakhst. 4 (1961) 64; Kovalevsk. in Opred. rast. Sr. Azii [Key to Plants of Mid. Asia] 3 (1972) 193; Claves pl. Xinjiang, 2 (1983) 298. —*A. narcissiflora* var. *protracta* Ulbr. in Bot. Jahrb. 37 (1906) 266; Fl. Sin. 28 (1980) 47. —*A. narcissiflora* var. *turkestanica* Schipcz. in Acta Horti Bot. Univ. Jurjev. 13, 2 (1912) 101, p.p. —*A. narcissiflora* auct. non L.: Forbes and Hemsl. Index Fl. Sin. 1 (1886) 11. —**Ic.:** Fl. SSSR, 7, Plate 17, fig. 4; Fl. Kazakhst. 4, Plate 8, fig. 4; Fl. Tadzh. 4 (1975), Plate 14, fig. 1-3.

Described from Mid. Asia. Type in Berlin (?B). Plate V, fig. 4. Map 2.

In grasslands, near brooks and along rock placers in subalpine belt.

IIA. Junggar: ***Tien Shan*** (Khonakhai, south of Kul'dzha, 1500-2100 m, June 16; Sumbe pass, 2100-2400 m, June 21; Dzhagastai mountains, Chapchal gorge, June 28 —1878, A. Reg.; Uch-Turfan, along brook between rocks, June 4, 1908—Divn.).

General distribution: Nor. and Cent. Tien Shan; Mid. Asia.

Note. On one hand, *A. protracta* is similar to Junggar *A. schrenkiana* and, on the other, to South Tibet-Himalayan *A. demissa*. It is characterised by 3-lobed leaf blades with central lobe on petiole up to 2 cm long; lobes narrow-cuneate at base, subobtuse; upward appressed pubescence of stem and leaf petioles as well as leaf undersurface; cuspidate perianth lobes, invariably pubescent outside.

12. **A. rivularis** Buch.-Ham. ex DC. Syst. 1 (1817) 211; Franch. Pl. David. 1 (1884) 18; Maxim. Enum. pl. Mong. 1 (1889) 10; Ulbr. in Bot. Jahrb. 37 (1906) 234; Gr.-Grzh. Opis. put. Zap. Kit. [Report of Expedition in West. China] 3 (1907) 455; Pamp. Fl. Carac. (1930) 111; Walker in Contribs. U.S. Nat. Herb. 28, 4 (1941) 615; Fl. Sin. 28 (1980) 22; Fl. Xizang. 2 (1985) 78. —*A. barbulata* Turcz. in Bull. Soc. natur. Moscou, 10 (1837) 149; Forbes and Hemsl. Index Fl. Sin. 1 (1886) 10; Hao in Bot. Jahrb. 68 (1938) 595. —*A. rivularis* var. *flore minore* Maxim. Fl. Tangut. 1 (1889) 6; Fl. Intramong. ed. 2, 2 (1990) 465. —*A. longipes* Tamura in Acta Phytotax. Geobot. (Kyoto) 15, 6 (1954) 192. —**Ic.:** Fl. Sin. 28, tab. 5, fig. 1-5; Fl. Xizang. 2, tab. 22, fig. 2; Fl. Intramong. ed. 2, 2, tab. 188, fig. 3-4.

Described from Himalayas (Nepal). Type in Paris (P). Isotype in St.-Petersburg (LE).

In moist grasslands, at sources and in scrubs along banks of brooks in subalpine and alpine mountain belts.

IA. Mongolia: ***East. Mong.*** (Oulachan, July—David; nor. slopes of Inshan' mountain range, July 7 and Aug. 8, 1871—Przew.).

IIA. Qinghai: ***Nanshan*** (South. Tetung mountain range, July 14, 1872; same site, Aug. 7, 1880—Przew.; [Pinfana region] July 24, 1875, Pias.; "Xining mountains, Myn'dan'sha river, June 14, 1890"—Gr.-Grzh., l.c.; "auf dem Gebirge ostlich vom see Kokonor, 3800 m, Hao"—Hao, l.c.; "Xining alps, 3000-3400 m, 1905"—Gr.-Grzh. l.c.; nor. Syamyngou gorge, along sand bed, July 20; Pinfan town, along irrigation canals, July 22—1909, Czet.; "Lien Ch'eng; T'u Er P'ing; Hsi Mi Yai—1923, Ching"—Walker, l.c.; 60 and 70 km south-east of Chzhan'e town, Matisy temple, 2600 m, Aug. 12; east. extremity of Nanshan, 25 km south of Gulan town, 2450 m, mountain steppe, Aug. 12 —1958, Petr.), ***Amdo*** ([Kha-Gomi area], 2100 m, June 9; Mudzhik river, 2700 m, June 16 —1880, Przew.).

IIIB. Tibet: ***Weitzan*** (Yangtze river basin, Donra area on Danechyu river, 4000 m, June 16; Mekong river basin, Tszechyu river, 3400-3700 m, Sept.; same site, Rechyu river, 3400-3700 m, Dec.—1900, Lad.), ***South.*** (Linzhou, Lhasa, Pulan'"—Fl. Xizang. l.c., "vicinity of Shigatsze and Lhasa"—Tamura, l.c.).

General distribution: China (Nor., Nor.-West., South-West., Cent.), Himalayas, Indo-Mal. (Ceylon island).

13. **A. schrenkiana** Juz. in Fl. URSS, 7 (1937) 738, 273; Fl. Kazakhst. 4 (1961) 65; Kovalevsk. in Opred. rast. Sr. Azii [Key to Plants of Mid. Asia] 3 (1972) 193; Claves pl. Xinjiang, 2 (1983) 299. —*A. narcissiflora* var. *turkestanica* Schipcz. in Acta Horti Bot. Univ. Jurjev. 13, 2 (1912) 101, excl. f. *protracta* Ulbr.; Fl. Sin. 28 (1980) 48. —*A. narcissiflora* var. *yuldussica* Schipcz. l.c. 100.

—*A. narcissiflora* auct. non L.: Maxim. Enum. pl. Mong. 1 (1889) 9, p.p. —**Ic.:** Fl. Sin. 28, tab. 13, fig. 1-3.

Described from Jung. Ala Tau. Type in St.-Petersburg (LE). Plate V, fig. 3.

On grassy slopes, in forests, their borders, alpine meadows in upper mountain belt.

IIA. Junggar: ***Tarb.*** (Saur mountain range, south. slope, Karagaitu river valley, Bain-Tsagan creek valley on right bank, larch and sparse forests, June 23, 1957—Yun. et al.; nor.-west of Khobsaira (Chagan Obo mountain), 2000 m, No. 10528, June 22, 1959—Lee et al.), ***Jung. Ala Tau*** (in Toli district, in mountain meadow, No. 993, 2621 and 1387, Aug. 6-8, 1957—Kuan), ***Tien Shan*** (old Mal. Yuldus valley [Kharety-gol river], on mounds, June 17; upper Tsanma river, 2100 m, June 18—1877, Przew.; Burkhan-tau, June 5; Sairam lake, July 23; Yuldus mountain range, Sept.—1878, Fet.; Borborogusun, 900-1200 m, April 27; upper Taldy river, 2400-2700 m, May 17; Kumbel', 3000 m, June 3; Bagaduslun river, east. tributary of Dzhin river, 2100-2700 m, June 4; Karagol', pass to Nilki, 3000 m, June 14; on Borborogusun river, 2700 m, June 15; Aryslyn, 2400 m, July 10—1879, A. Reg.; Oberfluss d. Musstamass, July 1907—Merzb.; on Syata-Ven'tsyuan' road, No. 1387, Aug. 13; mountains in Aksu region, on slope, No. 3492, Aug. 14—1957, Kuan; in Turfan, on shaded slope, 2500 m, near Kalangou village, No. 5783, June 23, 1958—Lee and Chu).

General distribution: Jung.-Tarb.

Note. South. Tien Shan plants (Kuan's collection in Aksu region, No. 3492) with appressed or upturned pubescence of stem and petioles represent intermediate forms to *A. protracta*.

14. **A. sylvestris** L. Sp. pl. (1753) 540; Franch. Pl. David. 1 (1884) 17; Maxim. Enum. pl. Mong. 1 (1889) 9; Ulbr. in Bot. Jahrb. 37 (1906) 254; Kryl. Fl. Zap. Sib. 5 (1931) 1160; Juz. in Fl. SSSR, 7 (1937) 266; Kitag. Lin. Fl. Mansh. (1939) 214; Fl. Kazakhst. 4 (1961) 64; Kovalevsk. in Opred. rast. Sr. Azii [Key to Plants of Mid. Asia] 3 (1972) 192; Fl. Sin. 28 (1980) 33; Grub. Opred. rast. Mong. [Key to Plants of Mongolia] (1982) 111; Claves pl. Xinjiang, 2 (1983) 298; Fl. Intramong. ed. 2, 2 (1990) 465; Gub. Konsp. Fl. Vneshn. Mong. [Conspectus of Flora of Outer Mongolia] (1996) 48. —**Ic.:** Fl. SSSR, 7, Plate 16, fig. 7 a and b; Fl. Kazakhst. 4, Plate 8, fig. 5; Fl. Sin. 28, tab. 6, fig. 7-9; Fl. Intramong. ed. 2, 2 tab. 187, fig. 3-5.

Described from West. Europe. Type in London (Linn.).

In larch forests and their borders, birch and willow thickets, scrubs, meadows and meadowy slopes, along banks of brooks.

IA. Mongolia: ***Khobd.*** (valley of Shur-Ulyastai-Gol river, larch grove, July 12, 1973—Banzragch et al.), ***Mong. Alt.*** (Dzhirgalanta river valley, larch forest, July 27, 1906—Sap.; Khan-Taishiri mountain range, south. slope in upper Shine-Usu, larch forest, June 18, 1971—Grub. et al.), ***East. Mong.*** (Inshan' [Muni-Ula], in birch groves, June 25, 1871—Przew.; Kharkhonte railway station, June 8, 1902—Litw.; Khalkhin-Gol river, near Khere-obo, June 10, 1925—Kazakevich; Khailar town, Sishan' mounds, June 7, 1951—B. Skvortzov; same site, No. 770, June 20, 1951—Lee et al.; Shiliin-Bogdo mountain, July 8, 1971—Dashnyam et al.; same site, July 12, 1985—Gub. (MW).

IIA. Junggar: ***Cis-Alt.*** (in Tuergen (=Koktogoi) settlement region, 1470 m, No. 10387, June 6, 1959—Lee et al.; Sanginin-Gol river basin (upper Chern. Irtysh), 25 km south-east of Dayan-Nur post, 1800-2100 m, July 16-17, 1988—Kam., Bud. et al.), ***Tarb.*** (nor.-west of Khobsaira (=Khobuk-Saira) (Chachan-obo mountain), 2200 m, No. 10519, June 22, 1959—Lee et al.), ***Jung. Alat.*** (mountains in Toli district, in mountain meadow, Barktok-Arba-Kezen', 1980 m, No. 1385, Aug. 8, 1957—Kuan), ***Tien Shan*** ([Koshety-Daban], on forest boundary, June 11, 1877—[Pot.]; between Kegen and Khorgos rivers, 600-900 m, April 30; Cent. Khorgos, 900 m, May 1; Upper Khorgos, 1500-1800 m, June—1878, A. Reg.; west of Sairam lake, July 20, 1878—Fet.; upper Taldy river, 2400-2700 m, May 5; Bagaduslun-Bainamyn, east. tributary of Dzhin river, 1500-1800 m, June 4—1879, A. Reg.; pass between Barkul' and Khami towns [south-east of Sunshitan village], forest, June 22, 1879—Przew.; bei Manass, July 5-17, 1908—Merzb.; Santai village, upper Shigou river valley, 2000 m, May 27, 1952—Mois.; Boro-Khoro mountain range, Ulan-Bulak mountains, Yangtze-khai river basin, 2900 m, above alpine meadows, May 29, 1953; upper right summit of Manas river, spruce development zone, 2200 m, May 26, 1954—Mois.; in Savan district, Datszymyao, No. 1278, July 8, 1957—Kuan; Manas river basin, midstream of Ulan-Usu river, at confluence with Koisu, left flank of valley, forest belt, July 17, 1957—Yun. et al.; Kizil-Zangi, south of Shichan, on shaded slope, No. 819, July 23; 6 km south of Shichan, in Savan district, 1700 m, No. 2302, July 26—1957, Shen-Tyan).

General distribution: Fore Balkh., Jung.-Tarb.; Europe, Caucasus, West. Sib., East. Sib., Far East, Nor. Mong., China (Dunbei).

15. **A. trullifolia** Hook. f. et Thoms. Fl. Ind. 1 (1855) 22; Ulbr. in Bot. Jahrb. 37 (1906) 243; Fl. Sin. 28 (1980) 44; Fl. Xizang. 2 (1985) 82. —*A. obtusiloba* subsp. *trullifolia* (Hook. f. et Thoms.) Bruhl in Ann. Bot. Gard. (Calcutta) 5 (1896) 77. —**Ic.:** Ann. Bot. Gard. (Calcutta) 5, pl. 106 A; Fl. Sin. 28, tab. 12, fig. 18.

Described from Himalayas (Sikkim). Type in London (K).

In alpine meadows, 3000-4600 m above sea level.

IIIB. Tibet: ***South.*** ("Lhasa, Nan'mulin"—Fl. Xizang. l.c.).

General distribution: China (South-West.), Himalayas (east.).

16. **A. vitifolia** Buch.-Ham. ex DC. Syst. 1 (1817) 211; Fl. Sin. 28 (1980) 29. —*A. japonica* (Thunb.) Sieb. et Zucc. Fl. Jap. 1 (1835) 15, non Houtt, 1778; Forbes and Hemsl. Index Fl. Sin. 1 (1886) 11; Ulbr. in Bot. Jahrb. 37 (1906) 252. —*A. tomentosa* (Maxim.) Pei in Contr. Biol. Lab. Sci. Soc. China, Bot. ser., 9 (1933) 2; Fl. Sin. l.c. 29. —**Ic.:** Fl. Sin. 28, tab. 6, fig. 4-6; Fl. Xizang. 2, tab. 22, fig. 5.

Described from Himalayas. Type in Geneva (G).

On rocks, in mountains, 1500-3000 m alt.

IIIA. Qinghai: ***Nanshan*** (on rocks, Tetung river, 2300 m, Aug. 7, 1880—Przew.; in mountains west of Dzhamba river (Itel'-Gol), in shrubs, March 14, 1885—Pot.; "Yao Chich, No. 293, 1923—Ching"—Walker, l.c.).

General distribution: China (Nor.-West., Cent., East., South-West.), Himalayas (east.), Korean peninsula, Japan.

Note. *A. vitifolia*, apart from Qinghai, naturally grows in South-West. China and Himalayas and is cultivated or runs wild in other regions.

11. Pulsatilla Mill.

Gard. Dict. abr. ed. 4 (1754)

1. Flowers erect, wide-open, blossom before unfolding of leaves; if simultaneous, leaves pinnatisected...2.
+ Flowers campanulate; when completely open, leaf profile oblong. ....4.
2. Leaves 3-pinnatisected into lanceolate lobes, their lower segments on fairly long petioles, manifest after blossoming of flowers, rarely simultaneous, coarse and held long after wilting. Plant initially with compact, silky hairs, but balding later. Flowers large; perianth segments 30-40 mm long, straight...8. **P. turczaninovii** Kryl. et Serg.
+ Leaves 3-palmatipartite..3.
3. Lobes of leaf blade oblong-cuneate. Flowers yellow, violet outside in upper portion of perianth segments...5. **P. flavescens** (Zucc.) Juz.
+ Lobes of leaf blade linear, 1-3 mm broad. Flowers blue-violet..6. **P. multifida** (G. Pritz.) Juz.
4. Leaf blade profile oblong, narrow, with several lateral segments. Plants 5-25 cm tall, often forming turfs with many stems.............5.
+ Leaf blade profile ovoid, broad, with 2-4 (rarely more) lateral segments. Plants 25-30 cm tall with single stems....................................8.
5. Leaves pinnate with rhomboid trifid lobes. Plant usually small (5-10 cm tall) and compactly appressed-pilose. Flowers 10-15 mm long....2. **P. bungeana** var. **astragalifolia** (Pobed.) Grub. ex Gubanov.
+ Leaves bipinnatisected with linear-lanceolate or oblong lobes. Plant 5-25 cm tall and diffuse-pilose...6.
6. Flowers not large; perianth segments 12-15 mm long, 3-7 mm broad, often open..................................2. **P. bungeana** var. **bungeana**.
+ Flowers of moderate size; perianth segments 20-30 cm long, 6-10 mm broad, half open..7.
7. Flowers violet-coloured...
..2. **P. bungeana** var. **tenuiloba** (Turcz. ex Hayek) Grub. ex Gubanov.
+ Flowers yellow...7. **P. sukaczewii** Juz.
8. Leaves coarse, pinnatisected into cuneate lobes with 2 pairs of deeply partite segments. Pedicels short, somewhat curved; flowers pale-blue-violet, closed, of moderate size; perianth segments 23-30 mm long, 9-12 mm broad, straight, not declinate at tip. Nuts with long (about 50 mm) awn..4. **P. dahurica** (Fisch. ex DC.) Spreng.
+ Leaves tender, rapid-caducous after vegetation, bipinnatisected into oblong or lanceolate lobes with sessile lateral segments. Peri-

anth segments 25-35 mm long. Nuts with very short, 25-40 mm long, awn..9.

9. Flowers narrow-campanulate, drooping, with perianth segments declinate at tip. Stamens somewhat shorter than perianth segments. Leaf blade compact-pilose, with lanceolate subobtuse lobes. ....3. **P. campanella** Fisch. ex Regel et Til.

\+ Flowers broad-campanulate, erect or inclined, with perianth segments straight at tip. Stamens 1/3-1/2 as long as perianth segments. Leaf blades diffuse-pilose, with broad-lanceolate or oblong acute lobes....................................1. **P. ambigua** (Turcz. ex Hayek) Juz.

1. **P. ambigua** (Turcz. ex Hayek) Juz. in Fl. URSS, 7 (1937) 307; Kitag. Lin. Fl. Mansh. (1939) 221; Fl. Kazakhst. 4 (1961) 68; Grub. Opred. rast. Mong. [Key to Plants of Mongolia] (1982) 112; Y.Z. Zhao in Acta Sci. Natur. Univ. Intramong. 19, 4 (1988) 656; Gub. Konsp. Fl. Vneshn. Mong. [Conspectus of Flora of Outer Mongolia] (1996) 50. —*Anemone ambigua* Turcz. ex Hayek in Festschr. Feier. Siebzigsten Geburtstages P. Ascherson (1904) 466. —*Pulsatilla ambigua* Turcz. ex G. Pritz. in Linnaea, 15 (1841) 601, nom. in syn. *Anemone albana;* Fl. Sin. 28 (1980) 68; Claves pl. Xinjiang, 2 (1983) 301; Fl. desert Sin. 1 (1985) 506; Pl. Vasc. Helanshan (1986) 105; Fl. Intramong. ed. 2, 2 (1990) 473. —*A. regeliana* Maxim. Enum. pl. Mong. 1 (1889) 11; id. Fl. Tangut. 1 (1889) 10. —*Pulsatilla regeliana* (Maxim.) Pavl. in Bull. Soc. natur. Moscou, 38 (1929) 64; Kryl. Not. Syst. (Tomsk) 5-6 (1930) 2, comb. superfl.; Kryl. Fl. Zap. Sib. 5 (1931) 1167. —**Ic.:** Maxim. Enum. pl. Mong. tab. 3; Fl. SSSR, 7, Plate 29, fig. 8 a and b; Fl. Kazakhst. 4, Plate 8, fig. 9; Fl. Sin. 28, tab. 19, fig. 3-5; tab. 20, fig. 10-11; Fl. Intramong. ed. 2, 2, tab. 191, fig. 7.

Described from East. Siberia (Irkut river). Type and isotype in St.-Petersburg (LE).

In subalpine meadows, *Cobresia* groves, mountain steppes.

IA. Mongolia: ***Khobd.*** (Sailyugem mountain range, Shar-Gobi river valley, June 13; Ulei-Daban pass, June 14; Bairimen-Daban pass, June 20; Ulan-Daban mountain pass, June 22; south of Kharkira river, alpine belt, July 23—1879, Pot.; summit of Ulan-Daban pass, between Kobdo river and Ukhoi, July 4, 1898—Klem.; Turgen' mountain range, lower portion of slope, along forest border, July 17, 1971—Grub., Ulzij. and Dariima), ***Mong. Alt.*** (in Taishiri-Ula mountains, July 12 and 15, 1877—Pot.; upper Dzhargalanty river (Kolgutty), steppe slopes, July 8, 1905—Sap.; 2 km west of Bulgan somon, July 24, 1947—Yun.; Khasagtu-Khairkhan, nor. sløpe of Tsagan-Irmyk-Ula, in upper Khunkerin-Ama, *Cobresia* grove, 2700-3100 m, Aug. 23, 1972—Grub., Ulzij. et al.; Ikh-Dzhargalant river basin (right tributary of Bulgan river), 10-15 km nor.-west of Bulgan settlement, July 27, 1988—Kam., Gamalei et al. (MW), ***Gobi-Alt.*** (Gub. l.c.).

IIA. Junggar: ***Tarb.*** (south. slope of Saur mountain range, Karagaitu river valley, Bain-Tsagan creek valley, June 23, 1957—Yun. et al.; north of Dachen, No. 1563, Aug. 13, 1957—Kuan; nor.-west of Khob[uk]saira, Chagan-obo mountain, No. 10540, June 22,

1959—Lee et al.), *Jung. Alat.* (mountains in Toli district, No. 1043, 1192 and 2638, Aug. 6; same site, No. 2619, Aug. 7; Chzhaosu district, in Ven'tsyuan' region, No. 3456, Aug. 14—1957, Kuan), *Tien Shan* (Piluchi pass, 2100-2400 m, April 26, A. Reg.; Kunges river, in forest, May 11, 1877—Przew.; Sairam lake, July [8-12]; Talki gorge, 1800-2400 m, Aug. 16; Muzart river valley, 2700 m, Aug. 19; Dzhagastai, 1500-2100 m, Aug. [1]—1877, A. Reg.; Sarybulak near Kul'dzha, 1200-1800 m, April 22; Sharysu river, 2100-2400 m, June 26—1878, A. Reg.; Irtak-Sary (Sairam), July 20—1878, Fet.; Bogdo mountain, 2400-2700 m, July; Sairam lake, south-east. bank—1878, A. Reg.; upper Taldy river, 2400-2700 m, May 17 and 18, 1879—A. Reg.; [Koshety-Daban] in forest, May 25, 1879—Przew.; Borborogussun river, 2700 m, June 15; Dumbedan to Kumdaban, 2400-2700 m, June 28; Menguto, 3000-3300 m, July 4—1879, A. Reg.; oberes Gross Juldus Tal am Wege zum Karagaitasch Pass, May 22-26; Oberstes Dunde-Kelde Tal und oberstes Chustai Tal, July 5-6—1908, Merzb.; Manas river basin, left bank, valley of Ulan-Usu river, at confluence with Dzhartas, July 8, 1957—Yun. et al.; 6-7 km south of Danyu, No. 488, July 22; on Danyu river, No. 2199, July 23—1957, Kuan; Manas river basin, Danu-Gol river valley near right branch to Danu-Daban pass, alpine steppe, July 23, 1957—Yun. et al.; Mal. Yuldus, 2550 m, No. 6364, Aug. 2; nor. foothills of Narat mountain range gently descending into Tsanma valley—left tributary of Kunges, 2150 m, July 7; south. slope of Narat mountain range descending into Bol'sh. Yuldus basin, on road from Dasht pass to Bain-Bulak, mountain steppe, Aug. 8 —1958, Yun. and Yuan').

IIIA. Qinghai: ***Nanshan*** (Tashitu river, May 26, 1886—Pot.).

General distribution: West. Sib. (Altay), East. Sib., Nor. Mong. (Fore Hubs., Hent., Hang., Mong.-Daur.), China (Dunbei, Nor.-West.).

Note. It is not always easy to distinguish *P. ambigua* from *P. turczaninovii*. Quite possibly, it arose by the hybridization of the latter with *P. campanella*. Compared to *P. turczaninovii*, *P. ambigua* is characterised by much earlier leaf formation. Leaves, like stem, are less pubescent, tender, rapid-caducous at end of vegetation (leaves in *P. turczaninovii* are coarse and held for long); flowers less large but perianth segments also straight (unlike those of *P. campanella* in which they are revoluted at tip).

2. **P. bungeana** C.A. Mey. in Ledeb. Fl. alt. 2 (1830) 371; Kryl. Fl. Zap. Sib. 5 (1931) 1169; Grub. Opred. rast. Mong. [Key to Plants of Mongolia] (1982) 111; Gub. Konsp. Fl. Vneshn. Mong. [Conspectus of Flora of Outer Mongolia] (1996) 50. —*Anemone bungeana* (C.A. Mey.) G. Pritz. in Linnaea, 15 (1841) 603; Maxim. Enum. pl. Mong. 1 (1889) 12. —*Pulsatilla astragalifolia* Pobed. in Tr. Mong. komiss. AN SSSR, 19 (1935) 51. —*P. tenuiloba* (Turcz. ex Hayek) Juz. in Fl. URSS, 7 (1937) 299; Y.Z. Zhao in Acta Sci. Natur. Univ. Intramong. 19, 4 (1988) 658. —*Anemone tenuiloba* Turcz. ex Hayek in Festschr. Feier. Siebzigsten Geburtstages P. Ascherson (1904) 472. —*Pulsatilla bungeana* var. *astragalifolia* (Pobed.) Grub. ex Gubanov, l.c. (1996) 50. —*P. bungeana* var. *bungeana*; Grub. l.c. (1982) 111; Gub. l.c. (1996) 50. —*P. bungeana* var. *tenuiloba* (Turcz. ex Hayek) Grub. ex Gubanov, l.c. (1996) 50. —**Ic.:** Ledeb. Ic. pl. Fl. Ross. 2, 1 (1830) tab. 110; Tr. Mong. komiss. AN SSSR, 19, fig.; Fl. SSSR, 7, Plate 19, fig. 1 a and c; Grub. 1982, Plate 51, fig. 233.

Described from Altay (Chuya river). Type in St.-Petersburg (LE).

In rock fissures and ledges and rocky steppe slopes.

IA. Mongolia: ***Khobd.*** (between Ukha river and Khatu, July 10, 1870—Kalning; Boirimen-Daban pass, June 20; in mountains north-east of Uryuk-Nur lake, June 10; desert between Ulangom and Kharkira river, July 9—1879, Pot.; mountains between Kobdo river and Ukhoi river, July 4, 1898—Klem.; valley of Kharkira river, upland steppe, Sept. 4, 1931—Bar.), ***Mong. Alt.*** (valley of Tsitsirin-Gol river, July 7, 1877—Pot.; between Ikys-nur and Tunkul'nur, on bank of river bed, July 24, 1897; nor. slope of Mal. Ulan-Daban pass, between Dabasutei lake and Bodunchi river, July 19, 1898—Klem.; Buyantu river, July 8, 1903—Gr.-Grzh.; Tsagan-Gol river, near Tsagan-kobu crossing, June 28, 1905—Sap.; valley of Urtu-Gol river, on rocks, Aug. 17; Khara-Adzarga mountain, range valley of Sakhir-sala river, on east. slope of Imertsik mountain, Aug. 22; Khara-Tszarga mountains, valley of Borogol river, east. rocky slopes, No. 291, Aug. 24, 1930—Pob., holotypus *P. astragalifolia* Pobed. !; Kobdo-Toguryuk course, Aug. 11, 1930; waterdivide summit of Buyantu river and Khara-Usu lake, Aug. 22, 1931—Bar.; 15 km south of Kobdo town, hummocky area, Aug. 16, 1978—Ogureeva (MW); south-west of Tonkhil somon, Aug. 13, 1945; 12 km north of Tamchi somon, July 16; Tonkhil somon, pass through Dobtsag-Khuren-Nuru, slopes of hummocky area, Aug. 11; 6-7 km south of Tonkhil-Nur lake, July 16—1947, Yun.; same site, Sept. 7, 1948; south. slope of Khan-Taishiri mountain range, in upper Shine-usu, 2380-2450 m, June 19; valley of Dzuilin-Gola, south-east of Serbeng-Ula, pass at 2925 m, June 26; west. extremity of Tsetseg-Nur basin, Temetiin-Khukh-Ula, 2150-2200 m, June 26; around Tugurin-Gola estuary, gorge floor, June 27; Kobdo river basin, Tsagan-Nur road at intersection with road from somon, June 29; Buyantu river basin, west. spur of Bugu-ula, 2470 m, July 2 —1971, Grub., Ulzij. et al.; valley of Ulgiin-Barun-kholai, 1 km north of pass, Aug. 11; valley of Ulan-Ergiin-Gol near Must somon, granite Mu-Ulan-Tolgoi mound, Aug. 12—1979, Grub., Muld. et al.; Tsirtiin-Khuren-ula mountains, 20 km nor.-east of Bugat somon, 2543 m, July 5, 1984—Kam. and Dariima; 35 km south-west of Khalyun settlement, Aug. 14, 1984—Gub. (MW), ***Cent. Khalkha*** (between Tsakhir spring and Chin-Tologoi mountain, May 19, 1894—Klem.; on east. slopes of Bain-Khan mountains, Central Kerulen, 1899—Pal.; between source of Uber-Dzhargalanta river and Agit mountain, rubble beds, Aug. 11; near foothills of Agit mountain, Aug. 26—1925, Krasch. and Zam.; Kholt area, June 10 and July 26, 1926—Gus.; vicinity of Ikhe-Tukhum-Nor lake, Aug. 20, 1926—Zam.; Darkhan-Khan somon, Khailyasutuin-sume, April 24; Nyalga somon, Burul'zhin-Tala area, knoll summit, May 22—1944; Delger Tsogtu somon, June 12—1945, Yun.; 15 km nor.-east of Santu somon, July 24, 1951—Kal.; 16 km south of Munkh-Khan somon, July 29, 1962—Dashnyam; 30 km south-west of Munkh-Khan somon, June 20, 1971—Dashnyam, Isach. et al.; 35 km from Unzhul somon, nor. slope of Ar-Unzhul-ula, June 17, 1973—Sanczir; 46 km nor. of Unzhul somon, June 17, 1974—Dariima; Munkh-Khan-Ula, 1607 m, July 25, 1974—Golubk. et al.; vicinity of Tumen-Tsogt settlement, June 30, 1985—E. Alekseeva (MW), ***East. Mong.*** (in monte Batu-Chan, inter Kerulen et Dolon-Nor; inter Kulussutajewsk et Dolon-Nor—1870, Lom.; in mountains around Kalgan town, May 9; Suma-Khada, June 8—1871, Przew.; rocky slopes of Khukh-Khoto town, July 24, 1926—A. Bulle; between Baltasain-Shanda area and Shabar-Dzhargalant, Oct. 5, 1928—L. Shastin, Matad somon, Bayan-Khanga gorge, July 1, 1975—Zhurba (MW), ***Depr. Lakes*** (left bank of Kharkira river, in forest, July 11; Shiba river, Sept. 1—1879, Pot.; Shanda-Daban pass, July 1, 1894—Klem.; Dzun-Dzhirgalantu mountain range, south-west. slope, Ulyastu-Gola gorge, 1850-2800 m, June 22, 1971—Grub., Ulzij. et al.), ***Val. Lakes*** (Ubten-Daban pass, around Orok-Nur lake, Aug. 3; Tuya river, Sept. 22—1886, Pot.; between Khara-Tologoi lake, Togomus well and Tsagan-delli lake, July 30, 1893—Klem.), ***Gobi-Alt.*** (Baga Bogdo, deep canyon terrace and upper slopes, 1900 m, No. 218, 1925—Chaney; nor. slopes of Ikhe-Bogdo mountain upper belt, Aug. 23, 1926—Tug.; same site, Nokhoitin-Khundei upper creek valley, Sept. 14, 1943; same site, lower mountain belt, upper Sevsul creek valley, July 2, 1945—Yun.; same site, south-south-west. slope, near mouth of Narin-Khurimt gorge, 2400-2500 m, July 30, 1948—Grub.; same site, nor. macroslope, Uvt gorge, 2630 m, June 27, 1972—Banzragch, Karam. et al.; same site, south. slope, Dzhargalantyin-Ama gorge, 2600 m, July 9, 1973—Golubk.; south. slope of Ikhe-Bogdo mountain, 2950 m, Aug. 4, 1973—Isach. and Rachk.; Tugreg somon, Maint-Daba pass, July 21, 1975—Zhurba (MW), ***Alash. Gobi*** (Alashan'sk mountains, alp. meadow, May 19, 1873—Przew.; on road from Tsakharin-Tala area to Bain-Barat area, June 11; Bain-Barat area, June 12—1909, Czet.).

General distribution: West. Sib. (Altay), Nor. Mong.

Note. Plants of var. *astragalifolia* are found mostly in Khobdo and Mong. Altay regions and those of var. *bungeana* and var. *tenuiloba* all over the distribution range of the species.

3. **P. campanella** Fisch. ex Regel et Til. in Nouv. Mém. Soc. natur. Moscou, 11 (1859) 30; Hemsl. Fl. Tibet (1902) 162; Kryl. Fl. Zap. Sib. 5 (1931) 1168; Juz. in Fl. SSSR, 7 (1937) 306; Fl. Kirgiz. 6 (1955) 76; Fl. Kazakhst. 4 (1961) 68; Ikonnik. Opred. rast. Pamira [Key to Plants of Pamir] (1963) 117; Kovalevsk. in Opred. rast. Sr. Azii [Key to Plants of Mid. Asia] 3 (1972) 194; Fl. Tadzh. 4 (1975) 74; Fl. Sin. 28 (1980) 69; Grub. Opred. rast. Mong. [Key to Plants of Mongolia] (1982) 112; Claves pl. Xinjiang, 2 (1983) 301; Gub. Konsp. Fl. Vneshn. Mong. [Conspectus of Flora of Outer Mongolia] (1996) 50. —*Anemone albana* Stev. in Mém. Soc. Hist. natur. Moscou, 3 (1812) 264, p.p.; Maxim. Enum. pl. Mong. 1 (1889) 12; Pamp. Fl. Carac. (1930) 110. —**Ic.:** Ledeb. Ic. pl. Fl. Ross. 2, 1 (1830) tab. 109; Fl. SSSR, 7, Plate 19, fig. 6; Fl. Kirgiz. 6, Plate 8, fig. 4; Fl. Sin. 28, tab. 19, fig. 6.

Described from Far East. Type in St.-Petersburg (LE).

On rocky slopes, among rocks, boulders, moraine, juniper and spruce forests in upper belt of mountains.

IA. Mongolia: ***Khobd.*** (South. macroslope of Sailyugem mountain range, 70 km nor.-west of Ulan-Khus settlement, Khara-Dzhamat-Gol river basin, left tributary of Uigoryn-Gol river, July 10, 1988—Gub., Dariima et al. (MW), ***Mong. Alt.*** (Tamchi somon, Bus-Khairkhan mountain range, 3200 m, alpine steppe, No. 2025, July 17; Bulugun river basin, Kharagaitu-Khutul' pass, alpine meadow and rubbly placers with snow patches, July 24; upper Indertiin-Gol, alpine steppe, July 25—1947, Yun.; waterdivide of Buyantu and Bulugun, Akhuntyn-Daba, 3050 m, sedge-*Cobresia* waste land, July 2; upper Buyantu river, Chigirtei-Gol, nor. slope of Chigirtei-Ula, 2600-2800 m, recent larch grove along ravine, July 4—1971, Grub., Ulzij. Dariima; left tributary of Ulyastei-Gola, on rocks near upper forest boundary, June 28, 1973—Golubk., Tsogt; Uirtiin-Khuren-Ula mountains, 20 km north-east of Bugat somon and 5 km from pass, 2543 m, July 5, 1984—Kam. (MW).

IIA. Junggar: ***Tarb.*** (Saur mountain range, south. slope, valley of Karagaitu river, right bank creek valley of Bain-Tsagan, subalpine meadow, June 23, 1957—Yun. et al.; nor.-west. of Khobsair [=Khobuk-Sair], Chagan-Obo mountain, 2000 m, No. 10540, June 22, 1959—Lee et al.), ***Jung. Alat.*** (in Toli district, No. 2619 and 2638, Aug. 6 and 7; Chzhaosu district in Ven'tsyuan' town region (=Arasan), in ravine, No. 3456, Aug. 14—1957, Kuan), ***Tien Shan*** (Mal. Yuldus, 2300 m, May 29 and June 4, 1877, Przew.; Koktyube mountains, 2100-2700 m, June 23, 1978; nor. Borborogusun, 1500 m, April 24; near sources of Dzhargalan and Pilyuchi rivers, 1800 m, April 24; Taldy river, 1200-

2700 m, May 15—1879, A. Reg.; between Barkul' town and Khami town [under Koshet-Daban pass], June 3, 1879—Przew.; Naryn-Gol brook, Tsagan-usu, 1800-2400 m, June 10; Mengute, 2700 m, July 3—1879, A. Reg.; Dschanart Tal, June 14-17; im Terek Tal (Chalyk Tau) hoch oben in Polstern, May 19; Sary Dschass, zwischen Lager und Gletscherende, Aug. 2-8—1903, Merzb.; Manas river basin, Ulan-Usu river valley at confluence with Dzhartas, subalpine steppe, July 18; Danu-Gol river near right turn to Danu-Daban pass, alpine steppe, July 23—1957, Yun.; Danyu river, in ravine, No. 2199, July 23, 1957—Kuan; upper Ili river, nor. foothills of Narat mountain range, gently descending into Tsanma valley, 2150 m, subalpine meadow, Aug. 7; south. slope of Narat mountain range descending into B. Yuldus basin, on road from Dasht pass into Bain-Bulak, mountain steppe, Aug. 8—1958, Yun.), ***Jung. Gobi*** (east.: 130 km south-east of Bulgan somon, Ikh-Khavtag-Ula mountain range, 2500 m, Aug. 12, 1977 —Volk. and Rachk.).

IIIB. Tibet: ***Chang Tang*** (Keria mountain range, Kyzylsu river, 3200 m, July 24, 1885—Przew.).

IIIC. Pamir (Billuli pass, June 13, 1909—Divn.; Issyk-su river, 3000-3100 m, June 21; Pil'nen river gorge, 4500-5000 m, July 1; waterdivide between Atrakyr river and Tyuzutek, mossy tundra, 4500-5000 m, July 20; Kara-Dzhilga river, alpine pasture, 4000-5000 m, July 22—1942, Serp.).

General distribution: Jung.-Tarb., Nor. and Cent. Tien Shan, East. Pam.; Mid. Asia, West. Sib. (Altay), Nor. Mong. (Fore Hubs., Hang.), China (Altay).

4. **P. dahurica** (Fisch. ex DC.) Spreng. Syst. Veg. ed. 16, 2 (1825) 663; Juz. in Fl. SSSR, 7 (1937) 303; Kitag. Lin. Fl. Mansh. (1939) 222; Fl. Sin. 28 (1980) 67; Y.Z. Zhao in Acta Sci. Natur. Univ. Intramong. 19, 4 (1988) 656; Fl. Intramong. ed. 2, 2 (1990) 475. —*Anemone dahurica* Fisch. ex DC. Prodr. 1 (1824) 17. —**Ic.:** Fl. SSSR, 7, Plate 19, fig. 2 a and b; Fl. Intramong. ed. 2, 2, tab. 192, fig. 1-2.

Described from East. Siberia (Dauria). Type in Geneva (G).

On riverine pebble beds, sand and mountain meadowy slopes.

IA. Mongolia: ***East. Mong.*** (Manchuria station, Beishan' mountain, on mountain slope, No. 797, June 23; vicinity of Manchuria station, mountain slope, No. 842, June 24 —1951, Lee et al.; Khailar town, Sishan' mounds, meadow, July 6, 1951—B. Skvortzov and Lee).

General distribution: East. Sib., Far East, China (Dunbei).

5. **P. flavescens** (Zucc.) Juz. in Fl. URSS, 7 (1937) 296; Grub. Opred. rast. Mong. [Key to Plants of Mongolia] (1982) 111; Gub. Konsp. Fl. Vneshn. Mong. [Conspectus of Flora of Ourter Mongolia] (1996) 50. —*Anemone flavescens* Zucc. in Regensb. Zeit. 1 (1826) 371; G. Pritz. in Linnaea, 15 (1841) 585.

Described from West. Siberia (vicinity of Omsk). Type in ? Munchen (M). Map 3.

In larch forests, forest borders, arid steppe slopes.

IA. Mongolia: ***Mong. Alt.*** (nor. slope of Taishiri-Ula mountain range, July 12, 1945—Yun.; south. slope of Khan-Taishiri mountain range, in upper Shine-Usu (north of Khalyun somon), 2380-2450 m, June 19, 1971—Grub. et al.).

IIA. Junggar: ***Cis-Alt.*** (15-20 km nor.-west of Shara-Sume (on Kran river), June 7, 1959—Yun. and Yuan'; Sanginin-Gol river basin (upper Chern. Irtysh), 25 km south-east of Dayan-Nur post, 1800-2100 m, July 16-17, 1988—Kam., Bud. et al. (MW).

General distribution: Europe (south-east. East Europe), West. Sib., East. Sib., Nor. Mong.

Note. This species is cited for the first time for the Chinese territory. It is possible that plants from Kran river represent hybrids with *P. turczaninovii* since the entire plant is pubescent while midlobe of leaf is noticeably petiolate.

6. **P. multifida** (G. Pritz.) Juz. in Fl. URSS, 7 (1937) 296; Fl. Kazakhst. 4 (1961) 67; Grub. Opred. rast. Mong. [Key to Plants of Mongolia] (1982) 111; Claves pl. Xinjiang, 2 (1983) 301; Gub. Konsp. Fl. Vneshn. Mong. [Conspectus of Flora of Outer Mongolia] (1996) 50. —*Anemone patens* var. *multifida* G. Pritz. in Linnaea, 15 (1841) 581; Fl. Sin. 28 (1980) 68. —*Pulsatilla patens* subsp. *multifida* (G. Pritz.) Zam. in Acta Horti Bot. Univ. Latv. 1 (1926) 98; Kryl. Fl. Zap. Sib. 5 (1931) 1166; Kitag. Lin. Fl. Mansh. (1939) 222. —*P. patens* var. *multifida* (G. Pritz.) S.H. Li et Y.H. Huang in Fl. Pl. Herb. Chin. Bor.-Or. 3 (1975) 163; Y.Z. Zhao in Acta Sci. Natur. Univ. Intramong. 19, 4 (1988) 655. —**Ic.:** Curtis' Bot. Mag. tab. 1994; Zam. l.c. tab. 8 and 10 b; Fl. SSSR, 7, Plate 18, fig. 7; Fl. Sin. 28, tab. 20, fig. 7.

Described from Siberia. Type in Munchen (M).

On mountain-steppe slopes.

IA. Mongolia: ***Khobd.*** (Kharkira mountain range, Namyur river, July 18, 1903—Gr.-Grzh.; south. macroslope of Sailyugem mountain range, 70 and 90 km nor.-west of Ulan-Us settlement, Khara-Dzhamat-Gol river basin, left tributary of Uigoryn-Gol river, alpine grassland, July 10, 1988—Gub., Bud. and Dariima (MW), ***Mong. Alt.*** (upper Kobdo river, Dayan-Nur, south. extremity, Yamantau mountain, rocky summit of nor. slope, 2550 m, rocks, July 10, 1971—Grub., Ulzij. et al.; south. macroslope of Sailyugem mountain range, 70 km nor.-west of Ulan-Khus settlement, Khara-Dzhamant-gol river basin (left tributary of Uigoryn-Gol river), July 10, 1988—Gub., Bud. and Dariima; 10-15 km west of Dayan-Nur post near Khara-Nur lake, July 20, 1988—Bud., Gub. et al.).

IIA. Junggar: ***Cis-Alt.*** (south-west. Shara-Sume town, on road to Karamai, on mountain trail, No. 10221, April 29, 1959—Lee et al.).

General distribution: Arct., Europe (Volga-Kam. and Trans-Volga regions of East. Europe), West. Sib., East. Sib., Far East, Nor. Mong. (Hang., Mong.-Daur.), China (Dunbei), Nor. Amer.

7. **P. sukaczewii** Juz. in Fl. URSS, 7 (1937) 301; Fl. Sin. 28 (1980) 70; Y.Z. Zhao in Acta Sci. Natur. Univ. Intramong. 19, 4 (1988) 658; Fl. Intramong. ed. 2, 2 (1990) 479. —? *P. bungeana* subsp. *mongolica* Zimmerman in Feddes Repert. 70, 1-3 (1965) 147. —**Ic.:** Fl. Intramong. ed. 2, 2, tab. 191, fig. 1-6.

Described from East. Siberia. Type in St.-Petersburg (LE).

On rubble slopes, mounds and ravines.

IA. Mongolia: ***East. Mong.*** ("bei Mandschouli II"—Feddes Repert. l.c.; "cent. portion of Shilingol'sk ajmaq—administrative territorial unit in Mongolia: west. spur of Datsinshan' mountain range"—Fl. Intramong. l.c.).

General distribution: East. Sib., China (Dunbei).

Note. Cited for Mongolia (Fl. Intramong. l.c.—widely distributed).

8. **P. turczaninovii** Kryl. et Serg. in Not. Syst. (Tomsk) 5-6 (1930) 1; Kryl. Fl. Zap. Sib. 5 (1931) 1167; Juz. in Fl. SSSR, 7 (1937) 294; Kitag. Lin. Fl. Mansh. (1939) 222; Fl. Kazakhst. 4 (1961) 67; Fl. Sin. 28 (1980) 69; Grub. Opred. rast. Mong. [Key to Plants of Mongolia] (1982) 112; Claves pl. Xinjiang, 2 (1983) 301; Fl. desert Sin. 1 (1985) 506; Pl. Vasc. Helanshan (1986) 105; Y.Z. Zhao in Acta Sci. Natur. Univ. Intramong. 19, 4 (1988) 657; Fl. Intramong. ed. 2, 2 (1990) 477; Gub. Konsp. Fl. Vneshn. Mong. [Conspectus of Flora of Outer Mongolia] (1996) 50. —*P. turczaninovii* f. *albiflora* Y.Z. Zhao, l.c. 657; Fl. Intramong. ed. 2, l.c. 477. —*Anemone pulsatilla* auct. non L.: Maxim. Fl. Tangut. 1 (1889) 10. —**Ic.**: Fl. Sin. 28, tab. 19, Fig. 11; Grub. Plate 52, fig. 236; Fl. desert Sin. 1, tab. 186, fig. 1-3; Fl. Intramong. ed. 2, 2, tab. 193, fig. 1-3.

Described from East. Siberia (Dauria). Type in Tomsk (TK).

On mountain-steppe, often sandy and rocky slopes, coastal pebble beds, pine groves, arid forest borders, sandy and forb-cereal grass steppes.

IA. Mongolia: ***Khobd.*** (east of Ulan-Daban, June 22, 1879—Pot.), ***Cent. Khalkha*** (nor. slope of mountains surrounding Tsilinam creek valley from south, May 17; between Tsakhir spring and Chintologoi mountain, May 19—1894, Klem.; Sorgol-Khairkhan mountain, in crevices of granite rocks, May 16, 1941; Mandal somon, Sakhil'te-Ula, May 20, 1944—Yun.; nor slope of Bayan-Khan-Ula, 54 and 75 km nor.-west of Khutliin-Khuduk, July 14 and 15, 1971—Dashnyam, Karam., Safronova; 35 km north of Unzhul somon, Ar-Unzhul-Ula, June 17; 26 km nor.-east of Unzhul somon, east. slope of Sorgol-Khairkhan-Ula, June 12—1973, Sanczir; 22 km nor.-west of Unzhul somon, June 11; 35 km north of Unzhul somon, Ar-Unzhul-Ula mountain trail, June 22—1974, Dariima; Choiren-Ula mountain, July 9, 1975—Zhurba (MW), ***East. Mong.*** (Oulachau occid. July 1866—David; Muni-Ula mountains, April 26, 1872—Przew.; Kulun-buir-nor plain, Angirtu lake, May 28; same site, Mandybai lake, May 30; same site, Dzhirim lake, May 31; same site, Elisyn-Khuduk collective, June 5—1899, Pot. and Sold.; Manchuria station, steppe, June 5, 1902—Litw.; Darkhan-Ula, 15 km east of Darkhan-Khan somon, April 27; Zodol-Ula mountain, 50 km west of Yugodzyra, May 14—1944, Yun.; Khailar town, Sishan' mound, No. 533, June 7, 1951—B. Skvortz.; 20 km north of Khukh-Khoto town, pass through Datsin'shan' mountain range, 1900 m, June 4, 1958—Petr.; Shilin-Khoto town, 1959—Ivan.; 47°25' N, 118°55' E, Khalkhin-Gol river basin, upper Altyn-Bulak brook, 800-850 m, July 18, 1987—Gandbold, Gub. et al. (MW).

IIA. Junggar: ***Cis-Alt.*** (15-20 km nor.-west of Shara-Sume (on Kran river), June 7, 1959—Yun. and Yuan'; south-west of Shara-Sume town, on road to Karamai, on mountain trail, No. 10221, July 29, 1959, Lee et al.).

IIIB. Tibet: ***Weitzan*** (upper course of Huang He river, 3000-3300 m, May 3, 1880—Przew.).

General distribution: West. Sib., East. Sib., Far East, Nor. Mong., China (Dunbei, Nor.).

***

The following species have also been cited for Central Asian territory:

P. cernua (Thunb.) Bercht. et Presl. Rostl. Ranuncul. (1820) 2.

IA. Mongolia: ***East. Mong.*** (Datsinshan).

P. chinensis (Bge.) Regel in Mém. Ac. Sci. St. Pétersb. sér. 7, 4, 4 (1861) 5.

IA. Mongolia: ***East. Mong.*** **and** ***Ordos.***

12. **Hepatica** Mill.

Gard. Dict. Abr. ed. 4 (1754)

1. **H. falconeri** (Thoms.) Steward in Rhodora, 29 (1927) 53; Juz. in Fl. SSSR, 7 (1937) 284 (cum auct. comb. Juz.); Fl. Kazakhst. 4 (1961) 66; Kovalevsk. in Opred. rast. Sr. Azii [Key to Plants of Mid. Asia] 3 (1972) 193; Karnisonova in Byull. Glavn. bot. sada SSSR, 91 (1974) 51; Claves pl. Xinjiang, 2 (1983) 299. —*Anemone falconeri* Thoms. in Hook. Ic. Pl., new ser. 5 (1852) pl. 899; Ulbr. in Bot. Jahrb. 37 (1906) 272. —**Ic.:** Hook. tab. 899; Fl. Kazakhst. 4, Plate 8, fig. 1.

Described from Kashmir. Type—sketch in first description. Plate I, fig. 3. In forest borders in upper forest boundary.

IIA. Junggar: ***Tien Shan*** (Talki brook, 900-1500 m, April; upper Khorgos river, 1500-1800 m, June; Altyn-Emel' near Koyan-kuz, 1200-1500 m, June 1; Khanakhai mountains, 1500-2100 m, June 16—1878; Nilki brook (from Kash side), 2100 m, June 8; Borgaty (Kash) brook, 1500 m, July 8—1879, A. Reg.; Manas river basin, left bank, valley of Ulan-Usu river at confluence with Dzhartas, upper forest boundary, subalpine meadow on northern slope, July 18, 1957—Yun. et al.).

General distribution: Jung.-Tarb., Nor. Tien Shan (Kungei Ala Tau and Trans-Ili Ala Tau); Mid. Asia (West. Tien Shan), Pam.-Al., Himalayas (west., Kashmir), Indo-Mal. (Cent. India).

Subtribe 2. **Clematidinae** DC.

13. **Clematis** L.[1]

Sp. pl. (1753) 543, s. str.

1. Plants with grassy fine-ribbed annually perishing, more rarely semi-evergreenstem ....2.
+ Shrubby lianas or subshrubs with more or less branched, lignified, round or subfaceted perennial stem ....5.
2. Sepals and stamens divaricated, flowers wide-open. Leaves branched-pinnatisected into narrow linear segments...6. **C. hexapetala** Pall.
+ Sepals and stamens erect or climbing, flowers campanulate, urceolate or cup-shaped. Leaves of different structure.......................3.
3. Leaves simple, entire, subsessile or short-petiolate...7. **C. integrifolia** L.

[1]Co-author—V.P. Serov.

+ All leaves, except leafy bracts, pinnately compounded, long-petiolate....................4.
4. Flowers urceolate, sepals recurved at tip; leaflets ovoid or subdeltoid, usually 3-lobed, with large, uneven teeth along margin ....13. **C. rehderiana** Craib.
+ Flowers narrow-campanulate, sepals erect; leaf segments trisected, generally into narrow linear or lanceolate lobes.................... ....1. **C. aethusifolia** Turcz.
5. All leaves entire or lower entire and middle and upper leaves pinnatisected....................6.
+ All leaves, except leafy bracts, once-twice pinnately compounded. ....7.
6. Flowers terminal. Leaves pinnatifid. Dwarf shrubs.................... ....10. **C. nannophylla** Maxim.
+ Flowers singly or more on twigs. Leaves entire or more or less incised-dentate, sometimes 3-lobed. Shrubs.......4. **C. fruticosa** Turcz.
7. Lower leaves entire, middle and upper leaves pinnatisected.......... ....2. **C. aspleniifolia** Schrenk.
+ All leaves entire, smooth-edged or sometimes serratodentate along margin....................14. **C. songorica** Bge.
8. Leaves pinnate, sometimes with ternate lower segments; leaflets large, ovoid or ovoid-lanceolate. Sepals obcuneate, obtuse............... ....3. **C. brevicaudata** DC.
+ Leaves once-twice tricompounded, usually with fine oblong smooth-edged or very large sinuate-serrate leaflets. Sepals ovoid-lanceolate, oblong, acute....................9.
9. Sepals glabrous inside....................10.
+ Sepals pubescent inside....................11.
10. Leaflets entire or more or less lobed but not dentate.................... ....8. **C. intricata** Bge.
+ Leaflets dentate along margin.........15. **C. tangutica** (Maxim.) Korsh.
11. Leaflets linear, not more than 3-5 mm broad....................12.
+ Leaflets not linear, usually considerably more broad....................13.
12. Leaflets long-cuspidate....................9. **C. ladakhiana** Grey-Wilson.
+ Leaflets not long-cuspidate, obtuse to short-cuspidate................ ....11. **C. orientalis** L.
13. Sepals glabrous outside, only sometimes with rather few stray hairs in upper portion....................14.
+ Sepals pubescent outside....................15.
14. Flowers usually single, sometimes in groups of not more than 3; sepals thick; anther filaments more than 2-5 times longer than anthers....................16. **C. tenuifolia** Royle.

\+ Flowers usually in groups of 5 or more, rarely 1-3 each; sepals thin; anther filaments less than double as long as anthers............... ....11. **C. orientalis** L.

15. Leaflets smooth-edged, sometimes with 1-2 teeth or lobed, glaucous. Flowers usually in dichasium...16.

\+ Leaflets dentate, green; flowers usually single...............................17.

16. Leaflets lanceolate-elliptical, usually entire, sometimes with 1-2 lobes of same form at base. Sepals more than 15 mm long............... ....5. **C. glauca** Willd.

\+ Leaflets usually lobed. Sepals not more than 15 mm long.................. ....11. **C. orientalis** L.

17. Dwarf subshrubs. Flowers campanulate with more or less straight sepals, usually terminal; peduncles 9-19 cm long........................ ....12. **C. pamiralaica** Grey-Wilson.

\+ Creeping or lodging plant. Flowers cup-shaped, with procumbent sepals, terminal or axillary; peduncles not more than 7 cm long. ....15. **C. tangutica** (Maxim.) Korsh.

1. **C. aethusifolia** Turcz. in Bull. Soc. natur. Moscou, 5 (1832) 181; Franch. Pl. David. 1 (1884) 12; Forbes and Hemsl. Index Fl. Sin. 1 (1886) 1; Maxim. Fl. Tangut. 1 (1889) 3; id. Enum. pl. Mong. 1 (1889) 5; Finet et Gagnep. in Bull. Soc. Bot. Fr. 50 (1903) 540; Krasch. in Fl. SSSR, 7 (1937) 315; Fl. Sin. 28 (1980) 115; Grub. Opred. rast. Mong. [Key to Plants of Mongolia] (1982) 112; Fl. desert Sin. 1 (1985) 483; Pl. Vasc. Helanshan (1986) 107; Y.Z. Zhao in Acta Sci. Nat. Univ. Intramong. 18, 2 (1987) 284; Fl. Intramong. ed. 2, 2 (1990) 528; Gub. Konsp. Fl. Vneshn. Mong. [Conspectus of Flora of Outer Mongolia] (1996) 49. —*C. aethusifolia* var. *latisecta* Maxim. Prim. Fl. Amur. (1859) 12; Kitag. Lin. Fl. Mansh. (1939) 216; Fl. Sin. l.c. 116; Fl. Intramong. l.c. 522. —**Ic.:** Fl. SSSR, 7, Plate 20, fig. 12; Fl. Sin. 28, fig. 10; Fl. desert Sin. 1, tab. 176, fig. 5-8; Fl. Intramong. 2, tab. 213, fig. 1-3, fig. 4 (var. *latisecta*).

Described from Inner Mongolia. Type in Kiev (KW). Isotypes in St.-Petersburg (LE). Plate VI, fig. 1.

In mountain gorges, valleys of mountain rivers, forests, steppe scrubs or in exposed sites between loose dune sands.

IA. Mongolia: ***East. Mong.*** (in rupestribus Mongoliae chinensis australis—1831 [Kusnetzov], 4 isotypi !; Mong. chin. adeundus Chinam—1840, Kirilow; montagnes de l'Ourato—June 1864, David; nor. slope of Muni-ula mountain range, June 30, 1871—Przew.; right bank of Huang He river, below Hekou, Aug. 7; Edzhin-khoro area, Aug. 18—1884, Pot.; Kalgan gorge, Aug. 23, 1898—Zab.; Shilin-Khoto town, 1959—Ivan.; sand massif at Moltsok-Elesu town, 12 km south of Dariganga, July 6, 1971—Dashnyam, Karam. et al.; Moltsog-Els sand, 45°15′ N, 113°50′ E, 1300 m, 12 km south-south-east of Dariganga settlement, July 7 and 10, 1985—Gub.); ***Alash. Gobi*** (in midportion of Alashan mountain range along south. slope, July 10, 1873—Przew.; Alashan mountain range, Tszosto gorge, May 15, 1908—Czet.), ***Ordos*** (5 km west of Ikochzhoumen town,

Aug. 3, 1957—Petr.); *Khesi* (road from Alashan to Kukunor lake, Loukhu-shan' mountain range plateau, midbelt, July 17, 1908—Czet.).

IIIA. Qinghai: ***Nanshan*** (in forests on mountain range along south. side of Tetung river, July 10, 1872; South. Tetung mountain range, forest belt, 2300 m, Aug. 6, 1880—Przew.; Chorten-tan temple, 2100-2400 m, Sept. 8, 1901—Lad.; 70 km south-east of Chzhan'e town, Tsinlyan'shan', Matisy temple, 2600 m, Aug. 12, 1958—Petr.).

General distribution: East. Sib.; Far East; China (Dunbei, Nor., Nor.-West., South-West.).

2. C. **aspleniifolia** Schrenk in Fisch. et Mey. Enum. pl. nov. 2 (1842) 68; Krasch. in Fl. SSSR, 7 (1937) 317; Fl. Kirgiz. 6 (1955) 78; Fl. Kazkhst. 4 (1961) 71; Claves pl. Xinjiang, 2 (1983) 304; Fl. desert Sin. 1 (1985) 479. — C. *songorica* var. *aspleniifolia* (Schrenk) Trautv. in Bull. Soc. natur. Moscou, 33 (1860) 56; Fl. Sin. 28 (1980) 151. —**Ic.:** Fl. SSSR, 7, Plate 20, fig. 3; Fl. Kazakhst. 4, Plate 9, fig. 3; Fl. Sin. 28, fig. 20 (sub C. *songorica* var. *aspleniifolia*); Fl. desert Sin. 1, tab. 174, fig. 6-7.

Described from East. Kazakhstan. Type in St.-Petersburg (LE). Plate VI, fig. 3.

On rocky mountain slopes, in foothills, lower and middle belts of mountains.

IA. Mongolia: ***Alash. Gobi*** (Urotov Land, Aug. 14, 1873—Przew.).

IB. Kashgar: ***Nor.*** (south. foothills of Taushkan-Dar'ya mountain range, 1500-1800 m, June 3, 1889—Rob.; Utsch-Turfan, June 1903—Merzb.; Uch-Turfan, May 27, 1908—Divn.; south. foothills of Tien Shan, east of Ak-su mountains, near Taz-Lyangar village, Aug. 11; same site, near Yakka-aryk village, in Muzart-Darya floodplain, Aug. 13—1929, Pop.; nor.-east. portion of Baisk basin, 23 km nor.-east of Kyzyl settlement, Sept. 1, 1958—Yun. and Yuan'; Zigdeichak in Kel'pin, No. 7471, Sept. 9, 1958—Lee and Chu; Kokshaal (Taushkan)-Dar'ya river valley, Sept. 18, 1958—Yun, and Yuan'; between Kucha and Takh'yar, No. 9842, July 9, 1959—Lee et al.), ***West.*** (Kizil-Su river basin (above Kashgar), Ulugchat mountains, July 2, 1929—Pop.; 120 km south of Kargalyk, Akkez Daban pass, June 5; west. foothills trail of King-Tau mountain range, 40 km south-west of Upal settlement, June 9, 1959—Yun. and Yuan'; on Kashgar-Irkeshtam road, 7-8 km east of Kensu, No. 9671, June 17, 1959—Kuan), ***East.*** (south of Khami, Sept. 3, 1875—Pias.; between Kuruk-Tag and Chol-Tag near Chokur village, Aug. 28, 1929—Pop.; Kiziltag mountain range, 25 km south-west of Kyumysh, 1400 m, Sept. 9, 1959—Petr.).

IIA. Junggar: ***Jung. Alat.*** (south. trails in Borotala valley, Archaty-Gol river, Aug. 17, 1957—Yun., Sh.I. Li, Yuan'), ***Tien Shan*** (Dzhirgalan river (tributary of lower Tekes), Aug. 1880—A. Reg.; vicinity of Turfan, Aug. 7, 1886—Krasn.; Savan, No. 3731, Oct. 4, 1956—Ching; 40 km south of Dzinkho, No. 4114, July 2, 1957—Kuan; Sin'yuan' town (=Aral-tyube), No. 3757, Aug. 22, 1957—Kuan; Ili valley, 42 km east of bridge on Kash river up along Ili-Kunges valley, Aug. 29, 1957—Yun., Sh.I. Li, Yuan'; 15 km nor.-nor.-west of Khari in Aksu, No. 8945, Sept. 29—1958, Lee and Chu), ***Jung. Gobi*** (on way from Kazvan in Tsinitsyuan', No. 1687, July 21, 1957—Kuan), ***Dzhark.*** (Suidun, July 8, 1877—A. Reg.).

IIIC. Pamir (Tash-Kurgan, July 25, 1913, Knorr.; in Arpalyk-Dar'ya river gorge, 10 km from Kyzyl-Davan, 3000 m, July 6; Posarbat area, 2500-3000 m, Aug.-Sept. —1941, Serp.).

General distribution: Fore Balkh., Jung.-Tarb., Nor. and Cent. Tien Shan; Mid. Asia, Fore Asia (Afghanistan).

Note. C. *aspleniifolia* differs from C. *songorica* in characteristic middle and upper leaves that are deeply lobed, with widely spaced lobes, sometimes tapered into petiolules.

3. **C. brevicaudata** DC. Syst. 1 (1817) 138; Prodr. 1 (1824) 3; Franch. Pl. David. 1 (1884) 14; Forbes and Hemsl. Index Fl. Sin. 1 (1886) 3; Maxim. Fl. Tangut. 1 (1889) 4; id. Enum. pl. Mong. 1 (1889) 5; Krasch. in Fl. SSSR, 7 (1937) 320; Kitag. Lin. Fl. Mansh. (1939) 216; Fl. Sin. 28 (1980) 188; Pl. Vasc. Helanshan (1986) 106; [Drevesn. rast. Tsinkhaya—Woody Plants of Qinghai] (1987) 193; Y.Z. Zhao in Acta Sci. Nat. Univ. Intramong. 18, 2 (1987) 284; Fl. Intramong. ed. 2, 2 (1990) 522; Gub. Konsp. Fl. Vneshn. Mong. [Conspectus of Flora of Outer Mongolia] (1996) 49. —**Ic.:** Fl. SSSR, 7, Plate 20, fig. 5; [Drevesn. rast. Tsinkhaya—Woody Plants of Qinghai], tab. 129; Fl. Intramong. ed. 2, 2 tab. 211, fig. 1-3.

Described from Nor. China. Type in Geneva (G).

In forest and scrubs in lower belt of mountains.

IA. Mongolia: ***East. Mong.*** (Muni-Ula, July 14, 1871—Przew.), ***Alash. Gobi*** (west. slope of midportion of Alashan mountain range, July 12, 1873—Przew.; 30 km east of Bayan-Khoto, 2200 m, July 6, 1957—Petr.), ***Ordos*** (valley of Ulan-Morin river, Aug. 22, 1884—Pot.).

IIIA. Qinghai: ***Nanshan*** (South. Tetung mountain range, lower belt, Aug. 10, 1872; forest belt of South. Tetung mountain range, Aug. 7, 1880—Przew.; "on Datunkhe river, Gunkhe ?—[Drevesn. rast. Tsinkhaya—Woody Plants of Qinghai] l.c.), ***Amdo*** (Mudzhikkhe river, June 30, 1880—Przew.).

General distribution: Far East, Nor. Mong. (Cis-Hing.), China (Dunbei, Nor., Nor.-West., East., Cent., South-West.), Korean peninsula.

4. **C. fruticosa** Turcz. Bull. Soc. natur. Moscou, 5 (1832) 180; Franch. Pl. David. 1 (1884) 12; Maxim. Enum. pl. Mong. 1 (1889) 2; Fl. Sin. 28 (1980) 148; Grub. Opred. rast. Mong. [Key to Plants of Mongolia] (1982) 112; Fl. desert Sin. 1 (1985) 481; Pl. Vasc. Helanshan (1986) 106; Y.Z. Zhao in Acta Sci. Nat. Univ. Intramong. 18, 2 (1987) 284; Grub., Gub. and Dariima in Byull. Mosk. obshch. isp. prir., otd. biol. 95, 6 (1990) 72; Fl. Intramong. ed. 2, 2 (1990) 517; Gub. Konsp. Fl. Vneshn. Mong. [Conspectus of Flora of Outer Mongolia] (1996) 49. —C. *fruticosa* var. *viridis* Turcz. l.c. 180. —C. *fruticosa* var. *canescens* Turcz. l.c. 180. —C. *canescens* (Turcz.) W.T. Wang et M.C. Chang, Fl. Sin. l.c. 150; Fl. desert Sin. l.c. 181; Fl. Intramong. l.c. 517; Gub. l.c. 49. —C. *canescens* subsp. *viridis* W.T. Wang et M.C. Chang in Fl. Sin. l.c. 148; [Drevesn. rast. Tsinkhaya—Woody Plants of Qinghai] (1987) 194. —**Ic.:** Fl. Sin. 28, tab. 43, fig. 1-6; Fl. desert Sin. 1, tab. 175, fig. 1-2; [Drevesn. rast. Tsinkhaya—Woody Plants of Qinghai] tab. 128; Fl. Intramong. ed. 2, 2, tab. 209, fig. 1-3.

Along flanks of gorges, on rocks, rubbly and rocky slopes of lower belt of mountains.

Described from Inner Mongolia. Type in Kiev (KW). Isotypes in St.-Petersburg (LE). Plate VI, fig. 2.

IA. Mongolia: ***East. Mong.*** (in arenosis montosis Mongoliae chinenses, 1831—[I. Kusnetzov], 3 isotypi !—var. *viridis* Turcz.; in salsis Mongoliae chinensis, 1831—[I. Kusnetzov], 2 isotypi !—var. *canescens* Turcz.; Mong. chin. 1840-1842—Kirilow; Monghol. Bunge. m. Sartchy, June 1866—David, Ordos, on sand north of Junggar prince's site, Aug. 9, 1884—Pot.; between Boro-obo and Khutul-dolon, Aug. 7, 1898—Zab.), ***Val. Lakes*** (Tuin-Gol river, July 9, 1893—Klem.; nor.-west. fringe of Ongon-Elesu sand. Derisni khuduk, May 17, 1944—Yun.), ***Gobi Alt.*** (along nor. descent of Tost mountain range, near Koshel'-Usu well, Aug. 19; Khaptsyl area, on south. slopes of Bain-Tsaganche mountain range, Aug. 23; on foothills of Bain-Tsagan mountain range, Aug. 26—1866, Pot.; Tszolin mountains, Sept. 8, 1931—Ik.-Gal.; south. spur of east extremity of Artsa-Bogdo mountain range, Khaldzan-Khairkhan-Ula, rocky slopes, July 19; 25 km nor. of Neme-getu-ula in Barun-Boro-Obo town region and Khara-Tologoi-khuduk well, poplar-saxaul oasis in gorge, Aug. 3; Nemegetu-nuru mountain range, Khara-Obo town, Aug. 7; nor. slope of west. end of Neme-getu-nuru mountain range, Sept. 5, 1948—Grub.; Artsa-Bogdo town, 98 km nor.-west of Bulgan somon, July 11, 1970—Banzragch, Karam. et al.; Somon-Khairkhan-Ula mountain range, south. macroslope, in gorge, July 21, 1977—Volk. and Rachk.), ***East. Gobi*** (Baga-ude, ravine in Khara-ula mountain, Aug. 16, 1926—Lis.; Delger-Khangai mountain range, Sept. 8, 1930—E. Kuznetsov; Del'ger-Khangai town, July 31, 1931—Ik.-Gal.; Kalgan road near Baga-Ude collective, Bain-gote mountains, Aug. 21, 1931—Pob.; 35-36 km north-north-west of Sain-Shanda, Sept. 10, 1940—Yun.; Bayan-Dzak area, Mandal-Obo somon, Oct. 21, 1947; same site, July 17, 1948—Grub.; Ulan-osh basin, Kholt somon, Aug. 14, 1950—Lavr. and Kal.; Bailinmyao town, 1954—Ivan.; 90 km north of Saikhan-Dulan, Dzhargalant-Nur lake basin, July 16; 30 km east-south-east of Khuvsgel, July 27, 1971—Isach. and Rachk.; hummocky massif south-east of Khan-Bogdo somon, 10 km north-west of Bolony-Khuduk collective, July 20, 1974—Rachk. and Volk.; south of Khotan-bulak, nor. part of Khoryu-ula mountain, Aug. 5, 1989—Grub., Gub. and Dariima), ***West. Gobi*** (Trans-Altay Gobi, Burkhuntu-Bulak spring, Aug. 23, 1948—Grub.; 60 km north-west of Tszin'ta town, July 23; 93 km north of An'si, July 26—1958, Petr.), ***Alash. Gobi*** (mont. Alaschan, 1880—Przew.; mont. Alaschan australis, Aug. 25, 1880—Przew.; Kobden-Usu area, north of Gashyun-Nor lake, Aug. 14, 1886—Pot.; south of Tostu-Nuru mountain range, Mukhur-Shainda spring, Aug. 14, 1948—Grub.; 15 km west and 100 km south of Bayan-Khoto town, July 5 and 10, 1957; Tengri sand, Bayan-Khoto, Dzhanamu, Aug. 15, 1958—Petr.; trail of Khaldzan-Ula mountain range, July 31; 10 km south of Shuulin post, Aug. 1, 1989—Grub., Gub. and Dariima), ***Ordos*** (along sandy sites of Yellow river valley, 900-1200 m, Aug. 15; along flooded clayey-sandy sites of Yellow river, Aug. 18—1971, Przew.; Gun'-Dzhagatai [Sept. 6] 1884—Pot.; vicinity of Ikochzhoumen town, 30 km south of Dalatachi, Aug. 10 and 11, 1957—Petr.), ***Khesi*** (15 km north of Yunchan town, rocky slopes of Beidashan' mountains, June 28; 5 km south of Aksai, Chzhan'e, high foothills of Altyntag mountain range, 2300 m, Aug. 2; 55 km west of Tszyutsyuan' town, Yugemyn' town, arid river bed running from nor. mountain range, Aug. 6, 1958—Petr.).

IB. Kashgar: ***East.*** (Khami desert, Shiguza well, 1200 m, Aug. 11, 1895—Rob.).

IIIA. Qinghai ("Uishu, Gunkhe ?"—[Drevesn. rast. Tsinkhaya—Woody Plants of Qinghai], l.c.).

General distribution: endemic.

5. **C. glauca** Willd. Berl. Baumz. (1796) 65; Krasch. in Fl. SSSR, 7 (1937) 321; Fl. Kirgiz. 6 (1955) 80; Fl. Kazakhst. 4 (1961) 72; Kovalevsk. in Opred. rast. Sr. Azii [Key to Plants of Mid. Asia] 3 (1972) 197; Fl. Sin. 28 (1980) 143;

Grub. Opred. rast. Mong. [Key to Plants of Mongolia] (1982) 113; Claves pl. Xinjiang, 2 (1983) 304; Fl. desert Sin. 1 (1985) 485; Gub. Konsp. Fl. Vneshn. Mong. [Conspectus of Flora of Outer Mongolia] (1996) 49. —*C. orientalis* var. *glauca* Maxim. Fl. Tangut. 1 (1889) 3. —*C. orientalis* auct. non L. —Ic.: Willd. l.c. tab. 4, fig. 1; Fl. Kazakhst. 4, Plate 9, fig. 4; Fl. desert Sin. 1, tab. 177, fig. 1.

Described from the material from "East" and from Siberia ("... im Oriente und Sibirien"). Lectotype in Berlin (B).

On rocky slopes, along flanks of gorges, canyons, along banks of brooks, steppe thickets and coastal shrubs, poplar groves.

IA. Mongolia: ***Mong. Alt.*** (Khara-Adzarga mountain range, valley of Shutyn-Gol river, Aug. 27; Bombotu-Khairkhan mountains, plateau between El' and Khapchik rivers, Oct. 30—1930, Pob.), ***Depr. Lakes*** (vicinity of Khirgis-Nur lake and Telin-Gol river, Aug. 10; estuary of Telin-Gol river, on Dzabkhyn river, Aug. 10—1879, Pot.; 10 versts [1 verst = 1.067 km] from Ubsa lake, July 3, 1892—Kryl.; left bank of Dzabkhin river, July 11, 1894—Klem.; valley of Buyantu river, Aug. 27, 1930—Bar.; Shargain-Gobi desert, Bulyk spring, Sept. 10, 1930—Pob.; 3 km east of Bain somon, fringe of Shargain-Gobi valley, Sept. 4, 1948—Grub.; 16 km east of Davst somon, in Bor-Tologoi river valley, July 13, 1977—Karam. et al.), ***Val. Lakes*** (Dzapkhyn river valley, 2 km beyond centre of Taishiri somon, Aug. 8, 1979—Grub., Muld. and Dariima), ***West. Gobi*** (90 km south-west of Ekhiin-Gol oasis, Tsagan-Burgasny-bulak spring, Aug. 16, 1976—Rachk.).

IB. Kashgar: ***Nor.*** (Kuerchu village, west of Kurl'ya, No. 8641 and 8642, Aug. 28 and 29, 1958; around Tekenlyk lake in Chiklyk, Aug. 17; Aksu, No. 8917, Sept. 16—1958, Lee and Chu), ***East.*** (Turfan, Sept. 2, Algoi river, west of Turfan, 1500-1800 m, Sept. 12-14—1899, A. Reg.; 7 km nor.-east of Nan'khu in Pichan district, No. 6692, June 16; Alagao in Toksun, No. 7288, June 16; along banks of Tsinshuikhe river in Khomote, No. 7633, July 28; from Nyutsgen to Tsatsantunchao in Khomote, No. 7705, Aug. 15—1958, Lee and Chu).

IIA. Junggar: ***Cis-Alt.*** (Tsinhe, No. 710, July 24, 1956—Ching; on way from Sharasume to Karamai, No. 10811, Aug. 2, 1959—Lee et al.), ***Tarb.*** (between Lasta and Tumanda rivers, Aug. 8, 1876—Pot.; Urta-Ulasty river, northern foothill of Saur mountain range, June 18, 1908—Sap.; Archaty-Gol valley, No. 1621, Aug. 17, 1957—Yun., Sh.I. Li and Yuan'), ***Tien Shan*** (picket, No. 130—1878, Larionov; Ili river valley, 610 m, Aug. 28, 1876—Przew.; upper Ili river, Kash river bank, July 11, 1877—Przew.; Talki gorge, July 18, 1877—A. Reg.; Nilki brook (on Kash river), June 30; in Sin'yuan' town region, along bank of Sin'yuan' river, No. 1992, Aug. 14; south. portion of Tekes town, No. 1008, Aug. 18; Arasan area [=Ven'tsyuan'], No. 4560, Aug. 21; Ulastai area, No. 3816, Aug. 26; on way from Nilki to Karasu, No. 3885, Aug. 30; Nilki district, Yakou, No. 4046, Sept. 1; in Fukan area, No. 4191, Sept.—1957, Kuan), ***Jung. Gobi*** (nor.: Chern. Irtysh. [Aug. 26] 1876—Pot.; Shipati, No. 2332, Aug. 11; Barbagai, No. 2813, Sept. 11—1956, Ching; west.: Uch-Tyube, 1200 m, Aug. 1878—A. Reg.; south.: Guchensk area, Sontai—1889, Gr.-Grzh.; between Sondzhi and Dzhimisar, Aug. 17, 1898—Klem.; east.: Baitak-Bogdo mountain range, Ulyastu-gola gorge, Sept. 14; same site, Takhiltu-Ula, Ulyastu-gola gorge, Sept. 17, 1948—Grub.; lower course of Uipchi-gola, 15-20 km below winter camp, July 29, 1947—Yun.; nor. macroslope of eastern portion of Baitak-Bogdo mountain range, 5 km west of Olan-Bulak post, Aug. 4, 1988—Kam., Gub. and Dariima), ***Dzhark.*** (Kul'dzha, July 5 and Aug. 5; Suidun, July 16, 1877—A. Reg.; same site, July 1878—Aliyakhun).

General distribution: Jung.-Tarb., Nor. and Cent. Tien Shan; West. Sib., East. Sib., Nor. Mong. (Hang.).

Note. At places of simultaneous growth with *Clematis orientalis*—Mongolia (Khesi), East. Kashgar, Junggar (Tien Shan, Nor. Jung. Gobi, Dzhark.)—intermediate forms between these two species are found.

6. **C. hexapetala** Pall. Reise, 3 (1776) 735; Krasch. in Fl. SSSR, 7 (1937) 318; Kitag. Lin. Fl. Mansh. (1939) 217; Fl. Sin. 28 (1980) 156; Grub. Opred. rast. Mong. [Key to Plants of Mongolia] (1982) 112; Fl. desert Sin. 1 (1985) 482; Fl. Intramong. ed. 2, 2 (1990) 519; Gub. Konsp. Fl. Vneshn. Mong. [Conspectus of Flora of Outer Mongolia] (1996) 49. —*C. angustifolia* Jacq. Ic. pl. rar. 1 (1786): 11; Franch. Pl. David. 1 (1884) 14; Forbes and Hemsl. Index Fl. Sin. 1 (1886) 2; Maxim. Enum. pl. Mong. 1 (1889) 2; Finet et Gagnep. in Bull. Soc. Bot. Fr. 50 (1903) 536. —**Ic.:** Pall. l.c., tab. Q, fig. 2; Jacq. l.c., pl. 104 (sub. *C. angustifolia*); Fl. Intramong. ed. 2, 2, tab. 210, fig. 4-5; Fl. Sin. 28, tab. 48, fig. 1-4; Fl. desert Sin. 1, tab. 174; fig. 14-18.

Described from East. Siberia. Type in St.-Petersburg (LE).

In sandy steppes and steppe slopes, coastal sand, scrubs.

IA. Mongolia: ***Cent. Khalkha*** (Gub. l.c.), ***East. Mong.*** (Muni-Ula mountains, July 10, 1871—Przew.; Khailar railway station, June 10, 1902—Litw.; 3 km south-east of Khalkha-Gol somon on road to Bain-Tsagan, No. 14791, Aug. 5; Shavorte-Obo area, 55 km east of Eren-tsaba, Aug. 19—1949, Yun.; Khailar town, Sishan' hillocks, No. 725, 1951 —V. Wang; Manchuria railway station, Nanshan mountain, 650-900 m, June 24, 1954 —Yu.L. Chang, T.Ch. Chao et al.; Khailar, Sept. 11, 1958—Lavr. and L.T. Chen; same site, 1959—Ivan.; left bank of Khalkhin-Gola, 20 km south of Khamar-Daba, Aug. 15, 1970—Grub., Ulzij. et al.; Bayan-Khan-Ula town, 75 km north-west of Khutliin-Khuduk, July 15, 1971—Dashnyam, Karam. and Safronova).

General distribution: East. Sib., Far East; Nor. Mong. (Hent., Mong.-Daur., Cis-Hing.), China (Dunbei, Nor., East., South-West.), Korean peninsula.

7. **C. integrifolia** L. Sp. pl. 1753) 544; Maxim. Enum. pl. Mong. 1 (1889) 2; Kryl. Fl. Zap. Sib. 5 (1931) 1170; Krasch. in Fl. SSSR, 7 (1937) 312; Fl. Kazakhst. 4 (1961) 71; Kovalevsk. in Opred. rast. Sr. Azii [Key to Plants of Mid. Asia] 3 (1972) 196; Fl. Sin. 28 (1980) 89; Claves pl. Xinjiang, 2 (1983) 303. —**Ic.:** Fl. SSSR, 7, Plate 20, fig. 9, 10; Fl. Kazakhst. 4, Plate 9, fig. 8; Fl. Sin. 28, tab. 23.

Described from Europe and, evidently, from East. Kazakhstan ("in Hungaria et Tataria"). Type in London (BM).

On slopes among shrubs in lower and middle belts of mountains.

IIA. Junggar: ***Cis-Alt.*** (Kanadagatai river, Sept. 15, 1876—Pot.; Shara-Sume, No. 2407, Aug. 26, 1956—Ching; in Altay mountains, 1200 m, No. 10664, July 16, 1959 —Lee et al.).

General distribution: Fore Balkh. (Balkh.-Alak), Jung. Tarb.; Europe (West. and East.), Balk.-Asia Minor, Caucasus, West. Sib., East. Sib.

8. **C. intricata** Bge. Enum. pl. China bor. (1832) 1; id. in Mém. Ac. Sci. St.-Pétersb., Sav. Etr. 2 (1835) 75; Kitag. Lin. Fl. Mansh. (1939) 217; Fl. Sin. 28

(1980) 142; Grub. Opred. rast. Mong. [Key to Plants of Mongolia] (1982) 112; Fl. desert Sin. 1 (1985) 485; Pl. Vasc. Helanshan (1986) 106; Y.Z. Zhao in Acta Sci. Nat. Univ. Intramong. 18, 2 (1987) 284; Grey-Wilson in Kew Bull. 44, 1 (1989) 45; Fl. Intramong. ed. 2, 2 (1990) 531; Gub. Konsp. Fl. Vneshn. Mong. [Conspectus of Flora of Outer Mongolia] (1996) 49. —*C. orientalis* var. *intricata* (Bge.) Maxim. in Bull. Ac. Sci. St. Pétersb. 22 (1876) 210; Forbes and Hemsl. Index Fl. Sin. 1 (1886) 6. —*C. glauca* var. *purpurea* Y.Z. Zhao, Fl. Intramong. 2 (1978) 369, 242; Fl. Intramong. ed. 2, l.c. 533. —*C. intricata* var. *purpurea* Y.Z. Zhao, Fl. Sin. l.c. 369. —**Ic.:** Grub. Plate 52, fig. 237; Fl. desert Sin. 1, tab. 177, fig. 2-3; Fl. Sin. 28, fig. 15; Grey-Wilson, fig. 3 E-G; Fl. Intramong. 2, tab. 214, fig. 3-5.

Described from Nor. China. Type and isotypes in St.-Petersburg (LE).

On hummocky and saline sand, chee grass thickets, rocky sites along brooks and rivers, up to 2800 m.

IA. Mongolia: ***Mong. Alt.*** (nor. macroslope of Adzhi-Bogdo mountain range, 2300 m, July 27, 1977—Volk. and Rachk.), ***Cent. Khalkha*** (Ikhe-Tukhum-nor lake, June 1926 —Zam.), ***East. Mong.*** (in locis subarenosis Mongoliae Chinensis australis, 1831—Turcz.; Mong. chin.—Kirilow), ***Depr. Lakes*** (Kobdo river crossing on Khobdo-Ulangom road, Aug. 23, 1944—Yun.; east coast of Khara-Us-nur lake, Zhargalant-Khairkhan cape, July 24, 1972—Metel'tseva), ***Val. Lakes*** (25 km south-west of somon, July 6, 1941 —Yun.), ***East. Gobi*** (Gub. l.c.), ***Gobi-Alt.*** (Dzhirgalante-bulak spring, south of Noyan-Bogdo mountain range, Khasapche-Usu ridge, July 23-Aug. 4, 1926—Glag.; Bayan-Tukhum area, Aug. 3; hummocky sand between Tszolin and Bayan-Tsagan mountains, south of Bain-Tukhum lake, Aug. 6; road from Khalga pass to Bayan-Dzak, Sept. 25 —1931, Ik.-Gal.; intermontane valley between Tszolen and Deng mountain ranges, July 12, 1943—Yun.; south-east. part of basin of Bayan-Tukhum lake, Sept. 10, 1979—Grub. et al.; east. portion of Khurkhu mountain range, on road to Bayan-bulak, July 30, 1989, Grub., Gub. and Dariima), ***East. Gobi*** (Shabarokh Usu, 1925—Chaney; Ongiin-gola valley, 45-50 km south of Khushu-khida, July 18, 1943—Yun.; 75 km west.-south-west of Khotan-Bulak somon, July 23, 1974—Rachk. and Volk.), ***Alash. Gobi*** (mont. Alaschan, Aug. 6, 1873—Przew.; Alashan, Dyn-yuan'-in oasis (Bayan-Khoto), May 3; Alashan mountain range, Khote-gol gorge, June 17—1908, Czet.; 15 km south-east of Min'tsin town, in Gaotszyas village region, July 3, 1958; Alashan desert, Min'tsin, Aug. 11, 1959—Petr.; south of Khotan-Bulak, Khugde-ula, Aug. 4, 1989—Grub., Gub. and Dariima), ***Ordos*** (Ordos, 1871—Przew.; Ordos, 1884—Artzeler; 15 km west of Ushinchi town, Aug. 2; Dalotachin, Ḥuang He river, Aug. 10—1957, Petr.), ***Khesi*** (Dadzhin town, Aug. 22, 1880—Przew.; 90 km north of Lanzhou town, June 29, 1957—Petr.).

IIA. Junggar: ***Jung. Gobi*** (Gub. l.c.).

IIIB. Tibet: ***Chang Tang*** (Keria mountain range, Aug. 6, 1885—Przew.).

General distribution: China (Nor., Nor.-West.).

9. **C. ladakhiana** Grey-Wilson in Kew Bull. 44, 1 (1989) 49. —*C. orientalis* L. var. *acutifolia* Hook. f. et Thoms. in Fl. Ind. 1 (1855) 9 and in Fl. Brit. Ind. 1 (1875) 5. —**Ic.:** Grey-Wilson, fig. 1 N, P (fragm.).

Described from Himalayas (Kashmir). Type in London (K).

On rocks and shrubs along river valleys.

IIIB. Tibet: ***Chang Tang*** ("Pangong"—Grey-Wilson, l.c.).

General distribution: Himalayas (Kashmir).

10. **C. nannophylla** Maxim. in Bull. Ac. Sci. St. Pétersb. 23 (1877) 305; Forbes and Hemsl. Index Fl. Sin. 1 (1886) 6; Maxim. Fl. Tangut. 1 (1889) 2; Fl. Sin. 28 (1980) 147; Fl. desert Sin. 1 (1985) 481; [Drevesn. rast Tsinkhaya —Woody Plants of Qinghai] (1987) 193; Fl. Intramong. ed. 2, 2 (1990) 519; Borodina-Grabovskaya in Catalogue type specim. Centr. Asia (LE) (2000) 194. —*C. nannophylla* var. *foliosa* Maxim. in Acta Horti Petrop. 11 (1890) 5; Fl. Sin. l.c. 148. —**Ic.:** Maxim. Fl. Tangut. tab. 1, fig. 1-9; Fl. Sin. 28, fig. 18; Fl. desert Sin. 1, tab. 174, fig. 8-13; [Drevesn. rast. Tsinkhaya—Woody Plants of Qinghai] tab. 127; Fl. Intramong. ed. 2, 2, tab. 209, fig. 6-8.

Described from Qinghai (Nanshan) and Nor.-West. China. Lectotype and syntype in St.-Petersburg (LE). Plate VI, fig. 4.

On mountain slopes in gorges.

IA. Mongolia: ***Alash. Gobi*** (road from Inchuan to Bayan-Khoto, arid slopes of spurs of Alashan mountain range, July 5, 1957—Kabanov), ***Khesi*** (Machan-shan' mountain range, July 16, 1908—Czet.; Lanzhou town area, June 21; 67 km north of Lanzhou town, June 29—1957, Petr.).

IIIA. Qinghai: ***Nanshan*** (Alpes S a fl. Tetung., ad declivitates montium praeruptus apertos frequens, No. 161, June 25/July 7, 1872—Przew., lectotypus ! (Borodina-Grabovskaya, l.c.); in upper Huang He river, 2100 m, July 4; Jugum S a fl. Tetung, regio Sylv. inferior, Aug. 6—1880, Przew.; Itel'-gol river, June-Aug. 1885—Ssambartscha; Chorten-tan temple, 2100-2400 m, Sept. 8, 1901—Lad.; "Min'khe, Ledu, Khuchzhu"—[Drevesn. rast. Tsinkhaya—Woody Plants of Qinghai], l.c.).

General distribution: China (Nor.-West.).

11. **C. orientalis** L. Sp. pl. (1753) 543, s.l.; Franch. Pl. David. 1 (1884) 12; Gr.-Grzh. Opis. put. Zap. Kit. [Report of Expedition in West. China] 3 (1907) 475; Ostenf. and Pauls. in Hedin, S. Tibet, 6, 3 (1922) 83; Pamp. Fl. Carac. (1930) 110; Krasch. in Fl. SSSR, 7, (1937) 322; Fl. Kirgiz. 6 (1955) 79; Fl. Kazakhst. 4 (1961) 72; Ikonnik. Opred. rast. Pamira [Key to Plants of Pamir] (1963) 117; Kovalevsk. in Opred. rast. Sr. Azii [Key to Plants of Mid. Asia] 3 (1972) 197 (s. str.); Fl. Sin. 28 (1980) 140; Claves pl. Xinjiang, 2 (1983) 305; Fl. desert Sin. 1 (1985) 185; Grey-Wilson in Kew Bull. 44, 1 (1989) 36; Fl. Intramong. ed. 2, 2 (1990) 531; Gub. Konsp. Fl. Vneshn. Mong. [Conspectus of Flora of Outer Mongolia] (1996) 49. —*C. orientalis* var. *vulgaris* Trautv. in Bull. Soc. natur. Moscou, 33 (1860) 57; Maxim. Enum. pl. Mong. 1 (1889) 3. —**Ic.:** Fl. SSSR, 7, Plate 20, fig. 1; Fl. Kazakhst. 4, Plate 9, fig. 4; Fl. desert Sin. 1, tab. 177, fig. 4-5; Grey-Wilson, fig. 2 A-E; Fl. Intramong. ed. 2, 2, tab. 214, fig. 1-2.

Described from Asia Minor. Type in London (Linn.).

IA. Mongolia: ***West. Gobi*** (Gub. l.c.), ***Khesi*** (near Dadzhina town, [Aug. 11] 1880—Przew.; 15 km nor.-east of Tszinyn town, vicinity of Sangunchen village, July 17; 12 km south-west of An'si town, Sanchun sand, July 28-29—1958, Petr.).

IB. Kashgar: ***Nor.*** (Bugur town, between Kucha and Kurl'ya, Aug. 20, 1929—Pop.; Bai, 5 km south of Keinlak, No. 8262, Sept. 8; near Shakh'yar, No. 8789, Sept. 21—1958, Lee and Chu), ***West.*** (around Andinsan-sai bazar, Aug. 5, 1913—Knorr.; midcourse of Moin river, Aug. 23, 1913—Schischk. and Chenina; upper Kizil-su river, above Kashgar, near Min-yu village, July 5, 1929—Pop; Arpalyk in Artushe, No. 7509, Sept. 11, 1958—Lee and Chu; Sinkiang-Tibet highway, Maigaity village, No. 476, June 3, 1959—Lee et al.), ***South.*** (nor. foothill of Kun'-Lun', 1800 m, Sept. 4, 1889—Rob.; nor. slope of Russky mountain range, Mol'dzha river, 2400 m, July 15, 1890—Rob.; Khoshan oasis, Budnya settlement, Sept. 27, 1958—Yun. and Yuan'; Khotan region, on bank of Khotan river, No. 9610, May 12, 1959—Lee et al.; Karakash, in oasis, Sept. 22, 1959—Petr.), ***Takla-Makan*** (Tarim river valley, 20-25 km before confluence of Yarkend-Dar'ya with Khotan, Aral area, No. 1882, Sept. 29, 1957—Yun., Lee and Yuan'), ***Lob-Nor*** ("Bash-kurgan, SE Lob-nor, July 3, 1900—Hedin"—Ostenf. and Pauls. in Hedin, l.c.; north of Lob-nor lake, No. 10120, Sept. 21, 1959, Lee et al.), ***East.*** (nor.-west. Khami town, Lyaergou village, No. 122, July 6; Barbagai, No. 2966, Sept. 11—1956, Ching; nor.-east of Turfan, Putougou village, No. 5486, June 1; nor.-east of Toksun, No. 7335, June 19—1958, Lee and Chu).

IIA. Junggar: ***Cis-Alt.*** (Altay, 350 [0] m, No. 3372, Sept. 22, 1956—Ching), ***Jung. Alat.*** (Dzhair mountain range, Sept. end, 1951—Mois.), ***Tien Shan*** (San'dookhetsze, No. 352, July 4; 60 km south of Urumchi town, Chaivonu village, No. 476, July 11; midcourse of Manas river, [vicinity of] Sondzhi (=Chantszy), No. 1339, July 12; nor. bank of Sairam lake, No. 4776, Aug. 31—1957, Kuan; Urumchi-Karashar (=Yan'tsy), No. 6224, July 22—1958, Lee and Chu; 20 km south of Nyutsyuan'tsza, No. 223, Sept. 8, 1959—Lee et al.), ***Jung. Gobi*** (nor.-west.: in Savan district, Datsyuan'gou village, No. 1208, July 4; midcourse of Gurtu river, Aug. 31—1957, Kuan; west.: Ebi-Nur lake, 58 km east of Dzin'khe, south. fringe of Ebinursk depression Sept. 3, 1959—Petr.; south.: "Bei-lu, between Foukan town and Santai village, 1905—Gr.Grzh."—Gr.-Grzh. l.c.; Chipeizy, Dzhirgalan river (Kuitun), Sept. 8, 1947—Shumakov).

IIIB. Tibet: ***Chang Tang*** ("Mandarlik, 3400 m, 37°47' N, 90°47' E, July 13-19, 1900—Hedin"—Ostenf. and Pauls. in Hedin, l.c.).

IIIC. Pamir ("Mustagh-ata, Little Karakul, 2600 m, July 5, 1900—Hedin"—Ostenf. and Pauls. in Hedin, l.c., Ak-Sarai village on Charlym river, Aug. 4, 1909—Divn.).

General distribution: Aralo-Casp., Fore Balkh., Nor. and Cent. Tien Shan, East. Pam.; Europe, Caucasus, Mid. Asia, Asia Minor, Fore Asia, Himalayas (Kashmir, west.), Nor. Amer., Afr. (south).

12. **C. pamiralaica** Grey-Wilson in Kew Bull. 44, 1 (1989) 54. —*C. tangutica* (Maxim.) Korsh. in Bull. Ac. Sci. St. Petersb., sér. 5, 9 (1898) 399, p.p.; Krasch. in Fl. SSSR, 7 (1937) 322, p.p. —**Ic.:** Grey-Wilson, fig. 5.

Described from East. Pamir. Type in London (K). Isotype in St.-Petersburg (LE).

On coastal rocks and rock talus, exposed sites.

IIIB. Tibet: ***Chang Tang*** (nor. slope of Kun'-Lun', upper Tiznaf river, near Kyude settlement, 161 km from Kargalyk, on Tibet highway, June 4, 1959—Yun. and Yuan').

IIIC. Pamir ("Tagdumbash Pamir"; "Mugarb Dawan"; "mt. Kongur"—Grey-Wilson, l.c.).

General distribution: East. Pam.

Note. C. *pamiralaica* was confused for long time with C. *tangutica* s. str. although the two no doubt differ: the former is a dwarf plant with grassy stem (perishing to the ground level in winter), with short shoots bearing a single terminal flower; its sepals have fine silky hairs on both sides. This species is a derivative of C. *tangutica* in isolated high-altitude ecotope.

13. **C. rehderiana** Craib in Kew Bull. 1914 (1914) 150; Fl. Sin. 28 (1980) 108; Fl. Xizang. 2 (1985) 87; [Drevesn. rast. Tsinkhaya—Woody Plants of

Qinghai] (1987) 189. —**Ic.**: Fl. Sin. 28, tab. 7; [Drevesn. rast. Tsinkhaya—Woody Plants of Qinghai] tab. 123.

Described from South-West. China (West. Sychuan). Type in London (K).

On grassy slopes among shrubs, 3100-4200 m.

IIIA. Qinghai ("Yuishu, Golo—[Drevesn. rast. Tsinkhaya—Woody Plants of Qinghai], l.c.).

IIIB. Tibet: ***Weitzan*** ("Sosyan'"—Fl. Xizang.).

General distribution: China (South-West.), Himalayas (Nepal).

14. **C. songorica** Bge. Del. Sem. Horti Dorp. (1839) 8; Maxim. Enum. pl. Mong. 1 (1889) 3 (var. *integrifolia* Trautv. et var. *serratifolia* Trautv.); Kryl. Fl. Zap. Sib. 5 (1931) 1171; Krasch. in Fl. SSSR, 7 (1937) 316; Fl. Kirgiz. 6 (1955) 78; Fl. Kazakhst. 4 (1961) 71; Kovalevsk. in Opred. rast. Sr. Azii [Key to Plants of Mid. Asia] 3 (1972) 196; p.p.; Fl. Sin. 28 (1980) 151; Grub. Opred. rast. Mong. [Key to Plants of Mongolia] (1982) 112, Claves pl. Xinjiang, 2 (1983) 303; Fl. desert Sin. 1 (1985) 479; Fl. Intramong. ed. 2, 2 (1990) 519; Gub. Konsp. Fl. Vneshn. Mong. [Conspectus of Flora of Outer Mongolia] (1996) 49. —**Ic.**: Fl. SSSR, 7, Plate 20, fig. 2; Fl. Kazakhst. 4, Plate 9, fig. 5; Grub. Plate 52, fig. 239; Fl. desert Sin. 1, tab. 174, fig. 1-5; Fl. Intramong. ed. 2, 2 tab. 210, fig. 1-3.

Described from Junggar. Type not found.

On rocky slopes and trails of mountains, talus, on rocky and sandy banks of brooks, flanks and floors of gorges.

IA. Mongolia: ***Mong. Alt., Gobi-Alt., East. Gobi, West. Gobi*** (Gub. l.c.), ***Alash. Gobi*** (5 km south of Obotu-Khural, on road to Sogo-Nuru, July 27, 1943—Yun.; Khukhutologoin bulak, 21 km west of Obotu-Khural, Aug. 13; south of Tostu-nuru mountain range, in Mukhur-Shanda spring region, Aug. 14—1948, Grub.; between Khaltszan-Ula and Dubnich-Ula mountains, Sept. 9, 1950—Lavr. and Kal.; Bordzon-Gobi, near Birukhain-Gol spring, desert, July 27, 1970—Grub. and Ulzij.; 80 km south of Nomgon, July 16, 1972—Gur. and Rachk.; volcanic massif in Shilt-ula town, 120 km south-east of Nomgon somon, south of Khurkhe mountain range, July 15, 1974—Rachk. and Volk.; southern fringe of Bordzon-Gobi, near Khaldzantsyan, on road to Shulin post, July 30 and Aug. 1, 1989—Grub., Gub. and Dariima), ***Khesi*** (An'si town, No. 67, July 4, 1956—Ching).

IB. Kashgar: ***Nor.*** (15 km nor. of Daman'yatsyk in Bai area, No. 8196, Sept. 4, 1958—Lee and Chu; Kucha, Kuerchu settlement, 115 km west of Yan'chi town, Sept. 11, 1959—Petr.), ***East.*** (Algoi river, between Karashar and Turfan, 1500-1800 m, Sept. 13, 1879—A. Reg.; between Kuruktag and Choltag near Chokur village, Aug. 28, 1929—Pop.; nor. west. Toksun, No. 7351, June 20; Turfan area, between Tsichantsun' village and Kalangou, No. 3780, June 22—1958, Lee and Chu).

IIIA. Junggar: ***Tarb.*** (nor. foothill of Saur, Urta-Ulasty river, steppe, June 17, 1908—Sap.; between Saur mountain range and Semistei, 5-6 km nor.-west of Koshtologoi settlement on road to Khobuk, on pebble beds, No. 269, June 22, 1957—Yun., Lee, Yuan'; 40 km nor. of Kratakbash (Kheishan'tou), No. 10638, Sept. 7 1959—Lee et al.), ***Jung. Alat.*** (Dzhair mountain range, Tuz-agny ravine, No. 104 and 106, Sept. 1951—Mois.; Dzhair mountain range, 8 km south-west of Boguta river, on road to Chuguchak from Aktam, along rim of gorge, No. 1109, Aug. 3, 1957—Yun., Lee and Yuan'; in Otu region, No. 881 and 934, Aug. 3 and 4, 1957—Kuan), ***Tien Shan*** (picket, 1875—Larionov; [south of Koshety-daban pass, 3 km beyond Nan'-shan'-keu] June 6—1877, Pot.; Upper

Kunges river, 1000 m, June 27, 1877; same site, June 5, 1879—Przew.; Talki gorge, July 18, 1877; Kash river, 900-1200 m, July 3, 1879—A. Reg.; zwischen Aksu und Tschokak, May end, 1903; Kapsalyon Tal und Nebental Kisyl-sai, auch Plateau hohe von Kara-Dschon, Aug. 2-3, 1907; Kiukönik Tal, Südliches Unter- und Mittellauf, June 15, 1908 —Merzb.; 11 km north of Dzhagastai, No. 647, Aug. 7; Dzhagastai mine in Ili area, No. 3120, Aug. 7; on way to Karasu from Nilki, No. 3887, Aug. 30—1957, Kuan; on road to Karashar from Urumchi, 1570 m, No. 5956 and 6108, July 21 and 22; on banks of Tsinshuikhe river in Khomot, No. 7635, July 28; in Bagrashkul' lake region, No. 6869, July 29; Lotogou village, in Khetszi district (Bagrashkul' lake), No. 6249, Aug. 1; south of Bortu in Khomot, No. 6926, Aug. 2—1958, Lee and Chu), ***Jung. Gobi*** (cent.: from Tsitai 60 km nor.-east, No. 2320, Sept. 25, 1957—Kuan; nor.-west of Tien Shan laoba to Lyaoergou, No. 2369, Aug. 3, 1957—Kuan; nor. trails of Saur mountain range, 30 km west.-nor.-west of Kheisan-gou settlement, on road to Karamai from Burchum, No. 1665, July 10, 1959—Yun. and Yuan'; south.: 10 km south of Shikho, Chidachu oasis, Aug. 10, 1947—E.M. Shumakov; east.: Adak village, June 17, 1877—Pot.; [south. Baitak-Bogdo trail] on cliffs near Kyup spring; between Santa-khu and Dzhimuchi, Aug. 19, 1898—Klem.; in Nom region, No. 2262, Sept. 29, 1957—Kuan), ***Dzhark.*** (Bayandai, in Kul'dzha region, July 27, 1877—A. Reg.).

IIIC. Pamir: (Karom [village], Pas-rabat, July 3, 1909—Divn.; Tash-Kurgan, July 25, 1913—Knorr.).

General distribution: Jung.-Tarb., Nor. and Cent. Tien Shan; Mid. Asia.

15. **C. tangutica** (Maxim.) Korsh. in Bull. Ac. Sci. St. Pétersb., sér. 5, 9 (1898) 399; Gr.-Grzh. Opis. put. Zap. Kit. [Report of Expedition in West. China] 3 (1907) 475; Krasch. in Fl. SSSR, 7 (1937) 250, 322, p.p.; Fl. Kirgiz. 6 (1955) 79; Ikonnik. Opred. rast. Pamira [Key to Plants of Pamir] (1963) 117; Kovalevsk. in Opred. rast. Sr. Azii [Key to Plants of Mid. Asia] 3 (1972) 197; Fl. Sin. 28 (1980) 144; Grub. Opred. rast. Mong. [Key to Plants of Mongolia] (1982) 112; Claves pl. Xinjiang, 2 (1983) 304; Fl. desert Sin. 1 (1985) 483; Fl. Xizang. 2 (1985) 89; Pl. Vasc. Helanshan (1986) 106; [Drevesn. rast. Tsinkhaya—Woody Plants of Qinghai] (1987) 192; Grey-Wilson in Kew Bull. 44, 1 (1989) 53; Fl. Intramong. ed. 2, 2 (1990) 533; Gub. Konsp. Fl. Vneshn. Mong. [Conspectus of Flora of Outer Mongolia] (1996) 49. —*C. orientalis* var. *tangutica* Maxim. Fl. Tangut. 1 (1889) 3; id. Enum. pl. Mong. 1 (1889) 3; Hemsl. Fl. Tibet (1902) 161; Borodina-Grabovskaya in Catalogue type specim. Centr. Asia (LE) (2000) 194. —**Ic.:** Fl. Sin. 28, fig. 17; Grub. Plate 52, fig. 238; Fl. desert Sin. 1, tab. 176, fig. 1-4; Fl. Xizang. 2, tab. 25, fig. 7; [Drevesn. rast. Tsinkhaya—Woody Plants of Qinghai] tab. 125; Fl. Intramong. 2, tab. 215, fig. 1-3.

Described from Qinghai (Nanshan), Tibet (Weitzan) and Khesi. Lectotype and syntypes in St.-Petersburg (LE).

On rock and rubble slopes of knolls and mountains, along flanks and pebble floors of gorges, on rocks.

IA. Mongolia: ***Mong. Alt.*** (South. Altay, Dzusylyn river gorge, June 29; on Saksa river, July 7; Tsitsirin-Gol river, July 9—1877, Pot.; Khara-Adzarga mountains, Sakhirsala river valley, Aug. 22; Khara-Adzarga mountains, Boro-gol river valley, Aug. 24; nor. slope of Khara-Adzarga, vicinity of Khairkhan-Duru river, Aug. 25; same site, Shutyn-gol river valley, Aug. 27; Khasagtu-Khairkhan mountains, Ulyastyn-gol river valley,

Sept. 20—1930, Pob.; Tugurin-gola gorge, June 27, 1971—Grub., Ulzij. et al.; Adzhi-Bogdo mountain range, Ikhe-gol river 2200-2300 m, Aug. 22, 1979—Grub., Dariima et al.; same site, upper Ubur-Tszusylan-gol creek valley, Khara-Adzargiin-Nuru mountains, Urd-Ulyastain-Gol gorge, July 3, 1984—Kam. and Dariima), ***Cent. Khalkha*** (around Kholt area, near Sonchzhi, July 17, 1926—Gus.). ***East. Mong.*** (Gatun-Bologai, 1925 —Chaney), ***Depr. Lakes*** (south-west. slope of Dzun-Dzhirgalant mountain range, Ulyastyn-gol gorge, 1850-2800 m, June 28, 1971—Grub., Ulzij. et al.), ***Gobi-Alt.*** (along nor. descent of Tostu mountain, near Kotel'-Usu well, Aug. 19, 1886—Pot.; Dundu-Saikhan mountains, July 4, 1909—Czet.; Baga-Bogdo and Ihke-Bogdo, 1700-1800 m, 1925—Chaney; Ikhe-Bogdo mountain range, June 21, 1926—E. Kozlova; Dundu-Saikhan mountains, Aug. 16; Dzun-Saikhan mountains, Aug. 25—1931, Ik.-Gal.; Dundu mountain range ánd Dzun-Saikhan, July-Aug. 1933—Simukova; Dzun-Saikhan mountain range, west. extremity, June 18, 1945—Yun.; east. extremity of Gichigine-nuru mountain range, 3 km south-west of Atani-bulak spring, Aug. 27, 1948—Grub.; Dzun-Saikhan mountain range, Elyn-ama gorge, July 22, 1970—Grub., Ulzij. et al.; Ikhe-Bogdo mountain range, south. slope of Uvd creek valley, June 29, 1972—Banzragch et al.; 30 km east-nor.-east of Bayan-Dalai, Dzun-Saikhan-Ula mountain range, July 22, 1972—Gur. and Rachk.; Bayan-Tsagan-nuru mountain range, Dzhargalantu-gol river, July 8, 1973—Golubk. and Tsogt; Ikhe-Bogdo town, Aug. 4, 1973—Isach. and Rachk.), ***West. Gobi*** (south. slope of Adzhi-Bogdo mountain range, Indertiin-Gol river valley, Aug. 6; same site, upper Ubur-Tszuslangol creek valley, Aug. 7—1947, Yun.; Atas-Bogdo mountain range, nor. slope, Aug. 22, 1976—Rachk.), ***Khesi*** (Kansu province, vulgaris, July 13/25, 1875, acc. 1876, Dr. Piasezky, syntypus !; Beidashan' mountain slopes, 15 km nor. of Yunchan town, June 28, 1958—Petr.).

IB. Kashgar: ***Nor.*** (Uch.-Turfan, Airi gorge, June 4, 1908—Divn.), ***West.*** (Tokhta-Khon, July 24, 1890—Grombch.; Taret pass, June 19, 1909—Divn.), ***South.*** (Keriya, June 30, 1885—Przew.), ***East.*** (north-west of Toksun, Chernaya mountain, No. 7383, June 21; Turfan district, between Tsichantsun' village and Kalangou, No. 5780, June 22—1958, Lee and Chu).

IC. Qaidam: ***Mount.*** (Nor. Qaidam, Orogyn-Gol river, 3000 m, June 22, 1895—Rob.; East. Qaidam, Barun-tszasaka khyrma, 2700 m, June 4; same site, on Bain-Gol river, 2700-3000 m, June 15—1901, Lad.).

IIA. Junggar: ***Tien Shan*** (east. Tien Shan, on Nan-shan'-keu, June 6, 1877—Pot.; [Nan-shan'-keu picket] on south. foothill, June 5, 1879—Przew.; von Aksu nach Tschokok, May; Inyltschek Tal, Aug. 17-25—1903; im Dunde-Kelde und im obersten Manass (Chustai) Tal, July 5-6—1908, Merzb., Manas river basin, Ulan-Usu river valley at confluence with Dzhartas, July 18, 1957—Yun., Lee and Yuan'; in Bogdoshan mountains near Turfan, no. 5734, June 18; 20 km from Lotogou village, on Khetszin in Bargashkul' lake region, No. 6292, Aug. 1; mountainous road from Bortu to timber works in Khomot, No. 6986, Aug. 3—1958, Lee and Chu), ***Jung. Gobi*** (east.: Baitag-Bogdo mountain range, nor. macroslopoe 110 km south-south-west of Bulgan somon, Aug. 8, 1977 —Volk. and Rachk.; nor. macroslope of Baitag-Bogdo mountain range, Budun-Khargaityn-gol gorge, 15 km east of Baitag-Bogdo post, Aug. 1, 1988—Gub., Kam. et al.).

IIIA. Qinghai: ***Nanshan*** (Ad pedem declivit. borealis prov. Kansu et per totam provincia, praesertim circa domos vulgaris, No. 105, June 19/July 1, lectotypus (Borodina-Grabovskaya, l.c.) et isolectotypus !; secus fl. Tetunggol, No. 105, June 29/ July 10, syntypus !—1872, Przew.; China occidentalis, Regio Tangut, ad Hoangho super., alt. 8000-9000 ft, in ripa glareosa frequens, No. 171, May 19/31—1880, Przew., syntypus !; Gobi australis, ad fl. Itelgol, 1885—Ssambartscha [G.N. Potanin] syntypus !; "Xining alps, Kuku-Nor, Bukhain-Gol river valley, 3000-3500 m, June, Aug. 1905"—Gr.-Grzh. 1907, l.c.; Kukunor lake, on bank, 3600 m, Aug. 29, 1908—Czet.; same site, Sept. 1925 —Rock; 60 km south-east of Chzhan'e town, tall Nanshan foothills, 2200 m, July 12; 70 km south-east of Chzhan'e, Matisy temple, Tsinlyan-shan' mountains, 2600 m, July 12; 15 km south of Aksay settlement, 2800 m, Aug. 2—1958, Petr.).

IIA. Junggar: *Tien Shan* ("Sinkiang, in Gunlyu town region, 1600 m, along fringes of deep gorges with water, or on fringes, No. 77069, Wen-hu Hou—typus"—Bull. Bot. Res. (Harbin), l.c.).

General distribution: Jung.-Tarb., Cent. Tien Shan.

3. **A. macropetala** (Ledeb.) C.A. Mey. in Ledeb. Fl. Alt. 2 (1830) 376; Krasch. in Fl. SSSR, 7 (1937) 308; Serov in Bot. zhurn. 76, 7 (1991) 995 (cum auct. comb. Ledeb.). —*Clematis macropetala* Ledeb. Ic. pl. Fl. Ross. 1, 1 (1829) 5; Franch. Pl. David. 1 (1884) 14; Forbes and Hemsl. Index Fl. Sin. 1 (1886) 5; Kitag. Lin. Fl. Mansh. (1939) 218; Fl. Sin. 28 (1980) 138; Pl. Vasc. Helanshan (1986) 106; Y.C. Zhao in Acta Sci. Nat. Univ. Intramong. 18, 2 (1987) 284; [Drevesn. rast. Tsinkhaya—Woody Plants of Qinghai] (1987) 192; Fl. Intramong. ed. 2, 2 (1990) 525. —*C. alpina* (L.) Mill. var. *macropetala* Maxim. Enum. pl. Mong. 1 (1889) subvar. *rupestris* Maxim. l.c. (1889) 6. —**Ic.:** Ledeb. 1829, tab. 11; Fl. SSSR, 7, Plate 20, fig. 7; Fl. Intramong. ed. 2, 2, tab. 126, fig. 1-4; Fl. Sin. 28, ic. 14 (sub *C. macropetala*); [Drevesn. rast. Tsinkhaya—Woody Plants of Qinghai], tab. 124; Serov, 1991, fig. 2 (map), 3 fig. of fragm.).

Described from East. Siberia. Type in St.-Petersburg (LE).

IA. Mongolia: ***East. Mong.*** (Muni-Ula, in shrubs along lowlands, June 26, 1871 —Przew.).

IIIA. Qinghai ("Ledu, Khuchzhu, Datun, Mynytsyuan'"—[Drevesn. rast. Tsinkhaya —Woody Plants of Qinghai], l.c.).

General distribution: East. Sib., Far East, China (Dunbei, Nor.), Japan.

4. **A. moisseenkoi** Serov in Bot. Journ. 71, 8 (1986) 1128. —*A. sibirica* auct. non L.: Grub. in Bot. mat. (Leningrad) 19 (1959) 539. —*Clematis sibirica* auct. non Mill.: Fl. Sin. 28 (1980) 135, p.p.; Claves pl. Xinjiang, 2 (1983) 303, p.p. —**Ic.:** Serov in Bot. zhurn. 76, 7 (1991) fig. 1.

Described from Junggar (Boro-Khoro mountain range). Type in St.-Petersburg (LE).

In coniferous forest belt.

IIA. Junggar: *Tien Shan* (Dekanby river, June 2, 1876—Larionov; Nilki gorge, on Kash river, 1800-2400 m, June 18, 1879—A. Reg.; Tuste river basin, in Boro-Khoro mountain range, 1600 m, in the zone of development of Tien Shan spruce, deep and narrow valley of seasonal brook (flanks up to 100 m high)..., No. 9, June 13, 1953—Mois., holotypus !).

General distribution: endemic.

5. **A. speciosa** Weinm. in Bull. Soc. natur. Moscou, 23 (1850) 538; Serov et Jarvis in Taxon, 37, 1 (1988) 167. —*A. sibirica* L. Sp. pl. (1753) 543, nom. ambig.; Kryl. Fl. Zap. Sib. 5 (1931) 1172; Krasch. in Fl. SSSR, 7 (1937) 308; Fl. Kirgiz. 6 (1955) 77; Fl. Kazakhst. 4 (1961) 69; Grub. Opred. rast. Mong. [Key to Plants of Mongolia] (1982) 112; Gub. Konsp. Fl. Vneshn. Mong.

[Conspectus of Flora of Outer Mongolia] (1996) 48. —*Clematis sibirica* Mill. Gard. Dict. ed. 8 (1768) No. 12; Kitag. Lin. Fl. Mansh. (1939) 218; Fl. Sin. 28 (1980) 135, p.p.; Claves pl. Xinjiang, 2 (1983): 303, p.p. (cum auct. comb. (L.) Mill.); Y.Z. Zhao in Acta Sci. Nat. Univ. Intramong. 18, 2 (1987) 284 (cum auct. comb. (L.) Mill.). —*C. tianschanica* (N. Pavl.) N.R. Cui in Claves pl. Xinjiang, 2 (1983) 303. —*Atragene tianschanica* N. Pavl. Vestn. Kas. SSR, 4 (1951) 89; Krasch. in Fl. SSSR, 7 (1937) 308, p.p.; Fl. Kazakhst. 4 (1961) 70. —**Ic.:** Fl. Kazakhst. 4, Plate 9, fig. 2; Fl. Sin. 28, tab. 40; Grub. Plate 53, fig. 240; Serov in Bot. zhurn. 76, 7 (1991) fig. 2 (map), 3 (fig.).

Described from East. Europe (Perm district). Type not known.

In larch and mixed forests, forest borders, birch stands and scrubs, along banks of rivers, and slopes, on shaded rocks, rock talus in forest belt and lower portion of high mountains.

IA. Mongolia: ***Khobd.*** (Ulan-daban pass, June 23; Kharkira river, July 23—1879, Pot.; 30 km west-south-west of Ulangom, Mukhur-Ulyasy-Gol gorge, slope exposed nor.-west, forb-dryad tundra, July 7, 1977—Karam., Sanczir and Sum.), ***Mong. Alt.*** (Dzusylyn gorge [Adzhi-Bogdo mountain range], July 13, 1879—Pot.; Tsagan-Gola river upper valley, wooded slope opposite Prokhladnoi river, June 30, 1905—Sap.; Shutyn-Gol river valley, larch forest on nor. slope, Aug. 29, 1930—Pob.; south. slope of Khan-Taishir mountain range, in upper Ishne-Usu (nor. of Khamok somon), 2380-2450 m, on rocks in forest, June 18; south-west. slope of Dzun-Dzhirgalante mountain range, Ulyastyn-Gola gorge, 1850-2800 m, June 28; upper Buyantu river, Chigirtei-Gol, 12 km above lake, nor. slope of Chigirtei-Ula, 2600-2800 m, recent larch grove, July 4; upper Bogdo river, Dayan-Nur, nor. slope of Yamatyn-Ula, 2350-2500 m, larch forest, July 9 —1971, Grub., Ulzij. and Dariima; Khasagtu-Khairkhan, nor. slope of Tsagan-Irtyk-Ula, nor.-exposed slope in upper Khunkerin-Ama, *Cobresia* grove on rocks, 2980 m, Aug. 23, 1972—Grub. et al.; Dzhelta river basin, Korumty-Gol valley Aug. 9, 1979 —Gub. (MW); 10-15 km west of Dayan-Nur post, near Khara-Nur lake, July 20, 1988 —Bud., Ganbold, Gub. et al.), ***Cent. Khalkha*** (Gub. l.c.), ***Depr. Lakes*** (on Ulyastai river, July 9, 1879—Pot.), ***Gobi-Alt.*** (Bago Bogdo, in moist canyon heads, 1800-2700 m, No. 242, 1925—Chaney; Dundusaikhan town, upper portion of Ulan-Khunde creek valley, Aug. 17; rubbly slope of Tszunsaikhan mountain, Aug. 25—1931, Ik.-Gal.; Dundu mountain range and Tszun Saikhan, mountain slopes and gorge from trail to upper belt, July-Aug. 1933—Simukova; Dzun-Saikhan mountain range, nor. portion, No. 2938, June 19, 1945—Yun.; Ikhe-Bogdo-Ula mountain range, south-east. slope, Narin-Khurulet gorge, on rocks, 2900 m, July 28, 1948—Grub.; Ikhe-Bogd mountain, nor. macroslope, Uleete creek valley, July 1, 1972—Banzragch et al.).

IB. Kashgar: ***Nor.*** (Uch-Turfan, Airi gorge, along brook, June 4, 1908—Divn.); ***East.*** (Turfan district, 10 km nor. of Kalangou village, No. 5821, June 23, 1958—Lee and Chu).

IIA. Junggar: ***Cis-Alt.*** (on Kandagatai river, Sept. 14, 1876—Pot.; Borbogusun, 2400 m, June 1879—A. Reg.; Chern. Irtysh valley, Kran river, June 23, 1903—Gr.-Grzh.), ***Tarb.*** (Dachen, No. 1606, Aug. 13, 1957—Kuan); ***Jung. Alat.*** (in Toli district, No. 1009 and 1257, Aug. 6, 1957—Kuan), ***Tien Shan*** (on Mal. Yuldus high plateau, May 17, 1877 —Przew.; same site, June 12, 1877—Pot.; Talki gorge, 1800-2400 m, July 16, 1877 —A. Reg.; Urtas-Aksu, Kul'dzha region, June 17, 1878—Fet.; Khonakhai mountain, 1500-2100 m, south-west. Kul'dzha, June 16; Agias river, 2100-2400 m, June 26—1878, A. Reg.; between Barkul' [lake] and Khami, May 21, 1879—Przew.; Baga-Dzuslun —Bayanamyn, east tributary of Dzhin river, 900-1800 m, June 4; Borgaty pass, 2400-2700 m, June 7; Naryngol on Tsaganusu (Dzhin), 1800-2400 m, June 10—1879,

A. Reg.; bei Kutschii und Oi-Tatun, sowie Schaichle, June 19-21; Aufstieg von Schauhle zum Karaburo Pass, June 22—1903; Karagai Pass, vor dem Karakul See, auch viel in Kotyl Val und im Tschon Jailek Tal, June 12/13, 1908—Merzb.; 25 km south-east of Nyutsyuan'tsza, Nintszyakhe river bank, No. 52 and 81, July 17, 1957—Kuan; nor. slope of main mountain range, descent into Tekes upper valley; 6-7 km south-east of Aksu settlement, Aug. 24, 1957; nor. slopes to Kunges, Aug. 7, 1958—Yun. and Yuan'), *Zaisan* ("Burchum"—Claves pl. Xinjiang, l.c.).

IIIC. Pamir (Ulugtuz gorge, in Charlysh river basin, among junipers, abundant, June 27, 1909—Divn.).

General distribution: Jung.-Tarb., Nor. and Cent. Tien Shan; Europe (East.), West. Sib., East. Sib., Mid. Asia, Nor. Mong., China (Dunbei, Nor.).

Note. *Atragene tianschanica* has been cited for Tien Shan and western regions of Junggar.

Subtribe 3. **Kingdoniinae** Janchen

15. **Circaeaster** Maxim.

in Bull. Ac. Sci. St. Pétersb. 27 (1882) 556

1. **C. agrestis** Maxim. in Bull. Ac. Sci. St. Pétersb. 27 (1882) 557; Fl. Sin. 28 (1980) 241; Claves pl. Xinjiang, 2 (1983) 558; Fl. Xizang. 2 (1985) 93; Fl. Xinjiang. 2,1 (1994) 351. —**Ic.:** Fl. Sin. 28, tab. 81, fig. 1-4; Fl. Xizang. 2, tab. 26, fig. 4-7.

Described from Qinghai. Type in St.-Petersburg (LE). Plate III, fig. 4. Map 3.

In borders of mountain forests, about 2000 m alt.

IIA. Junggar: ***Tien Shan*** ("Tomuerchi, in forest, in undergrowth"—Claves pl. Xinjiang, l.c.; "Sinkiang"—Fl. Xizang. l.c.; "Wusu Xian (nor. slope of Mt. Tomort)"—Fl. Xinjiang, l.c.).

IIIA. Qinghai: ***Nanshan*** (China occidentalis, Regio Tangut (Prov. Kansu) Jugum S a fl. Tetung, regio sylvaticum alt. s.m. 7500 ft [2300 m], in humo agrarum sinensium, frequens [South. Tetung mountain range, forest belt, 2300 m, on humus soil, in field] July 27/Aug. 8, 1880—Przew., holotypus !).

IIIB. Tibet: ***Weitzan*** ("Leinyaotszi"—Fl. Xizang. l.c.), ***South.*** ("Gunbutszyanda"—Fl. Xizang. l.c.).

General distribution: China (Nor.-West.), Himalayas (west., east.).

Tribe 3. **Ranunculeae** DC.

16. **Halerpestes** Greene

in Pittonia, 4 (1900) 207

1. Leaf blade entire, serrated at tip..2.
\+ Leaf blade tripartite with linear-oblong lobes subobtuse at tip; sometimes bi-tripartite. Diminutive, 1-2 cm tall plant..................... ....4. **H. tricuspis** (Maxim.) Hand.-Mazz.

2. Flowers 1.5-2 cm in diam.; petals 7-12, slightly longer than sepals. Leaf blade thickish, semicoriaceous, ovoid or broad-oval, with 3-5 rounded teeth at tip. Plant up to 20 cm tall.. ....2. **H. salsuginosa** (Pall. ex Georgi) Greene.

\+ Flowers 0.6-1 cm in diam.; petals 5, almost no longer than sepals. Small, up to 10 cm tall plant..3.

3. Leaf blade orbicular-ovoid or orbicular-reniform, 6-15 mm in cross-section, with 3-8 rounded teeth or lobed at tip.................................... ....3. **H. sarmentosa** (Adams) Kom.

\+ Leaf blade narrow-lanceolate, about 10 mm long, 1.5-3 mm broad, sometimes tridentate at tip.....1. **H. lancifolia** (Bertol.) Hand.-Mazz.

1. **H. lancifolia** (Bertol.) Hand.-Mazz. in Acta Horti Gotoburg. 13 (1939) 136; Tamura in Acta Phytotax. Geobot. (Kyoto) 15, 6 (1954) 197; Fl. Sin. 28 (1980) 336; Fl. Xizang. 2 (1985) 113. —*Ranunculus lancifolius* Bertol. in Mém. Ac. Sci. Bologn. ser, 2, 3 (1862) 423. —*R. tricuspis* var. *lancifolius* (Bertol.) Hara, Fl. E. Himal. 3 (1975) 39. —**Ic.:** Acta Phytotax. Geobot. (Kyoto) 15, 6, fig. 4; Fl. Sin. 28, tab. 105, fig. 10-11; Fl. Xizang. 2, tab. 34, fig. 7-8.

Described from Himalayas. Type ?

In marshy meadows.

IIIB. Tibet: ***South.*** ("Pulan', Chzhunba"—Fl. Xizang. l.c.; vicinity of "Shigatsze"—Acta Phytotax. Geobot. (Kyoto) l.c.).

General distribution: Himalayas.

2. **H. salsuginosa** (Pall. ex Georgi) Greene in Pittonia, 4 (1900) 208; Kovalevsk. in Opred. rast. Sr. Azii [Key to Plants of Mid. Asia] 3 (1972) 200; Grub. Opred rast. Mong. [Key to Plants of Mongolia] (1982) 113; Gub. Konsp. Fl. Vneshn. Mong. [Conspectus of Flora of Outer Mongolia] (1996) 49. —*Ranunculus salsuginosus* Pall. ex Georgi, Bemerk. Reise Russ. Reich. 1 (1775) 222. —*R. ruthenicus* Jacq. Enum. Hort. Vindob. 3 (1776) 19; Kryl. Fl. Zap. Sib. 5 (1931) 1190; Kitag. Lin. Fl. Mansh. (1939) 225. —*R. plantaginifolius* Murr. in Nov. Comm. Gött. 8 (1777) 39; Franch. Pl. David. 1 (1884) 20; Maxim. Enum. pl. Mong. 1 (1889) 16. —*Halerpestes ruthenica* (Jacq.) Ovcz. in Fl. URSS, 7 (1937) 331; Fl. Kazakhst. 4 (1961) 78; Fl. Intramong. 2 (1978) 246; Fl. Sin. 28 (1980) 336; Claves pl. Xinjiang, 2 (1983) 321; Pl. Vasc. Helanshan (1986) 107; Y.Z. Zhao in Acta Sci. Nat. Univ. Intramong. 18, 2 (1987) 284; Fl. Intramong. ed. 2, 2 (1990) 489. —**Ic.:** Fl. SSSR, 7, Plate 34, fig. 5; Fl. Intramong. 2, tab. 128, fig. 1-4; Fl. Sin. 28, tab. 105, fig. 1-3; Fl. Intramong. ed. 2, 2, tab. 197, fig. 3-4.

Described from East. Siberia. Type in St.-Petersburg (LE). Plate II, fig. 1.

In moist solonetzic and marshy meadows, along banks of rivers and lakes, springs, brooks, often in water, in shoals.

IA. Mongolia: ***Khobd.*** (Kholoin-Khavtsal gorge, 18 km east of Tsagan-Nur somon, July 24, 1977—Karam.; 8 km before Bugat somon, Bayan-Gol gorge, July 5, 1984

—Kam. and Dariima), ***Mong. Alt.*** (valley of Tsagan-Derisu river, July 3; on Tsitsiriin-Gol river, July 9—1877, Pot.; 2-3 km south-east of Yusun-Bulak, July 13, 1947—Tuvanzhab and Davasurun; South. bank of Ikhes-Nur lake, June 22, 1971—Grub., Ulzij. et al.; Nyutsugun-Gol river, tributary of Uenchi-Gol, 2100 m, June 26, 1973—Golubk. and Tsogt; Ulan-Ergiin-Gol river, 3-4 km beyond Must somon, Aug. 12, 1979—Grub., Dariima et al.), ***Cent. Khalkha, East. Mong., Depr. Lakes, Val. Lakes, Gobi-Alt.*** (Bayan-Tukhum area, Aug. 5, 1931—Ik.-Gal.; Khongor somon, Tergetu-Khuduk, June 20, 1945—Yun.; south. trail of Artsa-Bogdo mountain range, Dzhirgalant well, July 20, 1948—Grub.; 25 km nor.-east of Bayan-Dalai, July 23, 1972—Gur. and Rachk.; nor.-east. bank of Bain-Tukhum lake, Sept. 9, 1979—Grub. and Dariima), ***East. Gobi*** (Dalan-Dzadagad town, June 17, 1945—Yun.; Bailinmyao town, 1959—Ivan.; 45 km nor.-east of Nomgon, submontane plain south of Yamat-Ula town, July 12, 1972—Gur. and Rachk), ***West. Gobi*** (Grub. l.c., Gub. l.c.), ***Alash. Gobi*** (Alashan mountain range, Yamata gorge, May 5; same site, Dartymto gorge, May 25; Dynyuan'in oasis, June 1—1908, Czet.; 45 km south of Inchuan', July 3, 1957; 70 and 75 km south of Bayan-Khoto town, south-west. foothill of Kholan'shan', June 15; Bayan-Khoto, Tengeri sand, Bayan-Nor, July 25 —1958, Petr.), ***Ordos*** (25 km south-east of Otokachi town, Khaolaitunao lake, Aug. 1; 80 km west of Ushinchi town, bank of Ulibunor lake, Aug. 2; 50 km west-north-west of Khanchinchi town, on bank of Yan'khaitsza, Aug. 6—1957, Petr.) ***Khesi*** (on Edzin river, June 27, 1886—Pot.; nor. foothills of Nanshan, Dachuan' spring, 1500 m, May 12, 1894; Sachzhou oasis, 1100 m, Aug. 16, 1895—Rob.).

IB. Kashgar: ***East.*** (south. fringe of Khami oasis, Bugas village, 500 m, Aug. 19, 1895—Rob.).

IIA. Junggar: ***Cis-Alt.*** (Koktogoi, No. 10402, June 7, 1959—Lee et al.), ***Jung. Gobi*** (east.: Bodonchi river floodplain, 2-3 km south of Bodonchi-Khure, July 19, 1947 —Yun.; Uienchi somon, Borotsonchzhi area, Sept. 13, 1948—Grub.; 95 km south of Ertai settlement,. July 15, 1959—Yun. et al.; nor.-east. part of Altay somon, June 30, 1973—Golubk. and Tsogt; nor. macroslope of Baitag-Bogdo mountain range, valley of Nariin-Khargaityn-Gol river 8 km east of Baitag-Bogdo post, July 31, 1988—Gub., Ganbold (MW).

General distribution: Jung.-Tarb. (Zaisan basin, Tarbagatai); West Sib. (Irt., Altay), East. Sib., Nor. Mong., China (? Altay, Dunbei, Nor.).

3. **H. sarmentosa** (Adams) Kom. in Kom. et Klob.-Alis, Key Plants Far East. USSR, 1 (1931) 550; Kovalevsk. in Opred. rast. Sr. Azii [Key to Plants of Mid. Asia] 3 (1972) 199; Fl. Tadzh. 4 (1975) 88; Grub. Opred. rast Mong. [Key to Plants of Mongolia] (1982) 113; Fl. Intramong. ed. 2, 2 (1990) 487; Gub. Konsp. Fl. Vneshn. Mong. [Conspectus of Flora of Outer Mongolia] (1996) 50. —*Ranunculus sarmentosus* Adams in Nouv. Mém. Soc. natur. Moscou, 9 (1834) 244; Kryl. Fl. Zap. Sib. 5 (1931) 1190. —*R. subsimilis* H. Printz, Veg. of Sib. Mong. frontiers (1921) 236; Ostenf. and Pauls. in Hedin, S. Tibet, 6, 3 (1922) 81. —*Halerpestes subsimilis* (H. Printz) Tamura in Kitamura, Fl. Agfhan. (1960) 127. —*Ranunculus cymbalaria* auct. non Pursh: Franch. Pl. David. 1 (1884) 20; Forbes and Hemsl. Index Fl. Sin. 1 (1886) 14; Maxim. Fl. Tangut. 1 (1889) 12; id. Enum. pl. Mong. 1 (1889) 17; Hemsl. Fl. Tibet (1902) 162; Kitag. Lin. Fl. Mansh. (1939) 223. —*Halerpestes cymbalaria* auct. non (Pursh) Greene: Fl. Intramong. 2 (1978) 246; Fl. Sin. 28 (1980) 335; Claves pl. Xinjiang, 2 (1983) 322; Fl. desert Sin. 1 (1985) 489; Fl. Xizang. 2 (1985) 111; Pl. Vasc. Helanshan (1986) 107; Y.Z. Zhao in Acta Sci. Nat. Univ. Intramong. 18, 2 (1987) 284. —**Ic.:** H. Printz, 1921, fig. 88 and tab. VI,

fig. 1; Fl. SSSR, 7, Plate 34, fig. 4 (sub. nom. *H. salsuginosa*); Fl. Sin. 28, tab. 105, fig. 4-7; Grub. 1982, Plate 53, fig. 243; Fl. desert Sin. 1, tab. 179, fig. 1-3; Fl. Xizang. 2, tab. 34, fig. 1-4; Fl. Intramong. ed. 2, 2, tab. 147, fig. 1-2.

Described from East. Siberia. Type in ?

In moist and marshy solonchak meadows, solonetzic banks of lakes, rivers, springs, brooks, from plains to upper belt of mountains.

IA. mongolia: ***Khobd.*** (pass between Kharkira river valley and Nyuttsugun river (tributary of Namyur river), July 25—1879, Pot.; Kholoin-Khavtsal gorge, 18 km east of Tsagan-Nur somon, July 24, 1977—Karam. et al.; 8 km before Bugat somon, Bayangol gorge, July 5, 1984—Kam. et al.), ***Mong. Alt.*** (on Urten-Gol river, July 2; bank of Saksa river, July 7, 1877—Pot.; Tsagan-Balun, Aug. 1, 1899—Lad.; Daingol lake, July 5, 1906—Sap.; Khara-Adzarga mountain range, valley of Sakhir-Sala river, Aug. 21; Khoitulyasutai river, Aug. 30—1930, Pob.; Khubtsiin-Nuru mountain range, west of Adzhi-Bogdo, Aug. 3, 1947—Yun.; centre of Bulgan somon, July 20, 1970—Sanczir; Ulan-Ergiin-Gol, 3-4 km beyond Must somon, 1990 m, Aug. 12, 1979—Grub., Dariima et al.), ***Cent. Khalkha*** (upper Kherulen, valley of river near Bain-Erkhetu, 1899—Pal.; Kholt area, June 19, 1926—Gus.; Uber-Dzhargalanta river, between Bolta and Achit mountains, near Dol'che-Gegen monastery, Aug. 26, 1926—Krasch. and Zam.; Choiren [station], July 22, 1928—Tug.; 5-7 km east of Choiren Ula, Gashyun-Gol brook, Aug. 23, 1940—Yun.; Kerulen valley, 5 km beyond Undurkhan, Aug. 9, 1989—Grub., Gub. et al.), ***East. Mong.*** (Khailar town, July 6, 1901—Lipsk.; Manchuria railway station, June 5, 1902—Litw.; Daringa, 22 km west of Baishintu-Sume, Sept. 10, 1931—Pob.; Khailar town, No. 362, 583, June 8, 1951; same site, No. 365, 3007 [without precise dates] 1954—Wang), ***Depr. Lakes*** (near Ulangom, July 2; Kirgiz-Nor, Gumbyurde, July 30; south. bank of Baga-Nor lake, Aug. 2; around Kirgiz-Nor lake, on bank of Dzabkhyn river, Aug. 7; flood-prone section of southern bank of Khara-Usu lake, Aug. 17—1879, Pot.; near Dzereg settlement, along bank of Tsagan-Nur lake, Aug. 3, 1974—Zhurba (MW), ***Val. Lakes*** (Orok-Nor lake basin, May 19, 1926—E. Khozlova; meadow along Ongin-Gol river, July 18, 1926—A. Bulle; Orok-Nor lake, Aug. 4, 1926—Tug.; Ongiin-Gol river floodplain, 30 km before Khushu-Khida, July 17, 1943—Yun.; in Khushu-Khid monastery, in Ongiin-Gol river floodplain, July 14, 1948—Grub.; nor. bank of Bon-Tsagan lake, west of Baidarik estuary, June 16, 1971—Grub., Ulzij. et al.), ***East. Gobi*** (Grub. l.c., Gub. l.c.), ***Gobi-Alt.*** (along Leg river, Aug. 28, 1886—Pot.; Legin-Gol river valley, Aug. 19-Sept. 1, 1927—Simukova; Bain-Tukhum area, Aug. 4, 1931—Ik.-Gal.; pass between Dundu- and Dzhun-Saikhan, July 22, 1943; Dalan-Dzadagad town, June 17; Khongor somon, Tergetu-Khuduk area, June 20, 1945—Yun.; 25 km east-nor.-east of Bayan-Dalai, July 23, 1972—Gur. and Rachk.), ***West. Gobi*** (Grub. l.c., Gub. l.c.), ***Alash. Gobi*** (midportion of Alashan mountain range, west. slope, in valley in foothills, July 5, 1873—Przew.; around Chzhun-vei town, July 2, 1957—Petr.; Bayan-Khoto town, 1959 —Ivan.; Chzhunvei, July 2; 45 km south of Inchuan' town, July 3—1957, Petr.), ***Ordos*** (right bank of Huang He river, Aug. 4, 1884—Pot.), ***Khesi*** (near Fui'itin town, June 5; Edzin river, June 20—1886, Pot.; Chzhan'e, 40 km north of Yunchan town, Ninyanlu settlement, July 1, 1958—Petr.).

IB. Kashgar: ***Nor.*** (Yarkend-Dar'ya 1500 m, June 9, 1889—Rob.; Uch-Turfan, Yamansu, June 9, 1908—Divn.; in Bugur village, Aug. 20, 1929—Pop.; (around) Bai, No. 8169, Sept. 2, 1958—Lee and Chu), ***West.*** (around Yangi-Gissar town, May 27, 1909—Divn.; Kashgar oasis, Yangi-Shar area, July 25, 1929—Pot.), ***South.*** (alongside Keriya-Chira highway, No. 120, Sept. 16, 1957—Kuan; 30 km west of Keriya on road to Chira, May 16, 1959—Yun. and Yuan'), ***East.*** (Khami, May 29, 1877—Pot.; Lyaodun in Khami, No. 454, May 21, 1957—Kuan; 1 km nor. of Pichan area, No. 5431, May 23; nor.-east of Toksun, No. 7339, June 19; on Bagrashkul' Lake, No. 6180, July 26—1958, Lee and Chu).

IC. Qaidam: ***Plain*** (around Barun-Tszasak site, 2400 m, May 30, 1901—Lad.), ***Mount.*** (Kurlyk area, 2700 m, May 8 and 15; Baga-Qaidamin-Nor lake, 3000 m, June 9, 1895, Dulankhit temple, May 30, 1901—Lad.).

IIA. Junggar: ***Cis-Alt.*** (Koktogoi, No. 10393, June 6, 1959—Lee), ***Tarb.*** (near Khobsair, No. 10481 and 10571, June 21 and 24—1959, Lee et al.), ***Jung. Alat.*** (Dzhair mountain range, road from Karamai to Chuguchak, Aug. 5, 1951—Mois.; Lyaergou village in Toli district, No. 2428, Aug. 4, 1957—Kuan), ***Tien Shan*** (Dzhagastai, Aug. 7, 1877; on Kash river, 900 m, Sept. 6, 1878—A. Reg.; am Piket Kukurtuk, June 26—July 2, 1903; Südliches Kiukönik Tal, beim Lager unter Tschon-Gailak Pass, June 15, 1908—Merzb.; vicinity of Sinyuan' town, No. 3773, Aug. 24, 1957—Kuan; nor.-east of Toksun, No. 7327, June 19, 1958—Lee and Chu), ***Jung. Gobi*** (south.: Urumchi—Khunyanchi, No. 589, June 7; east of Kuitun, No. 419, July 7—1957, Kuan; east.: in Barkul' lake region, Dakhechzhuan village, Sept. 26, 1957—Kuan; Nomyn-Gobi desert, 30 km west of Altay somon, July 27; south of Atas-Ula mountain, around Talyn-Bulak spring, July 29—1978; east. extremity of Baitag-Bogdo mountain range, at Olyn-Bulak post, Aug. 2, 1979; 35 km west of Bulgan settlement, on bank of Bulgan river, 1000 m, Aug. 14, 1982—Gub. (MW); nor. slope of Baitag-Bogdo mountain range, forest and alpine belt, Budun-Khargaityn-Gol gorge, 15 km east of Baitag-Bogdo post, Aug. 1, 1988—Gub., Kam. et al.), ***Dzhark.*** (Kuldscha, 1876—Golicke; north of Kul'dzha, May 3; Khoyur-Suman south of Kul'dzha, May 27; Suidun town, July 16, 1877; Bayandai near Kul'dzha, 600-1200 m, May 6, 1878—A. Reg.; Kul'dzha, 1906—Muromlyanskii).

IIIB. Tibet: ***Chang Tang*** ("33°32' N, 88°52' E, 5100 m, Aug. 1901—Hedin"—Ostenf. and Pauls. l.c.), ***South.*** ("Nan'mulin, Lankaza, Linzhou, Shigadze"—Fl. Xizang. l.c.).

IIIC. Pamir ("Little Kara-Kul, 1894—Hedin"—Ostenf. and Pauls. l.c.; Kenkol, Tom-Kara area, June 14, 1909—Divn., Tash-Kurgan, Aug. 25, 1913—Knorr.; in Tashkurgan, No. 299, June 13, 1959—Lee et al.).

General distribution: Aralo-Casp., Fore Balkh., Jung.-Tarb., Nor. and Cent. Tien Shan, East. Pam.; Mid. Asia (West. Tien Shan, Pamiro-Alay and plains adjoining it), West. Sib., East. Sib., Far East, Nor. Mong., China (Nor.), Himalayas, Japan, Indo-Mal.

4. **H. tricuspis** (Maxim.) Hand.-Mazz. in Acta Horti Gotoburg, 13 (1939) 135; Fl. Sin. 28 (1980) 336; Claves pl. Xinjiang, 2 (1983) 321; Fl. desert Sin. 1 (1985) 491; Fl. Xizang. 2 (1985) 111. —*Ranunculus tricuspis* Maxim. Fl. Tangut. 1 (1889) 12; id. Enum. pl. Mong. 1 (1889) 16; Deasy, Tibet, Chin. Turkest. (1901) 396; Hemsl. Fl. Tibet (1902) 163; Diels, Futturer durch Asien (1903) 10; Finet et Gagnep. in Bull. Soc. Bot. Fr. 51 (1904) 317; Gr.-Grzh. Opis. put. Zap. Kit. [Report of Expedition in West. China] 3 (1907) 476; Borodina-Grabovskaya in Catalogue type specim. Centr. Asia (LE) (2000) 200. —*Halerpestes filisecta* L. Liou in Fl. Sin. 28 (1980) 338, 362; Fl. Xizang. 2 (1985) 113. —**Ic.:** Maxim. Enum. pl. Mong., tab. IV, fig. 17-27; Fl. Sin. 28, tab. 105, fig. 12 (sub *H. filisecta*); Fl. desert Sin. 1, tab. 179, fig. 8-10; Fl. Xizang. 2, tab. 34, fig. 9 (sub *H. filisecta*).

Described from Qinghai. Lectotype and syntypes in St.- Petersburg (LE).

In moist solonetzic and marshy meadows, along banks of brooks and rivers, near springs.

IC. Qaidam: ***Plain*** (Barun-tszasaka khyrma region, Khatu gorge, 2400-2700 m, May 24 and June 15, 1901—Lad.), ***Mount.*** (Nor. Qaidam, Kurlyk area, 2700 m, May 25; same site, near Khabirga spring, 3300 m, June 3—1895, Rob.).

IIIA. Qinghai: ***Nanshan*** (alpes Nan-shan, regio infer., 8000 ft [2400 m], ad fontes, No. 274, June 28/July 10, 1879—Przew., lectotypus (Borodina-Grabovskaya, l.c.) et

2 isolectotypi *R. tricuspis* !; ad fl. Tschurmyn, 8000-9500 ft [2700-3000 m], in ripa argillosa, No. 54, April 28/May 10, 1880—Przew., syntypus et 2 isosyntypi !; ripa humida fl. Lonssyr, supra oppidum Li-yuan-in, [June 2] 1886—Pot.; valley of Sharagol'dzhin river, Psae area, June 12; Yamatynumru area, 3700-4000 m, June 13, 1894—Rob.; "valley of Datunkhe river, 3000 m, May 2, 1905—Gr.-Grzh."—Gr. Grzh. l.c.; "Kuku-Nor"—Diels, l.c.), ***Amdo*** (valle fl. Landsha-Lunwa, ripa arenosa humida, May 8, 1885—Pot., syntypus !).

IIIB. Tibet: ***Chang Tang*** (decliv. N, Jugi Russki, fauce fl. Moldsha, in glareosis ripae, parce, No. 31, May 5/17, 1885—Przew.; Keriya [June 3]—1885, Przew.; nor. slope of Russky mountain range, Bashbulak area in upper Aksu, 3700 m, June 1; same site, near Karasai village, 3000 m, June 13—1890, Rob.; "34°39′ N, 82°5′ E, 5100 m, July 27, 1896—Camp 23"—Deasy, l.c.; "Getszi (sub nom. *H. filisecta*), Zhitu, Shen'chzha"—Fl. Xizang. l.c.), ***Weitzan*** (nor. slope of Amnen-kor mountain range, Kuku-Bulak spring, June 5; nor. bank of Russkoe lake, 4100 m, June 20—1900, Lad.), ***South.*** ("Chzhada, Pulan', Lhasa"—Fl. Xizang. l.c.).

General distribution: China (? Nor., Nor.-West., South-West.), Himalayas (Kashmir, east.).

Note. Judging from the description and figure, we regard *H. filisecta* L. Liou in Fl. Sin. 28 (1980) 338, 362, described from solonchak meadows of Tibet (Chang Tang: Getszi), as identical to *H. tricuspis.* As a result of including *H. filisecta,* the distribution of *H. tricuspis* extended into Himalayas. In "Flora Kitaya" [Flora of China] and, immediately following it, in "Flora Tibeta" [Flora of Tibet], the concept of *H. tricuspis* (judging from fig. in Fl. Sin. 28, tab. 105, fig. 8-9 and Fl. Xizang. 2, tab. 34, fig. 5-6, and key) was incorrect since plants with unlaciniated leaf blades were included under it.

17. **Oxygraphis** Bge.

in Mém. Ac. Sci., St. Pétersb., Sav. Etr. 2 (1835) 556

1. **O. glacialis** (Fisch. ex DC.) Bge. in Mém. Ac. Sci. St. Pétersb., Sav. Etr. 2 (1835) 556; id. Suppl. Fl. alt. (1835) 46; Ostenf. and Pauls. in Hedin, S. Tibet, 6, 3 (1922) 81; Kryl. Fl. Zap. Sib. 5 (1931) 1174; Ovcz. in Fl. SSSR, 7 (1937) 326; Fl. Kirgiz. 6 (1955) 81; Fl. Kazakhst. 4 (1961) 76; Ikonnik. Opred. rast. Pamira [Key to Plants of Pamir] (1963) 118; Kovalevsk. in Opred. rast. Sr. Azii [Key to Plants of Mid. Asia] 3 (1972) 199; Fl. Tadzh. 4 (1975) 87; Fl. Sin. 28 (1980) 333; Grub. Opred. rast. Mong. [Key to Plants of Mongolia] (1982) 113; Claves pl. Xinjiang, 2 (1983) 320; Fl. Xizang. 2 (1985) 109; Gub. Konsp. Fl. Vneshn. Mong. [Conspectus of Flora of Outer Mongolia] (1996) 50. —*Ficaria glacialis* Fisch. ex DC. Prodr. 1 (1824) 44. —**Ic.:** Fl. SSSR, 7, Plate 34, fig. 1; Fl. Kazakhst. 4, Plate 10, fig. 4; Claves pl. Xinjiang, 2, tab. 53, fig. 241; Fl. Sin. 28, tab. 104, fig. 1-4; Fl. Xizang. 2, tab. 33, fig. 1-4.

Described from West. Siberia (Altay). Type in St.-Petersburg (LE).

On moist rocks, along banks of brooks, grasslands in alpine belt, in rubbly and rubbly-lichen mountain tundra, sometimes on snow patches; up to 4800 m.

IA. Mongolia: ***Khobd.*** (Bairimen-Daban, 2400 m, June 20, 1980—Pot.; south-east. spur of Turgen mountain range, south. macroscope of Kashgurt mountain, around brook, July 21, 1977—Karam. et al.). ***Mong. Alt.*** (between Dain-Gol and Ak-Korum lakes, June 29, 1903—Gr.-Grzh.; 10-15 km west of Dayan-Nur post, near Khara-Nur lake, July 20, 1988—Bud., Gamalei, Gub. et al.).

IIA. Junggar: ***Tien Shan*** (Mal. Yuldus, 2300-2700 m, May 24; Narat mountain range, 2700 m, June 5—1877, Przew.; Bogdo mountains, 3000 m, July 25, 1878; Irenkhabirga mountain range, Taldy gorge, 3000 m, May 24; Kum-Daban, 2700 m, May 29; Chungur-Daban pass, north of Borborogusun, [June 13]—1879, A. Reg.; Oberstes Dschanart Tal bis hinauf zum Pass, July 14-17, 1903; hohes Plateau eines Berges in oberem Koksu Gebiet, circa 3400 m, July 8-10, 1907; Kiukonik Pass, Nordseite, June 16, 1908—Merzb.; Taskhan canal, east of Khomota, No. 7138, Aug. 6, 1958—Lee and Chu).

IIIA. Qinghai: ***Nanshan*** (alp. meadow south of Tetung river, April 23, 1873—Przew.), ***Amdo*** (alp. belt of Dzhakhar-Dzhargyn mountains, on border of rocky belt, June 26, 1880—Przew.).

IIIB. Tibet: ***Chang Tang*** ("Zhitu"—Fl. Xizang. l.c.), ***Weitzan*** (20 km south-west of Alak-Nor-Gol, May 15; Odon'tala basin, May 24; upstream of Dzhagyn-Gol river, May 28 —1884, Przew.; Amnen-Kor mountain range, nor. and south. slopes, 2300-4700 m, June 4; Dzhagyn-Gol river, [July 2] 1900, Yantszytszyan river basin (Goluboi), Var-munchyu brook, on moist sandy-rocky banks, 4200 m, May 19; same site, on Gunte-Aa pass, on way to Khor-Gamdze from Chamdo, 4000 m, July 23, 1901—Lad.; "Sosyan', Bizhu"—Fl. Xizang. l.c.), ***South.*** ("Shuankhu"—Fl. Xizang. l.c.).

IIIC. Pamir ("Little Kara-Kul, 1894—Hedin"—Ostenf. and Pauls. l.c.; upper Pil'nen river, alpine mossy tundra near glacier, June 16, 1942—Serp.).

General distribution: Jung.-Tarb., Nor. and Cent. Tien Shan, East Pam.; Europe (East.), Arct., Mid. Asia (mountains), West. Sib. (Altay), East. Sib., Far East, Nor. Mong. (Fore Hubs., Hent., Hang.), China (Altay, Nor.-West., South-West.), Himalayas, Nor. Amer.

18. **Ranunculus** L.

Sp. pl. (1753) 548

1. Flowers initially pale pink, turning white later; calyx persists on fruit. Leaves proximated at tip of stem forming involucre surrounding flower.......................................32. **R. similis** Hemsl.

\+ Flowers yellow or whitish; calyx on fruit caducous or persistent. Leaves cauline and radical, not proximated on flowers.................2.

2. Nuts altogether flat or highly compressed laterally, often winged; calyx and corolla persist on fruit. Roots of 2 types: some elongated-tuberculate, cone-shaped or fusiform; rest thin and long; all aggregated into compact cluster. Land plants (section 2. *Ranunculastrum* DC.)...3.

\+ Nuts bulging or only slightly laterally compressed, not winged (sometimes, only with membranous border; calyx and corolla on fruit caducous. All roots monotypical—thin or thickened, fibrous, aggregated into cluster or form more or less developed rhizome. Water, marsh or dry land plants..6.

3. Glabrous plant. Radical leaves reniform or orbicular-cordate in profile, deeply tripartite into orbicular-cuneate segments obtuse above; these segments, in turn, divided or lacinated into 3 ovoid segments of second order, crenate along margin. Petiole of radical leaf narrowed at base. Nut 3-4 mm long........8. **R. gobicus** Maxim.

\+ Pubescent plant. Radical leaf 2-3-pinnatisected with long-petiolate segments of first order, in turn ternate- or pinnatisected into sessile or laciniated segments of second order; segments of latter order linear or lanceolate, cuspidate. Petiole of radical leaf enlarged at base into sheath. Nut 4-8 mm long...4.

4. Stylodium of nut very short (0.2-0.4 mm). Many nuts globose...9. **R. platyspermus** Fisch. ex DC.

\+ Stylodium of nut very long, 1-2 mm long. Many nuts terete..........5.

5. Leaf segments of latter order (terminal) narrow-lanceolate or linear.................................,...............................10. **R. regelianus** Ovcz.

\+ Leaf segments of latter order very broad—lanceolate or broad-lanceolate...11. **R. severzovii** Regel.

6. Plant with creeping or procumbent stems rooting at nodes (sometimes, only at lower nodes) ....7.

\+ Stems erect or even ascending, never rooting at nodes. Leaves entire, partite, laciniated or compound..14.

7. Leaves entire, elongated from ovoid-lanceolate to linear (section 5. *Flammula* (Webb ex Spach) Freyn...8.

\+ Leaves differently laciniated..10.

8. Small plant with procumbent-decumbent stems rooting all along at nodes; internodes arcuate. Leaves linear, imperceptibly transforming into petiole. Flowers small, 5-9 mm in diam. ....41. **R. reptans** L.

\+ Large plants with erect, thick , hollow stems, rooting in lower portion and creeping jointed rhizome. Leaves sessile, lanceolate. Flowers large, 1.6-4 cm in diam. ....9.

9. Leaves broad-lanceolate along margin with spaced fine teeth. Stem more or less glabrous or appressed-pubescent above. Flowers 3-4 cm in diam. Nuts glabrous..40. **R. lingua** L.

\+ Leaves linear-lanceolate, smooth-edged, with appressed setaceous hairs like stem. Flowers 1.6-2.5 cm in diam. Nuts with stray short hairs falling later...39. **R. amurensis** Kom.

10. Small pubescent plants. Leaf blades small, about 1 cm in diam., entire, deeply (often down to base) 3-5-palmatisected into segments...11.

\+ Very large glabrous plant. Leaf blade about 2-3 cm in diam., 3-5 lobed or partite into broad, subentire segments.............................12.

11. Aquatic floating or swamp-dry land plants. In land plants, stem simple, rooting at nodes; in aquatic plants, floating. Leaves 3-5-sected. Leaf segments linear or oblong, and in submerged plants filiform. Entire plant silky-haired. Nuts 1-1.2 mm long, many nuts 2-3 mm in diam. Plant of East. Mongolia. ......42. **R. gmelinii** DC.
+ Land plant with stem rooting at nodes. Leaves 3-sected; mid-segment entire, 3 (5)-lobed, obovoid; lateral segments bipartite with 2-3-lobed segments. Stems, leaf stalks and leaf blades with stiff hairs. Tibetan plant..............................18. **R. glareosus** Hand.-Mazz.
12. Leaf blade without notch at base, truncated or broad-cuneate. Stylodium of nut erect and very short (0.1 mm). Receptacle glabrous...43. **R. intramongolicus** Y.Z. Zhao.
+ Leaf blades with cordate base. Receptacle pubescent..................13.
13. Leaf blade entire, with large, obtuse teeth, 3-5-lobed or 3-partite (for 1/3-1/2 of length), coriaceous and glossy, up to 3.5 cm broad. Nuts 1.5-2 mm long, with erect uncinate stylodium at tip; many nuts 6-8 mm in diam. at head. Aquatic plant with floating stem and leaves. Found in Mongolia and China..44. **R. natans** C.A. Mey.
+ Leaf blade 3-lobed or 3-5-partite (for 1/2-2/3 of length) into orbicular, smoot-edged or broad-dentate lobes, not coriaceous, up to 2 cm broad. Nuts 1.2-1.4 mm long, with short beak; many nuts 5-7 mm in diam. at head. Land plant, sometimes aquatic with stems rooting at nodes. Found in west. part of Mongolia...45. **R. radicans** C.A. Mey.
14 (6). Annual or biennial plant, weakly rooting. Receptacle oblong-conical, hollow; many nuts oblong. Nuts small............................15.
+ Perennial plant with developed root system. Receptacle different form but not hollow. Nuts very large, with developed stylodium...16.
15. Plant divaricately stiff-haired. Leaves short-petiolate in profile, cuneate at base, deeply trifid into acute segments. Petals perceptibly longer than sepals. Nuts 2-3 mm long, glabrous...15. **R. chinensis** Bge.
+ Plant glabrous or, rarely, with soft appressed pubescence above. Leaves orbicular in profile, laciniated into 3-5 obtusilobous sessile segments. Petals barely longer than sepals. Nuts 1-1.3 mm long, rugose or pitted...46. **R. sceleratus** L.
16. Nuts somewhat laterally compressed, glabrous or rarely pilose, marginate, with perceptible oblong nerves. Lower leaves usually more apart than upper leaves. Rhizome very often developed (section 1. *Acris* Schur)......... ....17.

+ Nuts intumescent-convex, glabrous, smooth or pubescent, with single nerve, nearly confluent with dorsal portion, not marginate or with indistinct membranous border. Lower leaves usually less apart than upper leaves. Perennial plant, with cluster of fine fibrils (section 3. *Ranunculus*)..23.

++ Nuts obovoid, intumescent in upper portion, short-pubescent, with whitish membranous border. Radical and cauline leaves developed to equal extent. Perennial plant with cluster of thickened fibrils. Junggar plant (section 4. *Polyrhizos* Tzvel.)................................ ....38. **R. polyrhizos** Steph.

17. Leaf blades 3-5-partite down to base with often petiolate segments..18.

+ Leaf blades 3-partite deeply but not down to base (up to 1/7-1/5 of length) into cuneate-rhombic segments.............................. ....7. **R. submarginatus** Ovcz.

18. Stems rooting, ascending or procumbent in lower portion, glabrous like leaf petioles. Plant with creeping rooting surface shoots. Leaf blades ternate with petiolate segments, in turn deeply 3-sected into dentate segments. Sepals carinate, only imperceptibly declinate from petals..6. **R. repens** L.

+ Stems erect, rarely ascending, but never procumbent, glabrous or pubescent. Plant not forming creeping surface shoots. Leaf blades of different form..19.

19. Plants with well-manifest rhizome. Leaves deeply, but not down to base, 3- or more-partite. Rhizome elongated-procumbent. Stem with hairs bent down in lower portion and appressed hairs in upper. Nuts 3.5-4 mm long, with 0.5-0.8 (1) mm long beak.............. ....2. **R. grandifolius** C.A. Mey.

+ Plants without rhizome or rhizome highly contracted................20.

20. Receptacle pubescent..4. **R. polyanthemus** L.

+ Receptacle glabrous..21.

21. Leaf profile cuneate, 3-partite deeply, sometimes down to base, with lobes narrowed toward base.............1. **R. cuneifolius** Maxim.

+ Radical leaf profile orbicular, palmatipartite; partite not down to base of blade..22.

22. Stems glabrous or with rare appressed and semiappressed hairs, one- or few-flowered, 8-30 cm tall. Leaves predominantly radical on short petioles, pubescent with up-turned stiff hairs or glabrous; leaf blades glabrous or with scattered hairs.................................. ....5. **R. propinquus** C.A. Mey.

+ Lower stem and petiole of radical leaves divaricately hispid. Radical leaves on long petioles which, like their blades, diffusely, or sometimes compactly, appressed-hispid. Flowers generally

many, in corymbose inflorescence. Plant up to 60 cm tall........
....3. **R. japonicus** Thunb.

23(16). Leaf blades entire, of different form, from lanceolate to orbicular-reniform, smooth-edged or serrate-incised, rarely deeper than centre ....24.

\+ Leaf blades 3- or more-partite into narrow, linear or lanceolate lobes...36.

24. Radical leaf profile elliptical or lanceolate-elliptical....................25.

\+ Radical leaf profile orbicular..30.

25. Plant white-haired...26.

\+ Plant glabrous, subglabrous or stems with appressed hairs in upper portion..28.

26. All leaves linear....................................26. **R. membranaceus** Royle.

\+ Radical leaf ovoid-elliptical..27.

27. Radical leaf 3-lobed..23. **R. krasnovii** Ovcz.

\+ Radical leaf 5-7-lobed...............................19. **R. hetianensis** L. Liou.

28. Flowers up to 10 mm in diam., often with underdeveloped petals, almost no longer than sepals. Stem 2-3-forked or simple.................
....25. **R. longicaulis** C.A. Mey.

\+ Flowers 13-25 mm in diam. with large petals. Stems simple or with 1-3 unequal branches...29.

29. Radical leaf elliptical or lanceolate-elliptical; cauline leaf narrow-lanceolate; all leaves smooth-edged. Flowers 14-25 mm in diam. Stem with scattered dull, golden, semiappressed hairs in upper portion ....30. **R. pseudohirculus** Schrenk.

\+ Radical leaf elliptical or oblong-elliptical, serratodentate or serrately incised; cauline leaf more narrow, sessile, simple or 3-5-palmatisected into linear lobes. Flowers 13-17 mm in diam. Plant glabrous...31. **R. pulchellus** C.A. Mey.

++ Radical leaf broad-ovoid or reniform-orbicular, weakly cordate or broad-cuneate at base, 3-palmatilobed or 3-palmatipartite, with 3-5 orbicular segments above; cauline leaf more narrow, laciniated into lanceolate lobes. Flowers 5-20 mm in diam. Stem weakly pubescent.......................37. **R. yinshanensis** (Y.Z. Zhao) Y.Z. Zhao.

30(24). Radical leaf 1, rarely 2-3, orbicular or orbicular-reniform, mostly with narrow axil at base, entire or rarely with 3-7 broad lobes, dentate. Cauline leaf palmatisected into entire, linear segments. Stems glabrous. Fruit head subglobose. Nut short-pubescent. Flowers 14-18 mm in diam. Sepals subglabrous............
....27. **R. monophyllus** Ovcz.

\+ Characteristics different. Sepals pubescent.......................................31.

31. Sepals pubescent with dark brown or reddish brown stiff hairs...32.

+ Sepals pubescent with whitish, golden or reddish hairs............33.

32. Radical leaf with narrow-cuneate base, 3-5 dentate at tip, with narrow and long teeth..12. **R. altaicus** Laxm.

+ Radical leaf with cordate base, round-toothed all along length ....21. **R. indivisus** (Maxim.) Hand.-Mazz.

33. Radical leaf completely entire or 3-partite, crenate or round-toothed along margin, orbicular or narrow-emarginate at base................ ....17. **R. fraternus** Schrenk.

+ Radical leaf partite or laciniated to different extents, mostly 3-partite down to base into cuneately narrowed incised-denate segments. Segments of cauline leaves linear-lanceolate. Nuts glabrous..34.

34. Petals 5-7 (8), broad-obovoid...35.

+ Petals 8, oblong, thrice longer than breadth.. ....16. **R. chinghoensis** L. Liou.

35. Receptacle white-tomentose. Radical leaf profile suborbicular, deeply (down to base) 3-partite into broad-cuneate segments narrowed toward base..............................33. **R. songaricus** Schrenk.

+ Receptacle subglabrous. Radical leaf profile reniform-orbicular, 3-sected up to centre or slightly less into broad obovoid segments.... ....36. **R. trautvetterianus** Regel ex Ovcz.

36(23). Leaves deeply bipartite .... 37.

+ Leaves less laciniated..38.

37. Entire or only midsegment of leaf blade short-petiolate. Primary segments with orbicular or truncated base. Sepals green, white-haired, or with dull raddish hairs. Quite large (10-15 cm tall) Junggar plant..22. **R. karelinii** Czer.

+ Only midsegment of leaf blade short-petiolate, rarely petiole not developed. Segments narrow at base. Sepals purple, glabrous. Diminutive (2-3 cm tall) Tibetan plant.. ....35. **R. tibetanus** (Tolm. et Jurtz.) Borod.-Grabovsk.

38. Radical leaf multipartite; if 3-partite, segments deeply palmatifid or partite...39.

+ Radical leaf orbicular-reniform, deeply 3-partite or 3-sected into dentate, lobed segments or segments entirely cuneately narrowed toward base..40.

++ Radical leaf, at least some of them, cuneate at base, deeply partite into lanceolate segments or dentate, compactly pubescent. Small, 3-7 (10) cm tall plant...42.

39. Plant 10-20 cm tall, long, soft patent-haired or subglabrous. Radical leaves of 2 types: some orbicular, dentately incised to palmatisected, with very large midtooth; others palmatisected into

lanceolate lobes. Flowers 10-20 mm in diam. Nuts with short pubescence, rarely glabrous .... 28. **R. pedatifidus** Smith.

\+ Much larger, 10-30 cm tall plant compactly appressed-pilose. Radical leaf laciniated almost down to base into 3 oblong-cuneate segments. Nuts glabrous.. ....34. **R. tanguticus** (Maxim.) Ovcz. ex W.T. Wang.

40\. Roots thickened, fuseform, fascicled..........38. **R. polyrhizos** Steph.

\+ Roots thin, in contracted rhizome..41.

41\. Petals 5. Entire plant compactly pubescent....20 **R. hirtellus** Royle.

\+ Petals 8-14, set in 2-3 rows. Entire plant largely glabrous............... ....24. **R. lasiocarpus** C.A. Mey.

42\. Diminutive (1.2-4 cm tall), glabrous or faintly pubescent plant. Stem simple, monochromatic, glabrous. Leaves coriaceous, partite down to base into 3 entire broad-rhombic lobes............................... ....13. **R. banguoensis** L. Liou.

\+ Much larger (4-10 cm tall), compactly pubescent plant. Stem simple or branched, 1-2 or 4- or more-flowered. Leaves not coriaceous, of different form..43.

43\. Stem simple, 1-2-flowered, pubescent with appressed hairs, 3-7 (10) cm tall. Plant forms mat. Receptacle, ovary and nut glabrous..14. **R. brotherusii** Freyn.

\+ Stem branched, 4- or many-flowered, 4-6 cm tall, with compact tomentose, patent pubescence, single. Receptacle, ovary and nut pubescent..29. **R. popovii** Ovcz.

Section 1. **Acris** Schur

1\. **R. cuneifolius** Maxim. in Bull. Ac. Sci. St. Pétersb. 23 (1877) 306; id. Enum. pl. Mong. 1 (1889) 22; Fl. Intramong. 2 (1978) 255; Fl. Sin. 28 (1980) 314; Fl. desert Sin. 1 (1985) 495; Fl. Intramong. ed. 2, 2 (1990) 508; W.T. Wang in Bull. Bot. Res. (Harbin) 15, 3 (1995) 310. —**Ic.:** Maxim. 1889, tab. 1, fig. 1-7; Fl. Intramong. 2, tab. 133; Fl. Sin. 28, tab. 98, fig. 4-7; Fl. desert Sin. 1, tab. 181, fig. 3-5; Fl. Intramong. ed. 2, 2, tab. 206, fig. 1-2.

Described from Inner Mongolia (Ordos). Type and isotype in St.-Petersburg (LE). Map 4.

In meadows.

IA. Mongolia: ***Ordos*** (Terra Ordos, pratis vere inundatis secus fl. Hoangho, solo argilloso, frequens, No. 319, July 28/Aug. 9, 1871—Przew., holotypus ! et isotypus !; Ordos, meadow around Bain-Nor lake, Sept. 2; same site, Tszokul'-Chaidam area, Sept. 7—1884, Pot.).

IIIA. Qinghai: ***Nanshan*** (Tsinlyan'shan', Matisy temple, scrubs with grasslands in mountain valley, 2600 m, July 12, 1958—Petr.).

General distribution: China (Dunbei).

Note. Specimen from Nanshan differs from type plant in the smaller size of plant, reduced petals and nut with very long beak. It differs from other species of section in the characteristic form of leaves narrowed toward base.

2. **R. grandifolius** C.A. Mey. in Ledeb. Fl. alt. 2 (1830) 330; Kryl. Fl. Zap. Sib. 5 (1931) 1206; Ovcz. in Fl. SSSR, 7 (1937) 468; Fl. Kirgiz. 6 (1955) 101; Fl. Kazakhst. 4 (1961) 107; Kovalevsk. in Opred. rast. Sr. Azii [Key to Plants of Mid. Asia] 3 (1972) 215; Fl. Sin. 28 (1980) 317; Claves pl. Xinjiang, 2 (1983) 318; W.T. Wang in Bull Bot. Res. (Harbin) 15, 3 (1995) 308. —**Ic.:** Fl. Kazakhst. 4, Plate 13, fig. 6; Fl. Sin. 28, tab. 98, fig. 1-3.

Described from West. Siberia (Altay). Type in St.-Petersburg (LE).

In subalpine meadows and in forest belt, along banks of brooks and canals, forest meadows, scrubs.

IIA. Junggar: ***Jung. Alat.*** (Toli district, Albaksin town, No. 2563, Aug. 6, 1957—Kuan), ***Tien Shan*** (Sairam lake, July 1877—A. Reg.; Burkhantau, June 5, 1878—Fet.; from Borgata to Khapchagai, on nor. side of Kash river, 1800-2100 m, July 6; lower Aryslyn, 1700-1800 m, July 20—1879, A. Reg.).

General distribution: Jung.-Tarb., Nor. Tien Shan; Mid. Asia, West. Sib. (Altay), East. Sib.

3. **R. japonicus** Thunb. in Trans. Linn. Soc. (London) 2 (1794) 337; Kitag. Lin. Fl. Mansh. (1939) 224; Ovcz. in Fl. SSSR, 7 (1937) 470; Fl. Intramong. 2 (1978) 255; Fl. Sin. 28 (1980) 312; Grub. Opred. rast. Mong. [Key to Plants of Mongolia] (1982) 115; Claves pl. Xinjiang, 2 (1983) 318; Fl. desert Sin. 1 (1985) 497; Fl. Intramong. ed. 2, 2 (1990) 510; Gub. Konsp. Fl. Vneshn. Mong. [Conspectus of Flora of Outer Mongolia] (1996) 50; W.T. Wang in Bull. Bot. Res. (Harbin) 15, 3 (1995) 303, p.p. —*R. acris* L. var. *japonicus* (Thunb.) Maxim. Enum. pl. Mong. 1 (1889) 21. —*R. smirnovii* Ovcz. in Fl. URSS, 7 (1937) 745, 467; Kitag. Lin. Fl. Mansh. (1939) 226; Fl. Intramong. ed. 2, l.c. 508; W.T. Wang, l.c. (1995) 307. —*R. japonicus* var. *smirnovii* (Ovcz.) L. Liou in Fl. Sin. 28 (1980) 314. —**Ic.:** Fl. Intramong. 2, tab. 134; Fl. Sin. 28, tab. 97, fig. 1-4; Grub. Plate 54, fig. 248; Fl. desert Sin. 1, tab. 182, fig. 11-15; Fl. Intramong. ed. 2, 2, tab. 206, fig. 3-4 and 5.

Described from Japan. Type in Uppsala (UPS).

In larch and spruce forests, birch stands, forest borders, among shrubs on river banks, coastal meadows, along floors and moist slopes of creek valleys.

IA. Mongolia: ***East. Mong.*** (vicinity of Khailar, June 29, 1873—Fritshe; Muni-Ula mountains, alp. meadow, June 1871—Przew.; valley of Ergenten-Gol river, July 10, 1899—Pot. and Sold., Khailar town, June 8, 1951—S.H. Li et al.; same site, June 10, 1951—Kh. Wang; Shiliin-Bogdo-Ula, July 8, 1971—Dashnyam, Karam. et al.: Shiliin-Bogdo-Ula mount., 1650 m, July 12, 1985—Gub. (MW).

IIA. Junggar: ***Jung. Alat.*** (in Toli district, No. 1124 and 1260, Aug. 6 and 7, 1957—Kuan; 30 km west of Chzhaosu, Tszin'tsyuan' settlement, No. 3272, Aug. 12;

Ven'tsyuan'—Syata, No. 3536, Aug. 15, 1957—Kuan; Argaty river valley, Aug. 17, 1957—Yun. and Yuan'; 7 km north of Ven'tsyuan', No. 4586, Aug. 24; in Ven'tsyuan' region, No. 4613, Aug. 25—1957, Kuan), ***Tien Shan*** (Urumchi region, upper Tasenku river, Biangou area, Sept. 25, 1929—Pop.; 40 km south of Dzinkho, No. 104, Aug. 2; from Dzhagastai to Khaichzhou, No. 3191, Aug. 8—1957, Kuan; east. part of Ketmen' mountain range, Sarbushin pass on road to Kalmak-Kure from Ili, Aug. 23; floodplain of Tekes river near bridge on Aksu-Kalmak-Kure road, Aug. 26, 1957—Yun. and Yuan'), ***Jung. Gobi*** (south.: Urumchi—Yan'ervo, No. 566, May 31, 1957—Kuan).

General distribution: East. Sib., Far East, Nor. Mong. (Mong.-Daur., Cis-Hing.), China (Dunbei, Nor., Nor.-West., South-West, Taiwan), Korean peninsula, Japan, Indo-Mal. (Indochina).

4. **R. polyanthemus** L. Sp. pl. (1753) 554, "*polyanthemos*"; Kryl. Fl. Zap. Sib. 5 (1931) 1207; Ovcz. in Fl. SSSR, 7 (1937) 446; Fl. Kirgiz. 6 (1955) 100; Fl. Kazakhst. 4 (1961) 104; Kovalevsk. in Opred. rast. Sr. Azii [Key to Plants of Mid. Asia] 3 (1972) 215; Fl. Sin. 28 (1980) 324; Claves pl. Xinjiang, 2 (1983) 319; W.T. Wang in Bull. Bot. Res. (Harbin) 15, 3 (1995) 312. —**Ic.:** Fl. SSSR, 7, Plate 29, fig. 2 a-c; Fl. Sin. 28, tab. 102, fig. 1-3.

Described from Scandinavia. Type in London (Linn.).

Along arid meadows and slopes, forest borders.

IA. Mongolia: ***Khobd.*** (Ulangom, in forest, July 14 and Sept. 6, 1879—Pot.).

IB. Kashgar: ***Nor.*** (Muzart. Aug. [11] 1877—A. Reg.; Sakhe, Kurlya, No. 6837, July 20, 1958—Lee and Chu).

IIA. Junggar: ***Tarb.*** ("Dachen" [Chuguchak]—Claves pl. Xinjiang, l.c.), ***Jung. Alat.*** (Dzhagastai, July 26, 1877—A. Reg.; 2 km south-west of Dzhagastai, No. 721 and 3191, Aug. 8, 1957—Kuan; around Ven'tsyuan', No. 1534, 3536, 4586, 4613, Aug. 14, 15, 24 and 25—1957, Kuan), ***Tien Shan*** (Talki gorge, July 19 and Aug. 16; south-east. bank of Sairam lake, July 22; Tekes bank (midcourse), 1500 m, Aug. 13—1877, A. Reg.; Sairam lake, July 12, 1878—Fet.; nor. slope of Talkinsk mountain range on south-east. bank; Sairam-Nor lake, 2150 m, Aug. 18, 1953—Mois.; Chzhaosu region, No. 3272, 856, 886, Aug. 12, 13 and 14, 1957—Kuan; floodplain of Tekes river, No. 1588, Aug. 26, 1957—Yun. and Yuan'), ***Dzhark.*** (Kul'dzha, June 20, 1843—Schrenk; same site, 1906 —Muromskii).

General distribution: Jung.-Tarb., Nor. and Cent. Tien Shan; Europe, Balk.-Asia Minor (Balkan peninsula), West. Sib., East. Sib. (Ang.-Sayan).

5. **R. propinquus** C.A. Mey. in Ledeb. Fl. alt. 2 (1830) 332; Kryl. Fl. Zap. Sib. 5 (1931) 1204; Ovcz. in Fl. SSSR, 7 (1937) 461; Fl. Kazakhst. 4 (1961) 106; Gub. Konsp. Fl. Vneshn. Mong. [Conspectus of Flora of Outer Mongolia] (1996) 50. —*R. acris* var. *propinquus* (C.A. Mey.) Maxim. Enum. pl. Mong. 1 (1889) 22. —*R. japonicus* var. *propinquus* (C.A. Mey.) W.T. Wang in Bull. Bot. Res. (Harbin) 15, 3 (1995) 305.

Described from West. Siberia (Altay). Type in St.-Petersburg (LE).

On banks of brooks, in birch and larch forests, moist grasslands in alpine belt.

IA. Mongolia: ***Khobd.*** (Gub. l.c.), ***Mong. Alt.*** (Taishiri-Ula, July 15, 1877—Pot.; Ulasty river, July 1, 1908—Sap.; Khara-Adzarga mountain range, around Khairkhan-Duru river, Aug. 26, 1930—Pob.; Khan-Taishiri mountain range, south. slope in upper

Shine-Usu, 2380-2450 m, June 18, 1971—Grub., Dariima et al.; 30 km nor.-west of Bulgan, upper Altaal river, June 24, 1988—Kam., Bud., Gub.), *Cent. Khalkha* (between Nalaikha and Gagtsa-Khuduk [1840]—Tatarinov; Kharukha river, May 21, 1894—Klem.; Delger-Khuduk somon, Kerulen river valley, 1300 m, July 11, 1974—Golubk. and Tsogt), *East. Mong.* (Muni-Ula mountains, along banks, June; same site, in birch stand on mountain slope, June 14—1871, Przew.; around Khailar, June 29, 1873—Fritshe; valley of Ergenten-Gol river, July 10, 1899—Pot. and Sold.), *Depr. Lakes* (5 km north and 11 km south of Khirgis somon, July 15, 1973—Banzragch, Karam. et al.).

IIA. Junggar: *Cis-Alt.* (Kurtu river, 1903—Gr.Grzh.).

General distribution: West. Sib. (Altay), East. Sib., Nor. Mong. (Fore Hubs., Hent., Hang., Mong.-Daur.), China (Dunbei, Nor.).

6. **R. repens** L. Sp. pl. (1753) 554; Forbes and Hemsl. Index Fl. Sin. 1 (1886) 15; Maxim. Enum. pl. Mong. 1 (1889) 23; Kryl. Fl. Zap. Sib. 5 (1931) 1208; Ovcz. in Fl. SSSR, 7 (1937) 423; Kitag. Lin. Fl. Mansh. (1939) 225; Fl. Kirgiz. 6 (1955) 100; Fl. Kazakhst. 4 (1961) 104; Kovalevsk. in Opred. rast. Sr. Azii [Key to Plants of Mid. Asia] 3 (1972) 214; Fl. Intramong. 2 (1978) 251; Fl. Sin. 28 (1980) 326; Fl. Intramong. ed. 2, 2 (1990) 512; Gub. Konsp. Fl. Vneshn. Mong. [Conspectus of Flora of Outer Mongolia] (1996) 51; W.T. Wang in Bull. Bot. Res. (Harbin) 15, 3 (1995) 316. —**Ic.:** Fl. Kazakhst. 4, Plate 13, fig. 2; Fl. Intramong. 2, tab. 131, fig. 1-6; Fl. Sin. 28, tab. 96, fig. 5-7; Fl. Intramong. ed. 2, 2, tab. 207, fig. 3-4.

Described from Europe. Type in London (Linn.).

On banks of rivers, lakes, in moist meadows.

IA. Mongolia: *Khobd.* (Gub. l.c.), *Mong. Alt.* (10-15 km west of Dayan-Nur post, near Khara-Nur lake, July 20, 1988—Ganbold, Gub. et al. (MW), *Cent. Khalkha* (Gub. l.c.), *East. Mong.* (Khailar town, Nunlin'tun' village, June 10, 1951—Kh. Wang).

IIA. Junggar: *Jung. Alat.* (Toli district, Albaksin town, No. 2682, Aug. 6, 1957 —Kuan), *Tien Shan* (south-east. bank of Sairam lake, July 29, 1878—A. Reg.), *Dzhark.* (Kul'dzha, May 15, 1877—A. Reg.).

General distribution: Aralo-Casp., Fore Balkh., Jung.-Tarb., Nor. and Cent. Tien Shan; Europe, Mediterr., Balk.-Asia Minor, Fore Asia, Caucasus, Mid. Asia, West. Sib. (Altay), East. Sib., Far East, Nor. Mong. (Hent., Hang., Mong.-Daur., Cis-Hing.), China (Dunbei, Nor.), Korean peninsula, Japan, Nor. Amer. (introduced).

7. **R. submarginatus** Ovcz. in Fl. URSS, 7 (1937) 745, 439; Fl. Sin. 28 (1980) 317; Claves pl. Xinjiang, 2 (1983) 319; Fl. Intramong. ed. 2, 2 (1990) 512; W.T. Wang in Bull. Bot. Res. (Harbin) 15, 3 (1995) 308. —**Ic.:** Fl. SSSR, 7, Plate 29, fig. 3 a-b; Fl. Sin. 28, tab. 100, fig. 4-6; Fl. Intramong. ed. 2, 2, tab. 207, fig. 1-2.

Described from West. Siberia (Altay). Type in St.-Petersburg (LE).

In mixed, pine-larch and poplar forests in floodplains.

IA. Mongolia: *East. Mong.* ("Khulunbuir ajmaq—administrative territorial unit in Mongolia"—Fl. Intramong. ed. 2, l.c.).

IIA. Junggar: *Jung. Alat.* (south. trail of Archata river valley, poplar forest in floodplain, Aug. 17, 1957—Yun. and Yuan'), *Tien Shan* (40 km south of Dzinkho, No. 104, July 2; mountains east of Syata [=Santai], No. 1337, Aug. 11—1957, Kuan).

General distribution: West. Sib., China (Dunbei).

Note. "Flora Sin'tszyana" [Flora of Sinkiang] (1983) cites only the name without reference to location of find.

Section 2. **Ranunculastrum** DC.

8. **R. gobicus** Maxim. Enum. pl. Mong. 1 (1889) 15. —**Ic.:** Maxim. tab. IV, fig. 8-16.

Described from Sinkiang (Jung. Gobi). Type in St.-Petersburg (LE). Plate VII, fig. 2. Map 4.

On wet sand and clay.

IIA. Junggar: ***Jung. Gobi*** (east.: Gobi, Montes Schan-chanssian, in arenoso limosis humidis, parce, No. 54, May 12/24, 1879—Przew., holotypus !).

General distribution: endemic.

Note. New species for China.

9. **R. platyspermus** Fisch. ex DC. Prodr. 2 (1824) 37; Kryl. Fl. Zap. Sib. 5 (1931) 1183; Ovcz. in Fl. SSSR, 7 (1937) 505; Fl. Kazakhst. 4 (1961) 118; Kovalevsk. in Opred. rast. Sr. Azii [Key to Plants of Mid. Asia] 3 (1972) 221; Fl. Sin. 28 (1980) 330; Claves pl. Xinjiang, 2 (1983) 320; W.T. Wang in Bull. Bot. Res. (Harbin) 15, 3 (1995) 319. —**Ic.:** Maxim. Enum. pl. Mong. 1 (1889) tab. IV, fig. 1-7; Fl. SSSR, 7, Plate 33, fig. 2; Fl. Kazakhst. 4, Plate 15, fig. 4; Fl. Sin. 28, tab. 103, fig. 4-6.

Described from West. Kazakhstan (Inder lake). Type in Geneva (G). Map 4.

On sandy and loamy soil in deserts and semideserts.

IIA. Junggar: ***Dzhark.*** (Almaty valley, nor.-west of Kul'dzha, 900-1200 m, April 12 and May 22; Sarybulak river, 1200-1800 m, April 23—1878, A. Reg.).

General distribution: Aralo-Casp., Fore Balkh., Jung.-Tarb., Nor. Tien Shan (foothills), West. Siberia (Alt.), Mid. Asia (desert regions).

10. **R. regelianus** Ovcz. in Bull. Soc. natur. Moscou, 44 (1935) 267; Ovcz. in Fl. SSSR, 7 (1937) 506; Fl. Kirgiz. 6 (1955) 106; Fl. Kazakhst. 4 (1961) 120; Kovalevsk. in Opred. rast. Sr. Azii [Key to Plants of Mid. Asia] 3 (1972) 221; Fl. Sin. 28 (1980) 330; Claves pl. Xinjiang, 2 (1983) 320; W.T. Wang in Bull. Bot. Res. (Harbin) 15, 3 (1995) 319. —**Ic.:** Fl. SSSR, 7, Plate 33, fig. 1; Fl. Kazakhst. 4, Plate 14, fig. 4.

Described from East. Kazakhstan (Nor. Tien Shan). Type in St.-Petersburg (LE). Map 3.

On arid, turf-covered slopes, among steppe and meadow vegetation, from foothills to upper belt of mountains.

IIA. Junggar: ***Tien Shan*** (Kungess Tal, May 1-5, 1908—Merzb.), ***Dzhark.*** (Kul'dzha, April 20, 1877: Suidun river, 1200-1800 m, May 1, 1878—A. Reg.: Kul'dzha, April 8,

1878; Pilyuchi pass, 900-1200 m, April 22 and 24, 1879—A. Reg.; Kul'dzha, 1906—Muromskii; "Ili"—Claves pl. Xinjiang, l.c.).

General distribution: Jung.-Tarb., Nor. and Cent. Tien Shan; Mid. Asia (West. Tien Shan, Pam.-Al.—Alaisk mountain range).

11. **R. severzovii** Regel in Acta Horti Petrop. 5 (1877) 221, p.p.; Ovcz. in Fl. SSSR, 7 (1937) 500; Fl. Kirgiz. 6 (1955) 106; Fl. Kazakhst. 4 (1961) 117; Kovalevsk. in Opred. rast. Sr. Azii [Key to Plants of Mid. Asia] 3 (1972) 222. —**Ic.:** Fl. Kazakhst. 4, Plate 15, fig. 2.

Described from Mid. Asia (Karatau mountain range). Type in St.-Petersburg (LE). Map 4.

In rocky sections, along arid river beds and banks of mountain brooks and streams in lower belt of mountains.

IIA. Junggar: ***Tien Shan*** (Borborogussun, 900-1200 m, April 27, 1879—A. Reg.).

General distribution: Mid. Asia.

Note. New species for China.

Section 3. **Ranunculus**

(*Ranunculus* subgen. *Auricomus* Spach; *R.* sect. *Auricomus* Schur)

12. **R. altaicus** Laxm. in Novi Comment. Ac. Sci. Petrop. 18 (1773) 533; C.A. Mey. in Ledeb. Fl. alt. 2 (1830) 325; Kryl. Fl. Zap. Sib. 5 (1931) 1196; Ovcz. in Fl. SSSR, 7 (1937) 400; Fl. Kirgiz. 6 (1955) 95; Fl. Kazakhst. (1961) 97; Kovalevsk. in Opred. rast. Mong. [Key to Plants of Mongolia] (1982) 114; Claves pl. Xinjiang, 2 (1983) 315; W.T. Wang in Bull. Bot. Res. (Harbin) 15, 3 (1995) 289; Gub. Konsp. Fl. Vneshn. Mong. [Conspectus of Flora of Outer Mongolia] (1996) 50. —*R. sulphureus* Soland. var. *altaicus* Trautv. in Bull. Soc. natur. Moscou, 33 (1860) 69; Maxim. Enum. pl. Mong. 1 (1889) 19. —**Ic.:** Fl. Kazakhst. 4, Plate 12, fig. 4; Fl. Sin. 28, tab. 87, fig. 8.

Described from West. Siberia (Altay). Type in St.-Petersburg (LE).

On bald peaks, wet slopes and talus, moraine, grassland and banks of brooks in alpine belt.

IA. Mongolia: ***Khobd.*** (south. macroslope of Sailyugem mountain range, waterdivide of Uigoryn-Gol river and its left tributary Khara-Dzhaman-Gol, Obot-Shira mountain, July 10, 1988—Kam. and Ganbold), ***Mong. Alt.*** (Urmogaity pass, June 27, 1903—Gr.-Grzh.; Onkattu lake, Chingis-tei sentry post, alp. tundra, June 25, 1906—Sap.; Bulugun river basin, Kharagaitu-Khutul' pass, alpine meadow and rubble talus, July 24; same site, upper course of Ketsu-sairin-Gol river, on moraine and slopes toward glacier, in alpine meadow, July 26—1947, Yun.; upper Kobdo river, Danyagiin-Khara-nur lake, 2500 m, meadow steppe, July 7, 1971—Grub., Ulzij. et al.; south. bank of Dayan-Nur lake, July 26, 1974—Zhurba (MW); mountains along south. bank of Dayan-Nur lake, July 26, 1977—Karam., Sanczir et al.; pass from basin of Dayan-Nur lake to Irtysh basin, 15-20 km south of Dayan-Nur settlement, alpine meadow and dwarf birch thickets, 2400-2500 m, July 15; 10-15 km west of Dayan-Nur settlement, Khara-Nur lake, July 20; Ult-Daba pass from basin of Sagsai river to Chern. Irtysh basin (in Elt-Gol river valley), 2700-2750 m, July 22—1988, Gub., Kam. et al.).

IIA. Junggar: *Cis-Alt.* (Kandagatai river, Sept. 21, 1876—Pot.; in Timulbakhan region, Altay mountains, on swamp, 2650 m, No. 10711, July 19, 1959—Kuan).

General distribution: Jung.-Tarb., Nor. Tien Shan; Europe, West. Sib. (Altay), East. Sib. (south), Far East, Nor. Mong. (Fore Hubs., Hent., Hang.), Japan.

13. **R. banguoensis** L. Liou in Fl. Sin. 28 (1980) 362, 293; Fl. Xizang. 2 (1985) 102; W.T. Wang in Bull. Bot. Res. (Harbin) 15, 2 (1995) 177. —**Ic.:** Fl. Sin. 28, tab. 88, fig. 9-10; Fl. Xizang. 2, tab. 28, fig. 1-2.

Described from Tibet. Type in Beijing (PE).

In coastal meadows, 5200 m alt.

IIIB. Tibet: *Chang Tang* ("Banguo, 5200 m, No. 9494, June 23, 1976, K.Y. Lang, typus"—Fl. Sin. l.c.; "Ban'ge"—Fl. Xizang. l.c.).

General distribution: endemic.

14. **R. brotherusii** Freyn in Bull. Herb. Boiss. 6 (1898) 885; Ovcz. in Fl. SSSR, 7 (1937) 392; Fl. Kirgiz. 6 (1955) 94; Fl. Kazakhst. 4 (1961) 97; Kovalevsk. in Opred. rast. Sr. Azii [Key to Plants of Mid. Asia], 3 (1972) 210; Fl. Sin. 28 (1980) 295; Claves pl. Xinjiang, 2 (1983) 316; Fl. Intramong. ed. 2, 2 (1990) 496; W.T. Wang in Bull. Bot. Res. (Harbin) 15, 3 (1995) 276. —**Ic.:** Fl. SSSR, 7, Plate 24, fig. 4; Fl. Intramong. ed. 2, 2, tab. 200, fig. 1-2.

Described from Kirghiz (Tersk-Ala Tau). Type in Geneva (G). Isotype in St.-Petersburg (LE).

In alpine meadows, on rocks, 1500-2700 m alt.

IIA. Junggar: *Tien Shan* (Mal. Yuldus, 2300-2700 m, in moist meadow, June 1, 1877—Przew.; Bagaduslun—Bayannamyn, east. tributary of Dzhin river, 1500-1800 m, May 23; Borgaty pass, 2400-2700 m, June 7; Naryan-gol on Tsagan-usu (Dzhin), June 10 —1879, A. Reg.; zwischen Kungess Tal und Klein Juldus Tal, also Fai-aschu Pass und Zenma Tal, May 5-10, 1908—Merzb.; Danu pass, between Ulan-Usu valley and Danu-Gol, July 19, 1957—Yun. and Yuan'; on bank of Danu river, No. 2122, July 21; 3 km south of Datu, No. 411, July 21—1957, Kuan; upper Danu-Gol river, at ascent to Se-daban pass, July 22, 1957—Yun. and Yuan'; 7-8 km south of Danu, No. 504, July 22, 1957—Kuan; on Urumchi-Kucha highway, 2300 m, No. 6073, July 21; Bain-Bulak, Khotun-Sumbul district, 2560 m, No. 6399, Aug. 6; vicinity of Khomot, No. 7128 and 7151, Aug. 6 and 8—1958, Lee and Chu).

General distribution: Nor. and Cent. Tien Shan; Fore Asia (Nor. Pakistan), Himalayas (nor.-west. Kashmir, east.).

15. **R. chinensis** Bge. Enum. pl. China bor. (1832) 3; id. in Mém. Ac. Sci. St. Pétersb., Sav. Etr. 2 (1835) 77; Ovcz. in Fl. SSSR, 7 (1937) 409; Kitag. Lin. Fl. Mansh. (1939) 223; Fl. Kirgiz. 6 (1955) 99; Fl. Kazakhst. 4 (1961) 102; Kovalevsk. in Opred. rast. Sr. Azii [Key to Plants of Mid. Asia] 3 (1972) 214; Fl. Intramong. 2 (1978) 249; Fl. Sin. 28 (1980) 327; Claves pl. Xinjiang, 2 (1983) 319; Fl. desert Sin. 1 (1985) 497; Fl. Xizang. 2 (1985) 107; Pl. Vasc. Helanshan (1986) 108; Fl. Intramong. ed. 2, 2 (1990) 513; W.T. Wang in Bull. Bot. Res. (Harbin) 15, 3 (1995) 317; Gub. Konsp. Fl. Vneshn. Mong. [Conspectus of Flora of Outer Mongolia] (1996) 50. —*R. pensylvanicus* L.

var. *chinensis* (Bge.) Maxim. Enum. pl. Mong. 1 (1889) 23. —*R. pensylvanicus* auct. non L.: Franch. Pl. David. 1 (1884) 20; Finet et Gagnep. in Bull Soc. Bot. Fr. 51 (1904) 309. —**Ic.**: Fl. Kazakhst. 4, Plate 13, fig. 1; Fl. Intramong. 2, tab. 129, fig. 1-5; Fl. Sin. 28, tab. 102, fig. 4-6; Fl. desert Sin. 1, tab. 182, fig. 1-5; Fl. Xizang. 2, tab. 32, fig. 1-3; Fl. Intramong. ed. 2, l.c. tab. 208, fig. 1-2.

Described from Nor. China. Type in Paris (P).

On marshy and silty banks of rivers, shoals, moist meadows, fringes of irrigation canals.

IA. Mongolia: ***Khobd.*** (Gub. l.c.), ***East. Mong.*** (on plain around Kuku-khoto town, along irrigation canals, July 18, 1884—Pot.), ***Depr. Lakes*** (Gub. l.c.), ***Ordos*** (Ordos, 1884—Pot.), ***Khesi*** (along Edzin river, above Gaotai, June 20 1884—Pot.).

IB. Kashgar: ***Nor.*** (Sakhe—Kurlya, No. 6837, July 20, 1958—Lee and Chu).

IIA. Junggar: ***Tien Shan*** (Urumchi—Yan'ervo, No. 566, May 31, 1957—Kuan), ***Dzhark.*** (*Suman*, south of Kul'dzha, May 29, 1877—A. Reg.).

IIIA. Qinghai: ***Nanshan*** (Lovachen town, along fringes of irrigation canals, July 26 1908—Czet.).

General distribution: Fore Balkh., Jung.-Tarb., Nor. Tien Shan submontane regions); Mid. Asia, East. Sib., Far East, Nor. Mong. (Hent., Hang., Mong.-Daur.), China (Dunbei, Nor., East., South-West.), Himalayas (east.), Korean peninsula, Japan, Indo-Mal. (Vietnam, Laos).

16. **R. chinghoensis** L. Liou in Fl. Sin. 28 (1980) 287, 361; Claves pl. Xinjiang, 2 (1983) 317; W.T. Wang in Bull. Bot. Res. (Harbin), 15, 2 (1995) 175.

Described from Sinkiang. Type in Beijing (PE).

On forest borders.

IIA. Junggar: ***Cis-Alt.*** ("Sinkiang: Chingho [Tsinkhe], Datsingou, Mutougou, No. 720524, June 13, 1972, C.Y. Yang, typus"—Fl. Sin. l.c.; "Tsinkhe" [=Chinchil'] —Claves pl. Xinjiang, l.c.).

General distribution: endemic.

17. **R. fraternus** Schrenk in Fisch. et Mey. Enum. pl. nov. 1 (1841) 103; Ovcz. in Fl. SSSR, 7 (1937) 402; Fl. Kirgiz. 6 (1955) 95; Fl. Kazakhst. 4 (1961) 98; Kovalevsk. in Opred. rast. Sr. Azii [Key to Plants of Mid. Asia] 3 (1972) 212; W.T. Wang in Bull. Bot. Res. (Harbin) 15, 3 (1995) 288. —*R. albertii* Regel et Schmalh. in Acta Horti Petrop. 5 (1877) 223; Ovcz. in Fl. SSSR, 7 (1937) 401; Fl. Kirgiz. 6 (1955) 95; Fl. Kazakhst. 4 (1961) 98; Ikonnik. Opred. rast. Pamira [Key to Plants of Pamir] (1963) 121; Kovalevsk. l.c. (1972) 212; Fl. Tadzh. 4 (1975) 104; Fl. Sin. 28 (1980) 277; Claves pl. Xinjiang, 2 (1983) 314; W.T. Wang in Bull. Bot. Res. 15, 3 (1995) 287. —*R. sulphureus* Soland. var. *fraternus* et var. *albertii* (Regel et Schmalh.) Maxim. Enum. pl. Mong. 1 (1889) 19. —**Ic.**: Fl. Kazakhst. 4, Plate 12, fig. 8; Fl. sin. 28, ic. 37.

Described from East. Kazakhstan (Jung. Ala Tau). Type in St.-Petersburg (LE).

Along banks of brooks, grassy and rocky slopes of alpine and subalpine belt; descends into forest belt.

IB. Kashgar: ***Nor.*** (canal into Aksu, No. 8471, Sept. 26, 1958—Lee and Chu), ***West.*** (Sarykol' mountain range, Bostan-Terek location, July 11, 1929—Pop.; King-Tau mountain range, 3-4 km south-east of Kosh-Kulak settlement, June 10; Turugart settlement, on boundary with USSR, June 20—1959, Yun.).

IIA. Junggar: ***Jung. Alat.*** (Kazan pass, 2700-3300 m, Aug. 11, 1878—A. Reg.), ***Tien Shan*** (Mal. Yuldus, May 29 and June 15; Narat pass, 2600-3000 m, June 17—1877, Przew.; sources of Taldy river, 2400 m, May 17; Taldy gorge, 2400 m, May 24; Taldy river, 2700 m, May 26; Tsagan-Tunge, 1500-1800 m, June 8; Chungur-Daban pass, nor. of Borborogusun [June 13]; Aryslyn gorge, 2400-2700 m, July 11; Aryslyn, 2700-3000 m, July 12—1879, A. Reg.; Mukhurdai pass, 3000 m, among placers, June 19, 1893—Rob.; Pässe zwischen Kin-su und Kurdai, July 3; about 3400 m hoes plateau eines Berges in oberehn Koksu gebiet, July 8-10—1907; Kungess Tal, May 1-5, 1908—Merzb.; upper stream of one of the right branches of Manas river, spruce belt, 2500 m, May 26, 1954 —Mois.; Nilki district, Dzinkho—Yakou, on slope, No. 4034, Sept. 1, 1957—Kuan; Narat mountain range, Dachit pass, 3100 m, *Cobresia* meadow, Aug. 6, 1958—Yun. and Yuan'; Narat mountains in Kunges, No. 6550, Aug. 7, 1958—Lee and Chu, Muzart river valley, near Oi-Terek area, alpine belt, marshy meadow, Sept. 10, 1958—Yun. and Yuan').

General distribution: Jung.-Tarb., Nor. and Cent. Tien Shan, East. Pam., Mid. Asia (Pamiro-Alay).

Note. In "Flora SSSR" [Flora of USSR], Ovczinnikov erroneously cites *R. albertii* for mountains in nor.-west. part of Mongolia.

18. **R. glareosus** Hand.-Mazz. Symb. Sin. 7, 2 (1931) 307; Fl. Sin. 28 (1980) 300; Fl. Xizang. 2 (1985) 105; W.T. Wang in Bull. Bot. Res. (Harbin) 15, 2 (1995) 173. —**Ic.:** Fl. Sin. 28, tab. 89, fig. 8-9; Fl. Xizang. 2, tab. 30, fig. 6-7.

Described from South-West. China (Yunnan). Type in Edinburgh (E).

IIIB. Tibet: ***Weitzan*** (south. section of Amnen-Kor mountain range, gorge on nor. slope, 4000-4300 m, June 6, 1901—Lad.; "Bizhu"—Fl. Xizang. l.c.).

General distribution: China (South-West.).

19. **R. hetianensis** L. Liou in Fl. Sin. 28 (1980) 360, 274; Claves pl. Xinjiang, 2 (1983) 314; W.T. Wang in Bull. Bot. Res. (Harbin) 15, 3 (1995) 292. —**Ic.:** Fl. Sin. 28, tab. 88, fig. 1-2.

Described from Sinkiang (Kashgar). Type in Beijing (PE).

In steppified meadows.

IB. Kashgar: ***Nor.*** (near Uch-Turfan, Karagailik gorge, June 20, 1908—Divn.), ***South.*** (Sinkiang: Khotan, steppe pasture, 3200 m, No. 750489, July 20, 1975, C.Y. Yang, typus"—Fl. Sin. l.c.; "Khotan"—Claves pl. Xinjiang).

IIIB. Tibet: ***Weitzan*** (nor. and south. slopes of Burkhan-Budda mountain range, 3350-3600 m, June 12, 1901—Lad.).

General distribution: endemic.

Note. This species differs from *R. altaicus* Laxm. in thickened roots; ovoid radical leaves, slightly cordate or truncated at base, 5-7-lobed along margin; much smaller flowers, peduncles, receptacle and sepals with white pubescence; oblong petals, 8 mm, slightly longer than sepals.

20. **R. hirtellus** Royle Ili. Bot. Himal. Mount. 1 (1839) 53; Fl. Sin. 28 (1980) 292; Fl. Xizang. 2 (1985) 104; W.T. Wang in Bull. Bot. Res. (Harbin) 15, 2 (1995) 175. —**Ic.:** Fl. Sin. 28, ic. 38.

Described from Himalayas. Type in Liverpool (LIVU).

In meadows on shaded slopes, 3800-4800 m.

IIIA. Qinghai ("Qinghai"—Fl. Xizang. l.c.).

IIIB. Tibet: ***South.*** ("Lhasa, Nan'mulin"—Fl. Xizang. l.c.).

General distribution: China (South-West.), Himalayas.

21. **R. indivisus** (Maxim.) Hand.-Mazz. in Acta Horti Gotoburg, 13 (1939) 145; Fl. Sin. 28 (1980) 272; Tamura in Acta Phytotax. Geobot. (Kyoto) 15, 6 (1954) 198; W.T. Wang in Bull. Bot. Res. (Harbin) 15, 3 (1995) 293. —*R. affinis* R. Br. var. *indivisus* Maxim. Fl. Tangut. 1 (1889) 14; Gr.-Grzh. Opis. put. Zap. Kit. [Report of Expedition in West. China] 3 (1907) 476; Borodina-Grabovskaya in Catalogue type specim. Centr. Asia (LE) (2000) 197. —**Ic.:** Fl. Sin. 28, tab. 86, fig. 3-5.

Described from Qinghai (Amdo) and Tibet (Weitzan). Lectotype and syntypes in St.-Petersburg (LE). Plate VII, fig. 3.

On clayey descents of soft mountains and ridges, on meadowy slopes covered by shrubs.

IA. Mongolia: ***Alash. Gobi*** (midportion of Alashan mountain range, west. slopes, alpine belt, July 5 and 17, 1873—Przew.).

IC. Qaidam: ***Mount.*** (Sarlyk mountains [Kurlyka region], Kuku-bulak area, 3300 m, on moist sand, May 30, 1895—Rob.).

IIIA. Qinghai: ***Nanshan*** ("Xining alps, 3000-3400 m, June 1890"—Gr.-Grzh., 1907), ***Amdo*** (Jugo Balekun (10,000 ft) [3000 m], secus rivulum in argillosis, frequens, [Balekun mountain range, upper Huang He], No. 211, May 25/June 6—syntypus !; Ad Hoangho super. (7000 ft) [2100 m] [NW] [Chagomi st.], No. 456, June 23/July 5—lectotypus ! (Borodina-Grabovskaya, l.c.)—1880, Przew.; prov. Kansu occidentalis [Lanchzha-lunva river, Shon-pyn village], May 14, 1885, Pot., syntypus !).

IIIB. Tibet: ***Chang Tang*** (south of Keriya river, along Kyuk-Egil' river, July 11, 1884—Przew.), ***Weitzan*** (Ripa sinistra fl. Yang-tze, 13,000 ft. [4000 m], No. 143, June 11/June 23—syntypus ! alpine belt of Konchyun-chyu river, 4000 m, June 30; Talachyu river, 1000 m, July 6—1884, Przew.; nor. slope of Burkhan-Budda mountain range, Nomogun gorge, 2350-2600 m, May 25; south. slope of Amnen-Kor mountain range, 4100 m, on arid clayey cliffs, June 8; Russkoe lake, at emergence of Huang He river from it, 4100 m, June 17, 1900; Yangtze river basin, Makhmukhchyu brook, 4100 m, May 21; south-east. bank of Russkoe lake, 4100-4200 m [without date]—1901, Lad.), South. ("vicinity of Lhasa"—Tamura, l.c.).

General distribution: China (South-West.).

Note. Plants with thickened roots that brings this species close to *R. polyrhizos* Steph.

22. **R. karelinii** Czer. Pl. Vasc. URSS (1981) 425. —*R. gelidus* Kar. et Kir. in Bull. Soc. natur. Moscou, 15 (1842) 133, non Hoffmgg. ex Reichb. 1832; Ovcz. in Fl. SSSR, 7 (1937) 406; Fl. Kirgiz. 6 (1955) 98; Fl. Kazakhst. 4 (1961)

101; Kovalevsk. in Opred. rast. Sr. Azii [Key to Plants of Mid. Asia] 3 (1972) 213; Fl. Sin. 28 (1980) 300; Claves pl. Xinjiang, 2 (1983) 318; W.T. Wang in Bull. Bot. Res. (Harbin) 15, 2 (1995) 178. —*R. gelidus* var. *dasycarpus* Trautv. et var. *leiocarpus* Trautv. in Bull. Soc. natur. Moscou, 33 (1860) 74. —**Ic.:** Fl. Kazakhst. 4, Plate 12, fig. 3; Fl. Sin. 28, tab. 93, fig. 1-2.

Described from East. Kazakhstan (Jung. Ala Tau). Type in St.-Petersburg (LE).

On rock talus, rocks and exposures in upper mountain brooks, moist banks and beds of brooks, near patches of thawing snow in alpine belt at 2000-4600 m alt.

IIA. Junggar: ***Cis-Alt.*** (nor.-west. Shara-Sume, on trail, No. 10222, April 29, 1959—Lee and Chu), ***Jung. Alat.*** (Borotala river basin, south. slope of Junggar Ala Tau, below Koketau pass, July 21, 1909—Lipsk.), ***Tien Shan*** (Muzart gorge, 3000-3500 m, Aug. [15-17] 1877; Kok-Tyube mountains, 2100-2700 m, June 23; Kokkamyr upland, 3000 m, July 27—1878; from Dumbedan to Kumdaban, 2700 m, May 28; Kumdaban, 2700 m, May 29; Irenkhabirga mountain range, Chun-gur-daban pass, June 13—1879, A. Reg.; im oberen Kensu Gebiet, circa 3400 m, July 8-10, 1907—Merzb.; Danu pass, 3600 m, between Ulan-Usu valley and Danu-Gol, July 19; Danu-Gol river valley, July 21 —1957, Yun., Lee and Yuan'; vicinity of Danyu, No. 355 and 536, July 21 and 22; south of Yakou, No. 1704, Aug. 31; in Yakou, Nilki district, No. 4061, Sept. 1—1957, Kuan).

IIIA. Tibet: ***Weitzan*** (upper stream of] Konchyun-La, 4400-4600 m, on talus, July 3, 1884—Przew.).

General distribution: Jung.-Tarb. (Jung. Ala Tau), Nor. and Cent. Tien Shan.

Note. In Tibet, we have placed only the collections of N. Przewalsky on Konchyun-La river in this species. Rest of specimens with largely reduced leaf blade and some other differences have been treated as *R. tibetanus*.

23. **R. krasnovii** Ovcz. in Fl. URSS, 7 (1937) 742, 376; Ikonnik. Opred. rast. Pamira [Key to Plants of Pamir] (1963) 119; Kovalevsk. in Opred. rast. Sr. Azii [Key to Plants of Mid. Asia] 3 (1972) 210; Fl. Tadzh. 4 (1975) 101. —*R. pulchellus* C.A. Mey. var. *tridentata* O. Fedtsch. in Acta Horti Petrop. 21 (1903) 250.

Described from East. Pamir. Type in St.-Petersburg (LE).

On clayey slopes, short-grass meadows along brooks in alpine belt.

IIIB. Tibet: ***Weitzan*** (Burkhan-Budda mountain range, nor. slope, Nomogun gorge, 2300 m, May 25, 1900—Lad.).

IIIC. Pamir (under Ucha pass, June 18, 1909—Divn.).

General distribution: East. Pam.; Mid. Asia (West. Pamir).

Note. Cited for the first time in China. According to S.S. Ikonnikov, this species is endemic within Tajikistan in Pamir.

24. **R. lasiocarpus** C.A. Mey. in Ledeb. Fl. alt. 2 (1830) 323; Kryl. Fl. Zap. Sib. 5 (1931) 1197; Ovcz. in Fl. SSSR, 7 (1937) 392; Gubanov, Kamelin et al. in Byull. Mosk. obshch. isp. prir., otd. biol. 95, 1 (1990) 119; Gub. Konsp. Fl.

Vneshn. Mong. [Conspectus of Flora of Outer Mongolia] (1996) 50. —**Ic.:** Ledeb. Ic. pl. Fl. Ross. 2, 1 (1830) tab. 112.

Described from West. Siberia (Altay). Type in St.-Petersburg (LE).

In moist meadows, swamps, moss-lichen tundra in alpine belt.

IA. Mongolia: ***Khobd.*** (valley of Turgen river, July 6, 1973—Banzragch, Karam. et al.; Sailyugem mountain range, Khara-Dzhamat-Gol river basin, July 10, 1988—Gub., Bud. and Dariima (MW), waterdivide of Uigoryn-gol river and its left tributary Khara-Dzhamat-Gol, Obot-Shira mountain, July 10, 1988—Kam., Ganbold (MW), ***Mong. Alt.*** (Gub. l.c.).

General distribution: West. Sib. (Altay), Nor. Mong. (Fore Hubs., Hang.).

25. **R. longicaulis** C.A. Mey. in Ledeb. Fl. alt. 2 (1830) 308; Ovcz. in Fl. SSSR, 7 (1937) 375; Fl. Kirgiz. 6 (1955) 92; Fl. Kazakhst. 4 (1961) 92; Fl. Tadzh. 4 (1975) 98; Fl. Sin. 28 (1980) 267; Grub. Opred. rast. Mong. [Key to Plants of Mongolia] (1982) 114; Claves pl. Xinjiang, 2 (1983) 313. —*R. longicaulis* var. *longicaulis*; Gub. Konsp. Fl. Vneshn. Mong. [Conspectus of Flora of Outer Mongolia] (1996) 50. —*R. nephelogenes* Edgew. Trans. Linn. Soc. (London) 20 (1846) 28; Fl. Intramong. ed. 2, 2 (1990) 493. —*R. pulchellus* C.A. Mey. var. *longicaulis* (C.A. Mey.) Trautv. in Bull. Soc. natur. Moscou, 33 (1860) 68; Maxim. Enum. pl. Mong. 1 (1889) 18; Kryl. Fl. Zap. Sib. 5 (1931) 1189; Ikonnik. Opred. rast. Pamira [Key to Plants of Pamir] (1963) 120. —*R. pulchellus* C.A. Mey. var. δ. *tibeticus* Maxim. Fl. Tangut. 1 (1889) 13. —*R. nephelogenes* var. *longicaulis* (Trautv.) W.T. Wang in Bull. Bot. Res. (Harbin) 7, 2 (1987) 110; id. Bull. Bot. Res. (Harbin) 15, 3 (1995) 283. —**Ic.:** Ledeb. Ic. pl. Fl. Ross. 2, 1 (1830) tab. 117; Fl. Kazakhst. 4, Plate 11, fig. 8; Fl. Sin. 28, tab. 86, fig. 6; Fl. Xizang. 2, tab. 27, fig. 1; Fl. desert Sin. 1, tab. 181, fig. 1-2; Fl. Intramong. ed. 2, 2, tab. 198, fig. 3.

Described from West. Siberia (Altay). Type in St.-Petersburg (LE).

In swampy meadows, on moist banks of brooks.

IA. Mongolia: ***Khobd.*** (on Tszusylan river, along brook and at springs, on lower forest margin, July 11; sources of Kharkira river, July 21—1879, Pot.), ***Mong. Alt., Val. Lakes*** (Gub. l.c.).

IIA. Junggar: ***Jung. Alat.*** (between Sumbe [pass] and Kasan, 1200-2400 m, June 22, 1878—A. Reg.; Toli district, Albaksin, No. 2680, Aug. 6, 1957—Kuan), ***Tien Shan*** (Turkyul' lake, May 16, 1877—Pot.), ***Jung. Gobi*** (Gub. l.c.).

IIIB. Tibet: ***Weitzan*** (Ad fl. Dshagyn-gol, in paludibus, parce, 13,700 ft [4100 m], No. 319, July 11/23, 1884—Przew., typus *R. pulchellus* var. *tibeticus* Maxim. !; Dzhagyn-gol river, rather shallow lakes and meadows, 4300 m, July 5, 1900—Lad.).

General distribution: Jung.-Tarb., Nor. and Cent. Tien Shan, East. Pam.; Mid. Asia (mountains), West. Sib. (Altay), East. Sib., Nor. Mong. (Fore Hubs, Hent., Hang.), China (Nor.), ? Himalayas (Kashmir).

26. **R. membranaceus** Royle, Ill. Bot. Himal. Mount. 1 (1839) 53; Fl. Sin. 28 (1980) 269; Fl. Xizang. 2 (1985) 98; Pl. Vasc. Helanshan (1986) 108; W.T. Wang in Bull. Bot. Res. (Harbin) 15, 3 (1995) 284. —*R. alaschanicus* Y.Z. Zhao in Bull. Bot. Res. (Harbin) 9, 1 (1989) 64; Fl. Intramong. ed. 2, 2 (1990)

493. —**Ic.:** Fl. Sin. 28, ic. 36; Fl. Xizang. 2, tab. 27, fig. 2-3; Fl. Intramong. ed. 2, 2, tab. 199, fig. 1-3 (*R. alaschanicus*).

Described from Himalayas. Type in Liverpool (LIVU).

On mountain slopes, pebble beds in alpine belt.

IA. Mongolia: ***Alash. Gobi*** (Pl. Vasc. Helanshan, l.c.).

IIIB. Tibet: ***Weitzan*** (alpine belt along Konchyunchyu river, 4000-4300 m, June 30, 1884—Przew.; Russkoe lake, at exit of Huang He river from it, 4100 m, June 2, 1901 —Lad.).

General distribution: Himalayas (west., Kashmir).

27. **R. monophyllus** Ovcz. in Not. Syst. (Petrop.) 3 (1922) 54; 18, in Fl. SSSR, 7 (1937) 382; Fl. Kazakhst. 4 (1961) 94; Kovalevsk. in Opred. rast. Sr. Azii [Key to Plants of Mid. Asia] 3 (1972) 210; Fl. Sin. 28 (1980) 274; Grub. Opred. rast. Mong. [Key to Plants of Mongolia] (1982) 115; Claves pl. Xinjiang, 2 (1983) 314; Fl. Intramong. ed. 2, 2 (1990) 496. —*R. auricomus* subsp. *sibiricus* Korsh. Fl. Vost. Evrop. Ross. 1 (1892) 89; Kitag. Linn. Fl. Mansh. (1939) 223; W.T. Wang in Bull. Bot. Res. (Harbin) 15, 3 (1995) 287; Gub. Konsp. Fl. Vneshn. Mong. [Conspectus of Flora of Outer Mongolia] (1996) 50. —*R. auricomus* auct. non L.: Maxim. Enum. pl. Mong. 1 (1889) 21. —*R. krylovii* auct. non Ovcz.: Fl. Sin. l.c. 287. —**Ic.:** Fl. SSSR, 7, Plate 24, fig. 2; Fl. Sin. 28, tab. 87, fig. 1-3; Fl. Intramong. ed. 2, 2, tab. 199, fig. 4-5.

Described from East. Europe (Bashkir). Type in St.-Petersburg (LE).

In forests, meadows, banks of rivers and brooks in forest zone and occasionally in alpine belt.

IA. Mongolia: ***Khobd.*** (on Tszusylan river, in forest, July 24, 1879—Pot.), ***Mong. Alt.*** (Tsagan-Gola valley, saddle in Kharasu, alpine tundra, June 29, 1905—Sap.).

General distribution: Jung.-Tarb., Nor. Tien Shan; Arct., Europe (East.: nor., nor.-east., east), West. Sib., East. Sib., Far East, Nor. Mong., China (Dunbei, Nor.).

28. **R. pedatifidus** Smith in Rees, Cyclop. 29 (1814) No. 72; Ovcz. in Fl. SSSR, 7 (1937) 388; Fl. Kirgiz. 6 (1955) 93; Fl. Kazakhst. 4 (1961) 95; Ikonnik. Opred. rast. Pamira [Key to Plants of Pamir (1963) 121; Kovalevsk. in Opred. rast. Sr. Azii [Key to Plants of Mid. Asia] 3 (1972) 210; Fl. Sin. 28 (1980) 285; Grub. Opred. rast. Mong. [Key to Plants of Mongolia] (1982) 115; Claves pl. Xinjiang, 2 (1983) 316; Fl. Intramong. ed. 2, 2 (1990) 500; W.T. Wang in Bull. Bot. Res. (Harbin) 15, 2 (1995) 170; Gub. Konsp. Fl. Vneshn. Mong. [Conspectus of Flora of Outer Mongolia] (1996) 50. —*R. affinis* R.Br. var. α. *typicus* Maxim. Fl. Tangut. 1 (1889) 13; id. Enum. pl. Mong. 1 (1889) 20. —*R. amoenus* Ledeb. Fl. alt. 2 (1830) 320; Kryl. Fl. Zap. Sib. 5 (1931) 1148. —*R. rigescens* Turcz. ex Ovcz. in Fl. URSS, 7 (1937) 389, nom. nud., descr. ross.; Fl. Kazakhst. 4 (1961) 96; Kovalevsk. l.c. (1972) 211; Fl. Intramong. 2 (1978) 253; Fl. Sin. l.c. 285; Claves pl. Xinjiang, l.c. 316; Fl. Intramong. ed. 2, l.c. 498; W.T. Wang, l.c. (1995) 170. —**Ic.:** Fl. Kazakhst. 4, Plate 11, fig. 11; Fl. Intramong. 2, tab. 131, fig. 7-9 (*R. rigescens*); Fl. Sin. 28, tab. 92, fig. 3-4

(*R. rigescens*); Grub. Opred. rast. Mong. [Key to Plants of Mongolia] Plate 54, fig. 246; Fl. Intramong. ed. 2, 2 tab. 201, fig. 1-3 and 4-6 (*R. rigescens* and *R. pedatifidus*).

Described from Siberia. Type in London (Linn.).

In subalpine meadows.

IA. Mongolia: ***Khobd.*** (Ula-Daban, June 22; Tszusylan river, in forest, July 12; springs along Kharkira river, 2400 m, July 22—1879, Pot.; south. macroslope of Sailyugem mountain range, Khara-Dzhamat-Gol river basin (left tributary of Uigoryn-Gol river), July 10, 1988—Gub., Dariima et al. (MW); same site, Obot-Shira mountain, in waterdivide of Uigoryn-Gol river and its left tributary Khara-Dzhamat-Gol, July 10, 1988—Kam., Ganbold (MW), ***Mong. Alt.*** (130 km south-east of Bulgan somon, Ikh-Khavtag-Ula mountain range, Aug. 12, 1977—Volk., Rachk.; 10-15 km west of Dayan-Nur post, near Khara-Nur lake, July 20, 1988—Bud., Gub. et al. (MW), ***East. Mong.*** (Gub. l.c.), ***Gobi Alt.*** ("Baga-Bogdo-Ula" "Ikh-Bogdo-Ula"—Grub. l.c.; Adzhi-Bogdo town, Aug. 23, 1973—Isach. and Rachk.).

IIA. Junggar: ***Tarb.*** (Saur mountain range, south. slope, valley of Karagaitu river, Bain-Tsagan creek valley, subalpine meadow, June 23, 1957—Yun. and Yuan'), ***Tien Shan*** (Mal. Yuldus, 2300-2400 m, June 13; same site [June 14]—1877, Przew.; Yugantash, 1800-2100 m, May 25, 1878; upper Taldy river, 2700-3000 m, May 20, 1879—A. Reg.; Urumchi region, Biangou area, upper Tasenku river, Sept. 25, 1929—Pop.; Manas river basin, valley of Ulan-Usu river at its confluence with Dzhartas, July 18, 1957—Yun. and Yuan'; from Nyutsyuan'tsza 20 km in south, No. 207, July 18, 1957—Kuan).

IIIA. Qinghai: ***Nanshan*** ([Kuku-Usu river] July 11, 1879—Przew.; nor. slope of Humboldt mountain range, Blagodatnyi spring, alpine meadow, 3000-3300 m, May 24; same site, Kuku-Usu area, 3600 m, June 9—1894, Rob.).

IIIB. Tibet: ***Chang Tang*** (30 km south-east of Isachan [=Achan], Cherchen district, in alpine belt, 4060 m, No. 9412, June 4, 1959—Lee et al.), ***Weltzan*** (Burkhan-Budda mountain range, 3000-3600 m, 1900 or 1901—Lad.).

General distribution: Jung.-Tarb., Nor. and Cent. Tien Shan; West. Sib. (Altay), East. Sib. (Ang.-Sayan., Daur.), Far East, Nor. Mong., China (Nor.), Nor. Amer.

29. **R. popovii** Ovcz. in Fl. URSS, 7 (1937) 741, 393; Fl. Kirgiz. 6 (1955) 94; Fl. Kazakhst. 4 (1961) 97; Kovalevsk. in Opred. rast. Sr. Azii [Key to Plants of Mid. Asia] 3 (1972) 210; Claves pl. Xinjiang, 2 (1983) 315; Fl. Intramong., ed. 2, 2 (1990) 495; W.T. Wang in Bull. Bot. Res. (Harbin) 15, 2 (1995) 180. —*R. popovii* var. *popovii*; W.T. Wang, l.c. (1995) 180. —**Ic.:** Fl. SSSR, 7, Plate 24, fig. 5; Fl. Kazakhst. 4, Plate 12, fig. 1; Fl. Intramong. ed. 2, 2, tab. 198, fig. 4-6.

Described from Sinkiang. Type in St.-Petersburg (LE). Plate VII, fig. 4.

On shaded slopes in alpine and subalpine meadows.

IB. Kashgar: ***West.*** (Sarykol mountain range, west of Kashgar, Bostan-terek area, No. 123, July 11, 1929—Pop., holotypus !), ***East.*** (Tsakhanur pass in Khotun-Sumbul, 3100 m, No. 6528, Aug. 15, 1958—Lee and Chu).

IIA. Junggar: ***Tien Shan*** (Oberes Gross Yuldus Tal, am Wege zum Karagaitasch Pass, May 25-26; südliches Kiukönik Tal, beim Lager unter Tschon Yailak Pass, June 15, 1908—Merzb.).

General distribution: Nor. (West. Ala Tau) and Cent. Tien Shan; China (South-West.), Himalayas.

Note. Differs from *R. brotherusii* in radical leaves that are more or less cuneate at base, lobed or spaced, as well as in much taller and compact tomentose pubescence of stems.

30. **R. pseudohirculus** Schrenk in Fisch. et Mey. Enum. pl. nov. 2 (1842) 65; Ovcz. in Fl. SSSR, 7 (1937) 375; Fl. Kirgiz. 6 (1955) 92; Fl. Kazakhst. 4 (1961) 92; Ikonnik. Opred. rast. Pamira [Key to Plants of Pamir] (1963) 120; Fl. Tadzh. 4 (1975) 100; Grub. Opred. rast. Mong. [Key to Plants of Mongolia] (1982) 114; Gub. Konsp. Fl. Vneshn. Mong. [Conspectus of Flora of Outer Mongolia] (1996) 50. —*R. pulchellus* C.A. Mey. var. *pseudohirculus* (Schrenk) Trautv. in Bull. Soc. natur. Moscou, 33 (1860) 68; Maxim. Enum. pl. Mong. 1 (1889) 18; Ostenf. and Pauls. in Hedin, S. Tibet, 6, 3 (1922) 81; Kryl. Fl. Zap. Sib. 5 (1931) 1189—*R. longicaulis* C.A. Mey. var. *pseudohirculus* (Schrenk) Gubanov in Fl. Khang. (1989) 110. —**Ic.:** Grub. Plate 54, fig. 247.

Described from East. Kazakhstan (Jung. Ala Tau). Type in St.-Petersburg (LE).

In swamps, marshy meadows, along banks of water reservoirs from foothills to high mountains.

IA. Mongolia: ***Khobd.*** (Dzusylan, on upper forest boundary, June 13; spring south of Kharkira river, July 11—1879, Pot.; Turgen river valley, moss-lichen dwarf birch thicket, 2600 m, July 8, 1973—Banzragch, Karam. et al.; Turgen-Ula mountain range, Bairimen-Daba pass, 2500 m, alpine marshy meadow with rock outcrop and spring, Aug. 29, 1984; south. macroslope of Sailyugem mountain range, 80 km nor.-west of Ulan-Khus settlement, Dzhamai-Gol river (left tributary of Uigoryn-Gol), July 12, 1988 —Gub.), ***Mong. Alt.*** (Kak-Kul' (Khorpur) lake between Tsagankol and Kobdo, June 22, 1906—Sap., east. part of Khara-Adzargiin-Nuru mountain, Urd-Ulyastain-Gol gorge, middle and upper portions, July 3; Shadzgat-nuru mountain range, basin of Khait-Dzhargalant-Gol river, Bayansala gorge, July 27—1984, Kam., Dariima), ***Gobi-Alt.*** (Ikhe-Bogdo mountain range, upper mountain belt, June 29, 1945—Yun.; same site, head of Narin-Khurimt gorge, 3500 m, July 29, 1948—Grub.; Baga-Bogdo-Yla mountain range, valley of Tsagan-Burgas river, 2800 m, July 1978—Ogureeva (MW).

IB. Kashgar: ***West.*** (Sarykol' mountain range, west of Kashgar, Boston-Terek locality, July 10, 1929—Pop.).

IC. Qaidam: ***Mount.*** (Ritter mountain range, upper Bagda-Khatyn-Gol, alpine meadow, 3600 m, June 26, 1894—Rob.).

IIA. Junggar: ***Jung. Alat.*** (Kazan pass, Aug. 11, 1878—A. Reg.; Toli district, No. 2680, Aug. 6, 1957—Kuan), ***Tien Shan*** (Turkyul' lake, Jung 16, 1877—Pot.; Mal. Yuldus, 2300-2700 m, June 1, 1877—Przew.; south. slope [under Nan'shan'kou pass], June 11, 1877—Pot.; upper Taldy river, May 19; midcourse of Taldy river, 2100 m, May 26, Taldy river, May 27; Kum-daban pass, 2700 m, May 29, Bagadu-slun, east. tributary of Dzhin river, 2700 m, June 4; south. tributary of Karagol, near pass beyond Nilki, 2700 m, June 16; Aryslyn, 2700-3000 m [July 10]—1879, A. Reg.; valley of Muzart river, June 18, 1886—Krasn.; Manas river basin, valley of Koisu river, at its confluence with Ulan-Usu, July 17; on bank of Koisu river, No. 120, July 17; upper Ulan-Usu valley, before ascent to Danu-Daban pass, alpine belt, July 19—1957, Yun. et al.; between Daban and Danyu, in moist sites, No. 1998, July 19; from Nyutsyuan'tsza 30 km south along bank of Ulan-Usu, No. 314, July 19, 1957—Kuan; valley of Danu-Gol river, at ascent to Se-Daban pass, July 21, 1957—Yun. et al.; nor. rim of B. Yuldus basin floor, 30-35 km south-west of Bain-Bulak settlement, Aug. 10; east. fringe of Mal. Yuldus inter-montane basin,

1-2 km west of Kotyl' pass, Aug. 15—1958, Yun. and Yuan'), *Jung. Gobi* (east.: nor. macroslope of Baitag-Bogdo mountain range, forest and alpine belt, Budun-Khargaityn-Gol gorge, 15 km east of Baitag-Bogdo post, Aug. 1, 1988—Gub., Kam., Bud.).

IIIA. Qinghai: ***Nanshan*** (alpine meadow on South. Tetung mountain range [Sodi-Suruksum mountains], July 31, 1872—Przew.; Sungi-Nur area, 3000 m, valley of Sharagol'dzhin river, June 9; nor. slope of Humboldt mountain range, marsh around Kuku-Usu river, 3000 m—1894. Rob.), ***Amdo*** (on Churmyn river, 2700-2900 m, clay bank, May 10, 1880—Przew.).

IIIB. Tibet: ***Chang Tang*** ("Ban'ge"—Fl. Xizang. l.c.), ***Weitzan*** (springs in upper courses of Huang He river, 2900-3300 m, April 24, 1880; left bank of Yangtze river, 4000 m, June 13; mountain range between Huang He and Yangtze rivers, July 12; nor. slope in midportion of Burkhan-Budda mountain range, 3500-4000 m, Aug. 2—1884, Przew.; Amnen-Kor mountain range, Kuku-Bulak spring, nor. slope, valley of Alyknorin-Gol river, marshy banks of lakes, 3800 m, May 5; nor. slope of Burkhan-Budda mountain range, Nomokhun river gorge, 4000-4400 m, May 19, 21 and 25—1900, Russkoe lake and Rerg-Chyu brook, basin of Yellow river, 4100 m, May 29; nor. slope of Amnen-kor mountain range, 4200 m, June 6; nor. and south. slopes of Burkhan-Budda mountain range, Khatu gorge, Nomokhun and Ikhi-Gol, 4000-4200 m, June 12—1901, Lad.), ***South.*** ("Pulan", Chzhunba, Lhasa, Nan'mulin, Dyantszy"—Fl. Xizang. l.c.).

IIIC. Pamir ("Mustagh-ata, Kamper-Kishlak, moist meadow on a glacier, 4500 m, July 29, 1894—Hedin"—Ostenf. and Pauls, in Hedin, l.c.; ascent to Billuli pass from Chumbus river side, June 12; Taret pass, June 19; Ulug-Tuz gorge, in Charlym river basin, June 29—1909, Divn.; Kokat pass, 3800-4000 m, June 16, 1942—Serp.: 1-2 km west of Tashkurgan, on debris cone, No. 305, June 16, 1959—Lee and Chu; valley of Tashkurgan, on west. fringe of Tashkurgan town, along irrigation ditches, on short-grass meadows, June 13, 1959—Yun. and Yuan').

General distribution: Jung.-Tarb., Nor. and Cent. Tien Shan, East. Pam.; Mid. Asia, West. Tien Shan, Pamiro-Alay), West. Sib. (Altay), East. Sib.(south), Nor. Mong. (Fore Hubs., Hent., Hang.), China (Altay, Nor., Nor.-West., South-West.), Himalayas.

31. **R. pulchellus** C.A. Mey. in Ledeb. Fl. alt. 2 (1830) 333; Forbes and Hemsl. Index Fl. Sin. 1 (1886) 15; Maxim. Enum. pl. Mong. 1 (1889) 18; Hemsl. Fl. Tibet (1902) 163; Pamp. Fl. Carac. (1930) 111; Kryl. Fl. Zap. Sib. 5 (1931) 1189; Ovcz. in Fl. SSSR, 7 (1937) 374; Tamura in Acta Phytotax. Geobot. (Kyoto) 15, 6 (1954) 198; Fl. Kirgiz. 6 (1955) 91; Fl. Kazakhst. 4 (1961) 90; Kovalevsk. in Opred. rast. Sr. Azii [Key to Plants of Mid. Asia] 3 (1972) 209; Fl. Sin. 28 (1980) 270; Grub. Opred. rast. Mong. [Key to Plants of Mongolia] (1982) 114; Claves pl. Xinjiang, 2 (1983) 314; Pl. Vasc. Helanshan (1986) 108; Fl. Intramong. ed. 2, 2 (1990) 491; W.T. Wang sp. in Bull. Bot. Res. (Harbin) 15, 3 (1995) 280. —*R. longicaulis* subsp. *pulchellus* (C.A. Mey.) Gubanov, Fl. Vost. Khang. (1983) 130. —*R. longicaulis* var. *pulchellus* (C.A. Mey.) Gubanov, Fl. Khangaya (1989) 110; Gub. Konsp. Fl. Vneshn. Mong. [Conspectus of Flora of Outer Mongolia] (1996) 50. —**Ic.:** Ledeb. Ic. pl. Fl. Ross. 2, 1 (1830) tab. 111; Fl. Kazakhst. 4, Plate 11, fig. 7, Fl. Sin. 28, tab. 86, fig. 1-2; Fl. Intramong. ed. 2, 2, tab. 198, fig. 1-2.

Described from West. Siberia (Altay). Type in St.-Petersburg (LE).

On moist and silty banks of rivers, riverine streams, back waters and meanders, bottomland deciduous forests, moist solonetzic and marshy meadow, swamps, in forest belt and lower portion of high-mountain zone.

IA. Mongolia: ***Khobd.*** (Sailyugem mountain range, Katu river, June 16; same site, Altyn-Khatysyn area, June 19—1879, Pot.; Suok (Boro-Burgazy) river valley, June 18, 1906—Sap.; 90 km north-west of Ulan-Khus settlement, basin of left tributary of Khara-Dzhaman-Gol, July 10, 1988—Shiirevdamba, Gamalei and Ganbold (MW); same site, 80 km nor.-west of Ulan-Khus settlement, left tributary of Uigoryn-Gol river, July 12, 1988—Gub., Kam.), ***Mong. Alt.*** (Taishiri-Ula mountain range, July 5, 1877—Pot.; Bugat somon, near Bayan-Gol river, poplar forest, July 22 (MW); near Tsetseg somon, Tsast-Bogdo-Ula, floodplain of brook, slopes and rocks, July 28—1975, Zhurba (MW); Adzhi-Bogdo mountain range, Ikhe-Gol river, gorge at 2500-2700 m, Aug. 22, 1979—Grub., Dariima et al.), ***Cent. Khalkha*** (Gub. l.c.), ***East. Mong.*** (Mongolia chinensis, 1831 [Kuznetzov] Turczaninov), ***Depr. Lakes*** ("Khara-Us-Nur", Grub. l.c.), ***Gobi-Alt.*** (Ikhe-Bogdo mountain range, rock placer, June 29, 1945—Yun.).

IIA. Junggar: ***Tien Shan*** (Dumbedan-Kumdaban, nor. slope of Irenkhabirga mountain range, 2700 m, May 28; Kumbel' [pass], 3000 m, June 3; Borgaty pass, nor. of Kash river valley, 1500 m, July 8—1879, A. Reg.; Datszymyao village, in Savan district, No. 1741, July 22, 1957—Kuan).

IIIB. Tibet: ***Weitzan*** (basin of Golubaya river, Chokabu-vrun area, July 10, 1900—Lad.), ***South.*** ("Lhasa, Tszyantszy-Dzyantsy"—Fl. Xizang. l.c.; "Shigatze"—Tamura, l.c.).

General distribution: Jung.-Tarb., Nor. and Cent. Tien Shan; Mid. Asia (West. Tien Shan), West. Sib. (Altay), East. Sib., Nor. Mong., China (Nor.-West., South-West.), Himalayas (Kashmir).

32. **R. similis** Hemsl. in Hook. Ic. Pl. (1899) tab. 2586, fig. B; id, Fl. Tibet (1902) 163; Pamp. Fl. Carac. (1930) 112; W.T. Wang in Bull. Bot. Res. (Harbin) 6, 1 (1986) 28; W.T. Wang, l.c. 15, 3 (1995) 295. —*R. involucratus* Maxim. Fl. Tangut. 1 (1889) 15, non Lapeyr. 1813; Hemsl. in J. Linn. Soc. (London) Bot. 30 (1894) 107; Hemsl. l.c. (1902) 162; Finet et Gagnep. in Bull. Soc. Bot. Fr. 51 (1904) 323; Ostenf. and Pauls. in Hedin, S. Tibet, 6, 3 (1922) 81; Fl. Sin. 28 (1980) 284; Fl. Xizang. 2 (1985) 104; Borodina-Grabovskaya in Catalogue type specim. Centr. Asia (LE) (2000) 199. —*R. maximowiczii* Pamp. in Bull. Soc. Bot. Ital. (1915) 66. —**Ic.:** Hook. Ic. Pl. l.c. tab. 2586, fig. B; Maxim. tab. 22, fig. 7-13; Fl. Sin. 28, tab. 89, fig. 3-5; Fl. Xizang. 2, tab. 30, fig. 1-3.

Described from Tibet. Type, sketch on plate in first description.

On stone placers in rock crevices, coastal pebble beds in alpine and nival belts.

IC. Qaidam: ***Mount.*** (nor. slope of Ritter mountain range, placers, June 20, 1894—Rob.).

IIIA. Qinghai: ***Nanshan*** (Dzurge-Gol river, 3700 m, June 12; nor. slope of Humboldt mountain range, Kuk-Usnyn-Daban pass, 4300 m, June 24, 1894—Rob.).

IIIB. Tibet: ***Chang Tang*** ("Arka-tag, 4980 m, June 29, 1901—Hedin"—Ostenf. and Pauls. in Hedin, l.c.; "Zhitu, Shen'chzha"—Fl. Xizang. l.c.), ***Weitzan*** (Tibet borealis [fl. Dzhagyn-gol], limo humido ad rivulos frequens, 14,000-14,500 ft. s.m., No. 37, May 28/ June 9, syntypus *R. involucratus* !; Tibet borealis, montibus, secus fl. Yang-tze, June 8/20 —1884, Przew., lectotypus *R. involucratus* ! (Borodina-Grabovskaya, l.c.); Amnen-Kor mountain range, nor. slope, 4600-4800 m, June 4; Russkoe lake and bank of Huang He river, 4100 m, June 14; Dzhagyn-Gol river, 4000-4300 m, July 2—1900; on bank of Yellow river at its emergence from lake and on Russkoe lake, 4100-4600 m, May 25; nor. slope of Amnen-Kor mountain range, Qaidam basin, 4300 m, June 6—1901, Lad.), ***South.*** ("Nan'mulin, Zhada, Dzhunba"—Fl. Xizang. l.c.).

General distribution: endemic.

Note. Flowers pale pink initially, turning white later. Plant found under extreme conditions and often covered by snow.

33. **R. songaricus** Schrenk in Fisch. et Mey. Enum. pl. nov. 2 (1842) 67, "*songoricus*"; Maxim. Enum. pl. Mong. 1 (1889) 19; Ovcz. in Fl. SSSR, 7 (1937) 402; Fl. Kirgiz. 6 (1955) 96; Fl. Kazakhst. 4 (1961) 99; Kovalevsk. in Opred. rast. Sr. Azii [Key to Plants of Mid. Asia] 3 (1972) 212; Fl. Sin. 28 (1980) 288; Claves pl. Xinjiang, 2 (1983) 317; W.T. Wang in Bull. Bot. Res. (Harbin) 15, 2 (1995) 169. —*R. songaricus* var. *lasiopetalus* Maxim. l.c. 19. —**Ic.:** Fl. SSSR, 7, Plate 25, fig. 4, a-b; Fl. Kazakhst. 4, Plate 12, fig. 2.

Described from East. Kazakhstan (Tarbagatai). Type in St.-Petersburg (LE).

In alpine meadows, sparse spruce forests, on arid river beds and on banks of mountain brooks, up to 1700-2700 m.

IIA. Junggar: ***Tarb.*** (Ssaur-Argaltin, May 20—June 28, 1876—Pevtsov), ***Jung. Alat.*** (in Toli district, No. 659, Aug. 6, 1957—Kuan), ***Tien Shan*** (Koshety-daban pass, Aug. 11; Sairam, July; Dzhagastai pass, 2400-3000 m, Aug. 11; Dunguren gorge, 2600-3500 m, Sept. 8—1877, A. Reg.; on Borokhudzhir river, June 4, 1878—Fet.; Burkhantau, June 5 and 6; Karagol, 3000 m, April 17; Dzhirgalan gorge and Pilyuchi, 1800 m, April 24; Borborogusun, 900-1200 m, April 27; Taldy gorge, 2100-2700 m, May 15—1879, A. Reg.; nor. slopes of Tien Shan, around Koshetydaban pass, June 3 and 5, 1879—Przew.; south. tributary of Karagol, 2400-2700 m, June 16; Nilki gorge, 1500-1800 m, June 19; Irenkhabirga mountain range, Mengute, Aug. 2—1879, A. Reg.; Borokhoro mountain range, Chichkan, 1899—Korol'kov; im Oberen Koksu Gebiet, 3400 m, Aug. 8-10, 1907 —Merzb.; in Savan district, Datszymyao village, No. 1286, July 8, south of Nyutsyuan'tsza, No. 591, July 16; 20 km south of Ulastai, No. 3837, Aug. 28; Barat—Yakou, No. 1918, Aug. 31; Fukan district, vicinity of Tyan'chi lake, No. 4329, Sept. 19—1957, Kuan).

General distribution: Jung.-Tarb., Nor. and Cent. Tien Shan; Mid. Asia (West. Tien Shan, Pam.-Al.).

34. **R. tanguticus** (Maxim.) Ovcz. ex W.T. Wang in Bull. Bot. Res. (Harbin) 15, 3 (1995) 277. —*R. affinis* R. Br. var. *tanguticus* Maxim. Fl. Tangut. 1 (1889) 14; Gr.-Grzh. Opis. put. Zap. Kit. [Report of Expedition in West. China] 3 (1907) 476; Borodina-Grabovskaya in Catalogue type specim. Centr. Asia (2000) 198. —*R. tanguticus* (Maxim.) Ovcz. in Fl. URSS, 7 (1937) 393, nom. inval.; Fl. Sin. 28 (1980) 295; Grub. Opred. rast. Mong. [Key to Plants of Mongolia] (1983) 115; Pl. Vasc. Helanshan (1986) 108; Fl. Intramong. ed. 2, 2 (1990) 500; W.T. Wang, l.c. (1995) 277; Gub. Konsp. Fl. Vneshn. Mong. [Conspectus of Flora of Outer Mongolia] (1996). —*R. brotherusii* var. *tanguticus* (Maxim.) Tamura in Acta Phytotax. Geobot. (Kyoto) 15, 6 (1954) 198, nom. inval.; Fl. Xizang. 2 (1985) 105. —*R. tanguticus* (Maxim.) Ovcz. var. *dasycarpus* (Maxim.) L. Liou in Fl. Sin. l.c. 297. —*R. tanguticus* var. *capillaceus* (Franch.) L. Liou in Fl. Sin. l.c. 297. —**Ic.:** Fl. Sin. 28, tab. 92, fig. 1-2; Fl. Xizang. 2, tab. 31, fig. 1-2 (sub nom. *R. brotherusii* var. *tanguticus*); Fl. Intramong. ed, 2, 2, tab. 202, fig. 1-2.

Described from Qinghai. Lectotype and syntypes in St.-Petersburg (LE).

In alpine meadows, scrubs around springs in alpine belt.

IA. Mongolia: ***Khobd.*** ("Kharkhira mountain range"—Grub. l.c.; Gub. l.c.).

IIIA. Qinghai: ***Nanshan*** (Terra Tangutorum (prov. Kansu). In vallibus jugi S a fl. Tetung siti, locis udis vulgaris et gregaria [on moist sites in valleys of South. Tetungsk mountain range], No. 283, July 14/16, 1872, lectotypus (Borodina-Grabovskaya, l.c.) et isolectotypus !; Alpes S a fl. Tetung, regio alpina in pratis et faucibus frequens [in meadows and gorges of alpine belt of South. Tetungsk mountain range], No. 207, 1/13, 1873, Przew., syntypus et isosyntypus !; on Yusup-Khatyma river, in alpine meadows, 2700-3000 m, July 24, 1880—Przew.; "Kuku-Nor lake, valley of Bukhain-Gol river, 3400 m, July 1890—Gr.-Grzh."—Gr.-Grzh., 1907), ***Amdo*** (Ad rivulum Mudshikche, in fruticetis, secus fontes in limosis frequens et gregarius [on Mudzhikkhe river, in shrubs around springs], No. 279, June 4/16, 1880—Przew., syntypus !).

IIIB. Tibet: ***Weitzan*** (on bank of Bychyu river, 4000 m, on rubble, June 26/July 8, 1884—Przew.), ***South.*** ("Pulan', Chzhunba, Nan'mulin, Lhasa, Tszyantszy, Tsyuishui"—Fl. Xizang. l.c.; "Lhasa"—Tamura, l.c.).

General distribution: Nor. Mong. (Hang.), China (Nor., Nor.-West., South-West.), Himalayas (east.—Sikkim).

35. **R. tibetanus** (Tolm. et Jurtz.) Borod.-Grabovsk. comb. et stat. nov. —*R. gelidus* Kar. et Kir. var. *tibetanus* Tolm. et Jurtz. in Not. Syst. (Leningrad) 22 (1963) 116. —*R. affinis* R. Br. var. *tibeticus* Maxim. Fl. Tangut. 1 (1889) 14, p.p.

Described from Tibet (Weitzan). Type in St.-Petersburg (LE). Plate VII, fig. 5 Map 3.

On sandy, clayey and rocky banks of brooks in alpine belt.

IB. Qaidam: ***Mount.*** (Dulan-khit temple, 1901—Lad.).

IIIB. Tibet: ***Weitzan*** (Tibet borealis, regio alti alpina montium secus fl. Yangtze ad latera juxta campos lapidosos frequens, June 20; ad fl. Bo-tschu (affl. Yangtze) 13,000 ft s.m., in ripa glareosa parce, July 8—1884, Przew.; rocky steppe on Yangtze river, June 8; Ritter mountain range, upper Baga Khaltyana river, alpine meadow, 3700 m, June 24; on Bychyu river, 4000 m, June 26—1894, Rob.; nor. slope of Burkhan-Budda mountain range, Nomokhun gorge, 4000-4300 m, on wet sandy-clayey banks of brooks and streams, May 21 and 23, June 7; Russkoe lake and Huang He (Yellow) river, on sandy banks, 4100 m, June 14—1900, Lad.; Huang He river basin, valley of Sergchyu river, 4100 m, May 27; south. slope of Amnen-Kor mountain range, in pass in Russkoe lake valley, 4300 m, on moist, sandy-rocky bed of brook, June 4—1901, Lad., holotypus !).

General distribution: endemic.

Note. Plant material available with K.I. Maximowicz is mixed, only some plants belonging to *R. tibetanus*.

36. **R. trautvetterianus** Regel ex Ovcz. in Fl. URSS, 7 (1937) 403; Fl. Kirgiz. 6 (1955) 97; Fl. Kazakhst. 4 (1961) 100; Fl. Tadzh. 4 (1975) 104; Fl. Sin. 28 (1980) 288; Claves pl. Xinjiang, 2 (1983) 317; W.T. Wang in Bull. Bot. Res. (Harbin) 15, 3 (1995) 169. —*R. songaricus* Schrenk var. *partitus* Rupr. in Mém. Ac. Sci. St.-Pétersb. sér. 7, 14, 4 (1869) 37. —**Ic.:** Fl. SSSR, 7, Plate 25, fig. 3, a, b; Fl. Kazakhst. 4, Plate 12, fig. 6.

Described from East. Kazakhstan (Trans-Ili Ala Tau). Type in St.-Petersburg (LE).

In moist subalpine and alpine meadows.

IB. Kashgar: *Nor.* and *West.*
IIA. Junggar: *Tien Shan, Dzhark.*
IIIC. Pamir.
General distribution: Nor. and Cent. Tien Shan; Mid. Asia (Pamiro-Alay).

37. **R. yinshanensis** (Y.Z. Zhao) Y.Z. Zhao in Bull. Bot. Res. (Harbin) 9, 1 (1989) 67; Fl. Intramong. ed. 2, 2 (1990) 498; W.T. Wang in Bull. Bot. Res. (Harbin) 15, 3 (1995) 171. —*R. pulchellus* var. *yinshanensis* Y.Z. Zhao in Fl. Intramong. 2 (1978) 255, 369. —**Ic.:** Fl. Intramong. 2, tab. 132, fig. 1-6; Fl. Intramong. ed. 2, 2, tab. 200, fig. 3-4.

Described from Inner Mongolia. Type in Hohhot (HIMC).

On shaded slopes of mountain spurs.

IA. Mongolia: *East. Mong.* ("Datsinshan' mountain range, vicinity of Hohhot"—Fl. Intramong. ed. 2, l.c.).
General distribution: endemic.

Section 4. **Polyrhizos** Tzvel.

38. **R. polyrhizos** Steph. iln Willd. Sp. pl. 2 (1799) 1324; Kryl. Fl. Zap. Sib. 5 (1931) 1202; Ovcz. in Fl. SSSR, 7 (1937) 390; Fl. Kirgiz. 6 (1955) 93; Fl. Kazakhst. 4 (1961) 97; Kovalevsk. in Opred. rast. Sr. Azii [Key to Plants of Mid. Asia] 3 (1972) 211; Claves pl. Xinjiang, 2 (1983) 316; Fl. desert Sin. 1 (1985) 493; W.T. Wang in Bull. Bot. Res. (Harbin) 15, 2 (1995) 170. —**Ic.:** Fl. Kazakhst. 4, Plate 7, fig. 9; Fl. desert Sin. 1, tab. 180, fig. 3-5.

Described from West. Siberia. Type in ? Berlin (B).

On solonetzic steppe sinkholes, sand, rocky slopes from plains to midbelt of mountains.

IIA. Junggar: *Cis-Alt.* (Kurtu river, June 19, 1903—Gr.-Grzh.), *Tien Shan* (Talki gorge, July 18, 1877; Sarybulak river, 1200-1800 m, April 23; Kokkamyr upland, 1800-2100 m, April 29—1878; Talkibash, 900-1500 m [April]; Tagusturau (Kul'dzha) [April]; Pilyuchi and Dzhirgalan, April 24—1879, A. Reg.; "Urumchi, Ili"—Claves pl. Xinjiang, l.c.).
General distribution: Jung.-Tarb., Nor. and Cent. Tien Shan; Europe (East.), Caucasus, West. Sib.

Section 5. **Flammula** (Webb ex Spach) Freyn

39. **R. amurensis** Kom. Acta Horti Petrop. 22 (1903) 294; Ovcz. in Fl. SSSR, 7 (1937) 363; Fl. Intramong. ed. 2, 2 (1990) 502; W.T. Wang in Bull. Bot. Res. (Harbin) 15, 3 (1995) 297. —**Ic.:** Fl. Intramong. ed. 2, 2, tab. 203, fig. 1-2.

Described from Far East. Type in St.-Petersburg (LE).

In marshy meadows, overgrowing meanders.

IA. Mongolia: *East. Mong.* ("Khulunbuir"—Fl. Intramong. ed. 2, l.c.).
General distribution: East. Sib., Far East, China (Dunbei), Korean peninsula.

40. **R. lingua** L. Sp. pl. (1753) 549; Maxim. Enum. pl. Mong. 1 (1889) 18; Ovcz. in Fl. SSSR, 7 (1937) 362; Fl. Kirgiz. 6 (1955) 91; Fl. Kazakhst. 4 (1961) 89; Kovalevsk. in Opred. rast. Sr. Azii [Key to Plants of Mid. Asia] 3 (1972) 209; Fl. Sin. 28 (1980) 306; Claves pl. Xinjiang, 2 (1983) 313; W.T. Wang in Bull. Bot. Res. (Harbin) 15, 3 (1995) 297; Gub. Konsp. Fl. Vneshn. Mong. [Conspectus of Flora of Outer Mongolia] (1996) 50. —**Ic.:** Fl. Kazakhst. 4, Plate 11, fig. 4; Fl. Sin. 28, tab. 95, fig. 1-3.

Described from Nor. Europe. Type in London (Linn.).

In marshy meadows.

IIA. Junggar: ***Jung. Gobi*** (nor.: "South. Altay, on Urunchu river, Aug. 9, 1876—Pot."—Maxim.; east.: within Mongolia—Gub. l.c.), ***Jung. Alat.*** ("Ala Tau"—Claves pl. Xinjiang (without precise location, l.c.).

General distribution: Aralo-Casp., Fore Balkh., Jung.-Tarb., Europe, Caucasus, West. Sib., East. Sib., Himalayas (Kashmir).

41. **R. reptans** L. Sp. pl. (1753) 549; Kryl. Fl. Zap. Sib. 5 (1932) 1187; Ovcz. in Fl. SSSR, 7 (1937) 364; Fl. Kazakhst. 4 (1961) 90; Kovalevsk. in Opred. rast. Sr. Azii [Key to Plants of Mid. Asia] 3 (1972) 209; Fl. Sin. 28 (1980) 308; Grub. Opred. rast. Mong. [Key to Plants of Mongolia] (1982) 114; Claves pl. Xinjiang, 2 (1983) 313; Fl. Intramong. ed. 2, 2 (1990) 504; W.T. Wang in Bull. Bot. Res. (Harbin) 15, 3 (1995) 297; Gub. Konsp. Fl. Vneshn. Mong. [Conspectus of Flora of Outer Mongolia] (1996) 51. —**Ic.:** Fl. Kazakhst. 4, Plate 11, fig. 6; Fl. Sin. 28, tab. 94, fig. 9-10; Grub. Plate 54, fig. 244; Fl. Intramong. ed. 2, 2, tab. 203, fig. 3-4.

Described from Nor. Europe. Type in London (Linn.).

On moist pebbly and silty banks of rivers and lakes, moist coastal meadows.

IA. Mongolia: ***Mong. Alt.*** (on marshy bank of island in lower Kobde lake (sources of Kobdo river), Aug. 2; Kobdo river, Aug. 4—1899, Lad.; Dain-gol lake, June 29, 1909—Sap.), ***Depr. Lakes*** (Gub. l.c.).

General distribution: Fore Balkh. (Zaisan region); Arct., Europe, West. Sib., East. Sib., Far East, Nor. Mong. (Fore Hubs., Hent., Mong.-Daur.), China (Dunbei), Japan, Nor. Amer.

Section 6. **Xanthobatrachium** (Prantl) L. Benson

42. **R. gmelinii** DC. Syst. 1 (1817) 303, "Gmelini"; Kryl. Fl. Zap. Sib. 5 (1931) 1192 p.p.; Ovcz. in Fl. SSSR, 7 (1937) 354; Kitag. Lin. Fl. Mansh. (1939) 223; Fl. Kazakhst. 4 (1961) 88; Kovalevsk. in Opred. rast. Sr. Azii [Key to Plants of Mid. Asia] 3 (1972) 209; Fl. Intramong. 2 (1978) 249; Fl. Sin. 28 (1980) 305; Grub. Opred. rast. Mong. [Key to Plants of Mongolia] (1982) 114; Fl. Intramong. ed. 2, 2 (1990) 504; W.T. Wang in Bull. Bot. Res. (Harbin) 15, 3 (1995) 296; Gub. Konsp. Fl. Vneshn. Mong. [Conspectus of Flora of Outer Mongolia] (1996) 50. —**Ic.:** I. Gmelin, Fl. Sib. 4 (1769) 204,

tab. 83 a, fig. B; Fl. SSSR, 7, Plate 23, fig. 3 a-d; Fl. Intramong. 2, tab. 129, fig. 6-11; Fl. Sin. 28, tab. 194, fig. 7-8; Fl. Intramong. ed. 2, 2, tab. 204, fig. 1-2.

Described from Siberia. Type, sketch of I. Gmelin (tab. 83 a, fig. B).

In stagnant water, shallow water or on moist sandy banks.

IA. Mongolia: ***East. Mong.*** (vicinity of Khailar town, moist sites, No. 740, June 19, 1951—Lee et al.; "Shilingol'sk ajmaq—administrative territorial unit in Mongolia"—Fl. Intramong. l.c.; "East. Mong."—Gub. l.c.).

General distribution: Jung.-Tarb.; Arct., Europe (East.: Nor.), West. Sib., East. Sib., Far East, Nor. Mong. (Fore Hubs., Hent., Mong.-Daur., Cis-Hing.), China (Dunbei), Japan, Nor. Amer.

43. **R. intramongolicus** Y.Z. Zhao in Bull. Bot. Res. (Harbin) 9, 1 (1989) 69; Fl. Intramong. ed. 2, 2 (1990) 507; W.T. Wang in Bull. Bot. Res. (Harbin) 15, 3 (1995) 297. —**Ic.:** Fl. Intramong. ed. 2, 2, tab. 205, fig. 2-3.

Described from Inner Mongolia. Type in Hohhot (XIMC).

In swamps, on banks of rivers and brooks.

IA. Mongolia: ***East. Mong.*** (" [lake] Khulunbuir, Shilingol'sk ajmaq—administrative territorial unit in Mongolia"—Fl. Intramong. ed. 2, l.c.).

General distribution: China (Dunbei).

44. **R. natans** C.A. Mey. in Ledeb. Fl. alt. 2 (1830) 315; Kryl. Fl. Zap. Sib. 5 (1931) 1192; Ovcz. in Fl. SSSR, 7 (1937) 357; Kitag. Lin. Fl. Mansh. (1939) 225; Fl. Kirgiz. 6 (1955) 90; Fl. Kazakhst. 4 (1961) 88; Ikonnik. Opred. rast. Pamira [Key to Plants of Pamir] (1963) 119; Kovalevsk. in Opred. rast. Sr. Azii [Key to Plants of Mid. Asia] 3 (1972) 209; Fl. Tadzh. 4 (1975) 97; Fl. Sin. 28 (1980) 303; Grub. Opred. rast. Mong. [Key to Plants of Mongolia] (1982) 114; Claves pl. Xinjiang, 2 (1983) 312; Fl. desert Sin. 1 (1985) 492; Fl. Xizang. 2 (1985) 106; Fl. Intramong. ed. 2, 2 (1990) 506; W.T. Wang in Bull. Bot. Res. (Harbin) 15, 3 (1995) 296; Gub. Konsp. Fl. Vneshn. Mong. [Conspectus of Flora of Outer Mongolia] (1996) 50. —*R. hyperboreus* Rottb. var. *natans* (C.A. Mey.) Regel in Bull. Soc. natur. Moscou, 34 (1861) 43; Maxim. Enum. pl. Mong. 1 (1889) 17; Hemsl. Fl. Tibet (1902) 162; Pamp. Fl. Carac. (1930) 112. —**Ic.:** Fl. SSSR, 7, Plate 23, fig. 6; Grub. Plate 54, fig. 245; Fl. Sin. 28, tab. 94, fig. 1-4; Fl. Xizang. 2, tab. 31, fig. 3-4; Fl. Intramong. ed. 2, 2, tab. 204, fig. 3-4.

Described from West. Siberia (Altay). Type in St.-Petersburg (LE).

In stagnant and slow water streams, moist banks of rivers, lakes and brooks.

IA. Mongolia: ***Khobd.*** (Gub. l.c.), ***Mong. Alt.*** (in Urten-Gol river water, July 2, 1877—Pot.; Khashatu, Aug. 9, 1899—Lad.; Bzau-kul' lake (Beri-nor), lowland everglade, July 11, 1906—Sap.; Bodkhon-Gola gorge 3 km below Mankhan somon, in water of brook, June 26, 1971—Grub., Ulzij. and Dariima; Uenchiin-gola basin, Khargaityn-Gol valley 3 km upstream from estuary, floodplain, in flowing water, Aug. 15, 1979—Grub., Dariima et al.; pass from basin of Dayan-Nur lake to Irtysh basin, 15-20 km south of Dayan-Nur settlement, alpine meadow and dwarf birch thickets, 2400-2500 m, July 15, 1988 —Kam., Gub. et al.), ***Cent. Khalkha*** (upper course of Uber-Dzhargalanta, between Botoga and Agit mountains, in brook, Sept. 11, 1925—Krasch. and Zam.), ***East. Mong.*** (Gub. l.c.), ***Depr. Lakes*** (valley of Kharkhira river, in water, July 21, 1879—Pot.; Borig-Del sand, south. bank of Bayan-Nur lake, July 22, 1974—Zhurba (MW), ***Gobi-Alt.*** (Dzun-Saikhan mountain range, Yalo creek valley, in Tsagan-Gol river backwaters, Aug. 26, 1931—Ik.-Gal.), ***West. Gobi*** (Gub. l.c.).

IC. Qaidam: ***Plains*** (Kurlyk, vicinity of Barun-Tszasak khyrma, Khatu gorge, along springs, 3200 m; same site, 2400-3200 m, Khatu, Nomokhun and Ikhe-Gol, June 25, 1901—Lad.), ***Mount.*** (Ichegyn-Gol river, along springs 3000 m, June 16, 1895—Rob.).

IIIA. Junggar: ***Tien Shan*** (in water of brook of saline Turkyul' lake [June 16] 1877 —Pot.; Talki gorge, July 18; south. bank of Sairam lake, July 20—1877; Bogdo mountains, 1700 m, Aug. 15, 1878; Mengute pass, 3000-3400 m, July 4, 1879—A. Reg.; Mal. Luitsigen, in Khomota, No. 7176, Aug. 9, 1958—Lee and Chu (transforming into *R. intramongolicus*); nor. vicinity of floor of Bol. Yuldus basin, 30-35 km south-west of Bain-Bulak settlement, wet meadow, in water, Aug. 10, 1958—Yun. and Yuan').

IIIA. Qinghai: ***Nanshan*** (valley of Sharagoldzhin river, Sungi-Nur area, swamp, 3000 m, June 9; Yayakhu area, in moist sandy meadow, 3000 m, June 13—1894, Rob.).

IIIB. Tibet: ***Chang Tang*** ("Zhitu"—Fl. Xizang. l.c.), ***Weitzan*** (Burkhan-Budda mountain range, Nomokhun-Gol gorge, 3700-4000 m, May 18, 1900—Lad.), ***South.*** ("Linzhou" —Fl. Xizang. l.c.).

General distribution: Fore Balkh., Jung.-Tarb., Nor. and Cent. Tien Shan, East. Pam.; Mid. Asia (Pamiro-Alay), West. Sib., East. Sib., Far East, Nor. Mong., China (Dunbei), Himalayas (Kashmir), Nor. Amer.

45. **R. radicans** C.A. Mey. in Ledeb. Fl. alt. 2 (1830) 316; Ovcz. in Fl. SSSR, 7 (1937) 356; Kitag. Lin. Fl. Mansh. (1939) 225; Fl. Sin. 28 (1980) 303; Grub. Opred. rast. Mong. [Key to Plants of Mongolia] (1982) 114; Claves pl. Xinjiang, 2 (1983) 313; Fl. Intramong. ed. 2, 2 (1990) 506; W.T. Wang in Bull. Bot. Res. (Harbin) 15, 3 (1995) 295; Gub. Konsp. Fl. Vneshn. Mong. [Conspectus of Flora of Outer Mongolia] (1996) 51. —*R. hyperboreus* Rottb. var. *radicans* (C.A. Mey.) Hook. f. et Thoms. Fl. Brit. Ind. 1 (1872) 18; Maxim. Enum. pl. Mong. 1 (1889) 17; Pamp. Fl. Carac. (1930) 112. —*R. gmelinii* DC. var. *radicans* (C.A. Mey.) Kryl. Fl. Sib. Occid. 5 (1931) 1193. —**Ic.:** Fl. SSSR, 7, Plate 23, fig. 4 a-d; Fl. Intramong. ed. 2, 2, tab. 205, fig. 1.

Described from West. Siberia (Altay). Type in St.-Petersburg (LE).

In water and on moist banks of rivers and brooks, in waterholes in swamps, forest zone.

IA. Mongolia: ***Khobd.*** ("Turgen-Gola valley"—Grub. l.c.), ***Mong. Alt.*** (Buyantu river near Kobdo town, in stream, Aug. 15, 1899—Lad.; lower Kobdoss lake, valley of Tyurgun' river, lowland everglade, July 2, 1906—Sap.; Tamchi somon, Nukhenyin-Bulak spring, Sept. 7; upper Ubchugiin-Gola, in water of brook, Sept. 9—1948, Grub.; Munkh-Khairkhan-Ula, Aug. 14, 1978—Ogureeva (MW); 30 km west of Kobdo town, valley of Buyant-Gola river, 1650 m, Aug. 24, 1983—Gub. (MW), ***Cent. Khalkha*** (Ulkhuin-Bulun [area], in water of several fresh water pools, Aug. 9, 1925—Gus.), ***Depr. lakes*** (Ulangom, July 2, 1879—Pot.; valley of Kharkira river, birch stand, Sept. 1, 1931—Bar. and Shukhardin; saline delta of Kharkhira river, 20 km north of Ulangom, moist sections, Aug. 24, 1978; south. bank of Khara-Us-Nur lake, Aug. 10, 1982—Gub. (MW), ***Gobi-Alt.*** (Grub. l.c.).

IIA. Junggar: ***Jung. Gobi*** (east.: within Mongolia—Gub. l.c.).

General distribution: West. Sib. (Altay), East. Sib. (south), Nor. Mong., China (Dunbei, Nor.-West., South-West.).

Section 7. **Hecatonia** (Lour.) DC.

46. **R. sceleratus** L. Sp. pl. (1753) 551; Franch. Pl. David. 1 (1884) 19; Forbes and Hemsl. Index Fl. Sin. 1 (1886) 16; Maxim. Enum. pl. Mong. 1 (1889) 23; Kryl. Fl. Zap. Sib. 5 (1931) 1208; Ovcz. in Fl. SSSR, 7 (1937) 408; Kitag. Lin. Fl. Mansh. (1939) 226; Fl. Kirgiz. 6 (1955) 98; Fl. Kazakhst. 4 (1961) 102; Kovalevsk. in Opred. rast. Sr. Azii [Key to Plants of Mid. Asia] 3 (1972) 214; Fl. Sin. 28 (1980) 310; Grub. Opred. rast. Mong. [Key to Plants of Mongolia] (1982) 114; Claves pl. Xinjiang, 2 (1983) 317; Fl. desert Sin. 1 (1985) 492; Fl. Intramong. ed. 2, 2 (1990) 502; W.T. Wang in Bull. Bot. Res. (Harbin) 15, 3 (1995) 300; Gub. Consp. Fl. Vneshn. Mong. [Conspectus of Flora of Outer Mongolia] (1996) 51. —**Ic.:** Fl. Kazakhst. 4, Plate 13, fig. 2; Fl. Intramong. 2, tab. 132, fig. 7-10; Fl. Sin. 28, tab. 96, fig. 1-4; Fl. desert Sin. 1, tab. 182, fig. 6-10; Fl. Intramong. ed. 2, 2, tab. 202, fig. 3-4.

Described from Europe. Type in London (Linn.).

On moist and marshy banks of lakes, rivers and springs, in marshy meadows.

IA. Mongolia: ***Khobd.*** (10 km east of Tsagan-Nur settlement, on bank of lake, Aug. 21, 1979—Gub. (MW), ***Mong. Alt.*** (Daingol lake, July 5, 1906—Sap.), ***Cent. Khalkha*** (right bank of Kerulen river, 5 km beyond Under-Khana town, Aug. 9, 1989—Grub., Gub. and Dariima), ***East. Mong.*** (Manchuria station, bank of brook, June 5 and 6, 1902—Litw.; around water, No. 1067, June 30—Lee et al.; Khailar town, Sishan' knolls, No. 550, July 6, 1951—B. Skvortzov; west. fringe of Ongon-Elis sand, south of Khongor pass, Tun-bulak spring, 850 m, July 31, 1981—Gub. (MW); valley of Kerulen river, 60 km north of Choibolsan, on swampy bank of stream, July 21, 1987—Gub., Kam. et al. (MW), ***Depr. Lakes*** (Ulangom, in swamp, July 3, 1879—Pot.), ***Val. Lakes*** (nor.-west. swampy bank of Tsagan-Nur lake, between mounds, June 28, 1924—Pavlov (MW); Orok Nor, edge of lagoon in water and on moist banks, No. 314, 1925—Chaney), ***Gobi-Alt.*** (Bain-Boro-Nuru mountain range, in water around spring, Sept. 10, 1931 —Ik.-Gal.).

IIA. Junggar: ***Jung. Alat.*** (in Arasan [=Ven'tsyuan'] region, on bank of brook, No. 4571, July 22, 1957—Kuan), ***Tien Shan*** (Dzhagastai pass, Aug. 7, 1877—A. Reg.; on way to Khaichzhou from Dzhagastai, on water, No. 31663, Aug. 8, 1957—Kuan; Bol. Yuldus, 30-35 km west of Bain-Bulak settlement, in stagnant pools, Aug. 10, 1958 —Yun. and Yuan'; Bol. Yuldus, No. 6481, Aug. 10, 1958—Lee and Chu), ***Jung. Gobi*** (south: 3-4 km north of St. Kuitun settlement, on Shikho-Manas road, swampy meadow, on water, June 30, 1957—Yun. and Yuan'; east.: nor. slope of Baitak-Bogdo-Nuru mountain range, Ulyastu-Gola gorge, 3-5 km from estuary, in water on bank, Sept. 17-18, 1948—Grub.; in Kuitun region, No. 246, June 26, 1957—Kuan; 35 km west of Bulgan settlement, on bank of Bulgan river, 1000 m, Aug. 14, 1982—Gub. (MW); vicinity of Ushigiin-Us spring and Budun-Ula mountains along bank of brook, July 29, 1988 —Gub. et al.), ***Dzhark.*** (along Ili river, May 16; near Kul'dzha, May [3]; Kul'dzha, July 5 —1877, A. Reg.).

General distribution: Aralo-Casp., Fore Balkh., Jung.-Tarb., Nor. and Cent. Tien Shan; Europe, Balk.-Asia Minor, Fore Asia (Iran), Caucasus, Mid. Asia, West. Sib., East. Sib., Far East, Nor. Mong., China, Himalayas (east., Kashmir), Indo-Mal. (Nor. India, Indo-China), Korean peninsula, Japan, Nor. Amer., Afr. (Nor.).

19. **Batrachium** (DC.) S.F. Gray

Nat. Arr. Brit. Pl. 2 (1821) 720
—*Ranunculus* sect. *Batrichium* DC. Syst. 1 (1817) 233

1. Plants with uniform leaves, submerged in water, tripartite many times into filiform segments..2.
+ Plants with different leaf types: floating on water surface, long-petiolate, with orbicular-reniform 3-5-lobed blade and submerged, short-petiolate blade of orbicular profile, 2-4-tripartite into divaricate filiform segments.......4. **B. mongolicum** (Kryl.) V. Krecz.
2. Leaves amplexicaul, orbicular to reniform in profile, 1-2 cm long, sessile or subsessile, perpendicular to stem, not more than 2-tripartite into rigidulous divaricate segments arranged in the same plane and not convergent outside water. Flowers on very long pedicels, large, 15-18 mm in diam. Nuts weakly bristly.....................
....1. **B. circinatum** (Sibth.) Spach.
+ Leaves not amplexicaul, from semiorbicular to obdeltoid in profile, 1-6 cm long, often with fairly long petioles, sessile at acute angle to stem, (1)-2-5-tripartite into segments arranged in different planes, converging or not converging outside water. Flowers 5-15 mm in diam...3.
3. Leaves large, 7-10 cm long, 4-5-tripartite, with terminal segment longer than 1 cm, obdeltoid in profile, convergent outside water; petiole enlarged into squamiform sheath. Flowers 12-15 mm long
....3. **B. kauffmannii** (Clerc) V. Krecz.
+ Leaves soft, not longer than 2-5 cm, 2-3-tripartite, with terminal segments shorter than 1 cm, reniform or flabellate in profile, not convergent or only partly convergent outside water.......................4.
4. Leaves thin and tender, light green, petiolate, flabellate-semiorbicular in profile, (1)-2-tripartite, 2-3 cm broad. Plant glabrous. Flowers 5-8 mm in diam. Nuts 8-15, glabrous. Small plant................
....2. **B. eradicatum** (Laest.) Fries.
+ Leaves rigid, dark green, sessile, with broad hairy sheath, reniform in profile (inner lobes shorter than lateral), 3-tripartite, 3-5 cm broad. Entire plant pilose above. Flowers 12-15 mm in diam. Nuts 25-30, highly bristly. Very large plant...
....5. **B. trichophyllum** (Chaix) Bosch.

1. **B. circinatum** (Sibth.) Spach, Hist. natur. vég. 8 (1839) 201; Fries, Nov. Fl. Suec. Mant. 3 (1842) 52; Grub. Opred. rast. Mong. [Key to Plants of Mongolia] (1982) 113; Gub. Konsp. Fl. Vneshn. Mong. [Conspectus of Flora of Outer Mongolia] (1996) 48. —*Ranunculus circinatus* Sibth. Fl. Oxon. (1794) 175. —*R. foeniculaceus* Gilib. Fl. Lith. 5 (1782) 261, nom. illeg. —*Batrachium*

foeticulaceum (Gilib.) V. Krecz. in Fl. URSS, 7 (1937) 338; Fl. Kazakhst. 4 (1961) 80; Fl. Sin. 28 (1980) 342; Claves pl. Xinjiang, 2 (1983) 322; Fl. Intramong. ed. 2, 2 (1990) 484. —**Ic.:** Fl. Kazakhst. 4, Plate 10, fig. 7; Fl. Sin. 28, tab. 106, fig. 1-2.

Described from Europe. Type in Oxford (OXF).

In slow streams, backwaters and meanders.

IA. Mongolia: ***East. Mong.*** (Gub. l.c.), ***Depr. Lakes*** ("lower Kobdo, Khara-Us-Nur" —Grub. l.c., Gub. l.c.).

IIA. Junggar: ***Cis-Alt.*** ("Altay"—Claves pl. Xinjiang).

General distribution: Aralo-Casp., Fore Balkh.; Europe (East.), West. Sib., East. Sib., Far East, Nor. Mong. (Hant., Mong.-Daur., Cis-Hing.), China (?), Nor. Amer.

2. **B. eradicatum** (Laest.) Fries in Bot. Not. (Lund) (1843) 114; Krecz. in Fl. SSSR, 7 (1937) 339; Fl. Kazakhst. 4 (1961) 80; Fl. Sin. 28 (1980) 344; Grub. Opred. rast. Mong. [Key to Plants of Mongolia] (1982) 113; Claves pl. Xinjiang, 2 (1983) 323; Fl. Intramong. ed. 2, 2 (1990) 482; Gub. Konsp. Fl. Vneshn. Mong. [Conspectus of Flora of Outer Mongolia] (1996) 48. —*Ranunculus aquatilis* var. *eradicatus* Laest. in Nova Acta Soc. Sci. Upsal. 11 (1839) 242. —**Ic.:** Fl. Kazakhst. 4, Plate 10, fig. 8; Fl. Sin. 28, tab. 106, fig. 3-4; Fl. Intramong. ed. 2, 2, tab. 195, fig. 3-4.

Described from Nor. Europe. Type in Uppsala (UPS).

In lakes, riverine backwaters and meanders.

IA. Mongolia: ***Khobd.*** (valley of Suok river (Boro-Burgazy), marshy pool, June 18, 1906—Sap.), ***Mong. Alt.*** (on and around water on moist sand of Tsagan-bulun river, Aug. 1, 1899—Lad.; Bulgan river, 30 km west of Bulgan settlement, 1000-1100 m, on water, Aug. 14, 1982—Gub. (MW), ***East. Mong.*** (Gub. l.c.), ***Depr. Lakes*** (on Khirgis-Nur lake and Dzabkhyn river, Aug. 8 and 9, 1879—Pot.; Khara-Nur lake, July 27, 1872 —Metel'tseva), ***Alash. Gobi*** ("Alashan—Bain-Khoto"—Fl. Intramong. ed. 2, l.c.).

IIA. Junggar: ***Cis-Alt.*** ("Burchum"—Claves pl. Xinjiang, l.c.).

General distribution: Aralo-Casp., Fore Balkh., Jung.-Tarb.; Europe, West. Sib., East. Sib., Far East, Nor. Mong. (Cis-Hing.), China (Dunbei), Nor. Amer.

3. **B. kauffmannii** (Clerc) V. Krecz. in Fl. URSS, 7 (1937) 343; Fl. Kazakhst. 4 (1961) 82; Fl. Sin. 28 (1980) 340; Claves pl. Xinjiang, 2 (1983) 323. —*Ranunculus kauffmannii* Clerc in Bull. Soc. Oural Sci. Nat. 4 (1878) 107; Kitag. Lin. Fl. Mansh. (1939) 225. —**Ic.:** Fl. SSSR, 7, Plate 21, fig. 5; Fl. Kazakhst. 4, Plate 10, fig. 10; Fl. Intramong. ed. 2, 2, tab. 195, fig. 5-6.

Described from Europe. Type ?

In slow rivers.

IIA. Junggar: ***Jung. Gobi*** (south.: "Urumchi"—Claves pl. Xinjiang).

General distribution: Jung.-Tarb.; Europe, West. Sib., East. Sib., Far East, Nor. Mong. (Hent.), China (Dunbei), Nor. Amer.

4. **B. mongolicum** (Kryl.) V. Krecz. in Fl. URSS, 7 (1937) 350; Fl. Kazakhst. 4 (1961) 83; Grub. Opred. rast. Mong. [Key to Plants of Mongolia] (1982)

113; Gub. Konsp. Fl. Vneshn. Mong. [Conspectus of Flora of Outer Mongolia] (1996) 49. —*Ranunculus aquatilis* subsp. *mongolicus* Kryl. Fl. Sib. Occid. 5 (1931) 1180. —*R. mongolicus* (Kryl.) Serg. in Kryl. Fl. Zap. Sib. 12, 2 (1964) 3293. —**Ic.**: Grub. Plate 53, fig. 242.

Described from West. Siberia. Type in St.-Petersburg (LE).

Grows in water of shallow lakes and brooks.

IA. Mongolia: ***Mong. Alt.*** (upper course of Bulugun river, valley of Ulagchin-Gol, on road to Budzhurtu, 2 km beyond Astrakhyn-Gol, pool on marshy floor of creek valley, July 2; upper course of Kobdo river, Khoton-Gol valley, 10 km beyond bridge, small swamp patch with water, July 8—1971, Grub., Ulzij. and Dariima), ***Depr. Lakes*** (Khirgis-Nur—Grub. l.c., Gub. l.c.), ***Val. Lakes*** (Dzapkhyn-Gol, 5 km before "Guilin" state farm, on water on bank, Aug. 25, 1972—Grub. et al.).

General distribution: Nor. Tien Shan; West. Sib. (Altay), East. Sib., Nor. Mong. (Hang.).

5. **B. trichophyllum** (Chaix) Bosch in Prodr. Fl. Batav. (1850) 7; Krecz. in Fl. SSSR, 7 (1937) 344; Fl. Kirgiz. 6 (1955) 83; Fl. Kazakhst. 4 (1961) 82; Fl. Intramong. 2 (1978) 244, cum auct. comb. F. Schultz; Fl. Sin. 28 (1980) 340; Grub. Opred. rast. Mong. [Key to Plants of Mongolia] (1982) 113; Fl. Intramong. ed. 2, 2 (1990) 486; Gub. Konsp. Fl. Vneshn. Mong. [Conspectus of Flora of Outer Mongolia] (1996) 49. —*Ranunculus trichophyllus* Chaix in Vill. Hist. Pl. Dauph. 1 (1786) 335. —*Batrachium divaricatum* auct. non Wimm.: Krecz. in Fl. SSSR, 7 (1937) 340; Fl. Kirgiz. 6 (1955) 8; Ikonnik. Opred. rast. Pamira [Key to Plants of Pamir] (1963) 119; Pl. Vasc. Helanshan (1986) 107; Fl. Sin. 28 (1980) 340. —**Ic.**: Fl. SSSR, 7, Plate 21, fig. 3 and 4 (sub *B. divaricatum*); Fl. Kazakhst. 4, Plate 18, fig. 9 and 9a; Fl. Intramong. ed. 2, 2, tab. 196, fig. 5-6.

Described from Europe (France). Type ?

Overgrowing pools, slow rivers, backwaters and meanders.

IA. Mongolia: ***Cent. Khalkha*** (on bank of Ugei-Nor, June 11, 1893—Klem.; depression in the second terrace of Kerulen river, Aug. 7, 1927—Zam.), ***East. Mong.*** (left bank floodplain of Khalkhin-Gol river, Aug. 15, 1970—Grub., Ulzij. et al.), ***Depr. Lakes*** (Buyantu-Gol, at exit from Khobdo basin, 2 km north of town, Oct. 2, 1948—Grub., 30 km west of Kobdo town, valley of Buyant-Gol river, 1650 m, on water of meander, Aug. 24, 1983—Gub.); ***Val. Lakes*** (spring entering Orok-Nor lake, June 16, 1926 —E. Kozlova; Orok Nor, shallow lagoons, 1000 m, 1925—Chaney), ***Alash. Gobi*** (Pl. Vasc. Helanshan, l.c.).

IB. Kashgar: ***East.*** (Khami oasis, June 13, 1879—Przew.).

IC. Qaidam: ***Mount.*** (Kurlyk area, 2700 m, May 3, 1895—Rob.; Tsagan-Nor lake and Dulan-Khit temple, 3000 m, Aug. 12, 1901—Lad.).

IIA. Junggar: ***Jung. Gobi*** (east.: "lower Bulgan-Gola"—Grub. l.c., Gub. l.c.; south.: "Urumchi, Manas"—Claves pl. Xinjiang).

IIIB. Tibet: ***Weitzan*** (south. bank of Russkoe lake, in pond, 4100 m, July 27; paleoplain around Alak-nor lake, 4300 m, Aug. 8—1884, Przew.; Russkoe lake and Ekspeditsii, 4100 m, June 29, 1900—Lad.).

General distribution: Nor. Tien Shan, East. Pam.; Europe, Caucasus, West. Sib., East. Sib., Far East, Nor. Mong. (Fore Hubs., Hent., Hang.), China (Dunbei, Nor.), Nor. Amer.

20. **Ceratocephala** Moench

Meth. pl. (1794) 218

1. **C. testiculata** (Crantz) Bess. Enum. pl. Volhyn. (1822) 70; Kovalevsk. in Opred. rast. Sr. Azii [Key to Plants of Mid. Asia] 3 (1972) 198; Fl. Tadzh. 4 (1975) 84. —*Ranunculus testiculatus* Crantz, Stirp. Austr. 2 (1763) 97. —*Ceratocephala orthoceras* auct. non DC.: Maxim. Enum. pl. Mong. 1 (1889) 24; Kryl. Fl. Zap. Sib. 5 (1931) 1175; Ovcz. in Fl. SSSR, 7 (1937) 324; Fl. Kirgiz. 6 (1955) 80; Fl. Kazakhst. 4 (1961) 75; Ikonnik. Opred. rast. Pamira [Key to Plants of Pamir] (1963) 118; Fl. Sin. 28 (1980) 344; Claves pl. Xinjiang, 2 (1983) 324; Fl. desert Sin. 1 (1985) 489. —**Ic.:** Fl. Kazakhst. 4, Plate 10, fig. 2; Fl. Sin. 28, ic. 41; Fl. desert Sin. 1, tab. 178, fig. 9-10.

Described from Europe. Type in Budapest (BP).

In steppes and deserts, in clayey and sandy soil, on arid slopes, pebbly and solonetzic sites.

IIIA. Junggar: ***Jung. Gobi*** (south.: in midcourse of shoals and on bank of Urumchi river, No. 499, May 29, 1957—Kuan; nor. foothills at Bogdo-Ula town, 10-15 km east of Urumchi, on road to Fukan, wormwood thickets on ridgy foothills, April 26; on road to sacred lake from Fukan, steppe belt, pea shrub thickets, on rocky slope of foothills, May 26—1959, Yun.; east.: in Shankhausyan' mountains, May 25, 1879—Przew.), ***Dzhark.*** (Khorgos, April 22; near Kul'dzha, May—1877, A. Reg.; Kul'dzha, 1906 —Muromskii).

General distribution: Aralo-Casp., Fore Balkh., Jung.-Tarb., Nor. and Cent. Tien Shan, East. Pam.; Europe, Balk.-Asia Minor, Fore Asia, Caucasus, Mid. Asia, West. Sib. (Irt., Altay), China (Altay).

Subfamily 4. **ISOPYROIDEAE** Tamura

21. **Paropyrum** Ulbr.

in Notizbl. Bot. Gart. Mus. Berl. 9 (1925) 218. —*Paraquilegia* J. Drumm. et Hutch. in Kew Bull. (1920) 154, 156, p.p. —*Isopyrum* L. Gen. pl. ed. 2 (1742) 245, p.p.

1. **P. anemonoides** (Kar. et Kir.) Ulbr. in Notizbl. Bot. Gart. Mus. Berl. 9 (1925) 218; Pakhom. in Opred. rast. Sr. Azii [Key to Plants of Mid. Asia] 3 (1972) 148; Kamelin, Gubanov et al. in Bot. zhurn. 74, 9 (1989) 1355; Gub. Konsp. Fl. Vneshn. Mong. [Conspectus of Flora of Outer Mongolia] (1996) 50. —*Isopyrum anemonoides* Kar. et Kir. in Bull. Soc. natur. Moscou, 15 (1842) 135; Forbes and Hemsl. Index Fl. Sin. 1 (1886) 18; Maxim. Fl. Tangut. 1 (1889) 19, id. Enum. pl. Mong. 1 (1889) 25; Franch. in J. Bot. (Paris) 11 (1897) 192; Kryl. Fl. Zap. Sib. 5 (1931) 1128; Fl. Kirgiz. 6 (1955) 27; Ikonnik. Opred. rast. Pamira [Key to Plants of Pamir] (1963) 113; Fl. Sin. 27 (1979) 468; Claves pl. Xinjiang, 2 (1983) 288. —*Paraquilegia anemonoides* (Kar. et Kir.) Schipcz. in Not. Syst. (Leningrad) 5, 4 (1924) 53; non Ulbr. in Feddes Repert. 12 (1922) 369; Schipcz. in Fl. SSSR, 7 (1937) 78; Fl. Kazakhst. 4

(1961) 23. —*P. uniflora* J. Drumm. et Hutch. Kew Bull. (1920) 158; Fl. Tadzh. 4 (1975) 21. —**Ic.:** Maxim. Fl. Tangut. tab. 8, fig. 8; Fl. Kazakhst. 4, Plate 2, fig. 3; Fl. Tadzh. 4, Plate 4, fig. 1, 2; Fl. Sin. 27, tab. 110, fig. 1-5 (sub. *I. anemonoides*).

Described from Junggar Ala Tau. Type in St.-Petersburg (LE).

In subalpine and alpine meadows, rocky slopes and talus, rock crevices, sometimes in mountain forests.

IA. Mongolia: ***Mong. Alt.*** (Kamelin et al. 1989, l.c.; Gub. l.c.).

IB. Kashgar: ***Nor.*** (Uchturfan, on shaded side of narrow Airi slit, under spruce, June 3, 1908; Ulug-Tuz gorge, in Charlym river basin, under shade of juniper, June 3, 1908; Ulug-Tuz gorge, in Charlym river basin, under shade of juniper, June 21, 1909—Divn.), ***West.*** (Kun'-Lun', nor. slope, Akkez-Daban pass, 110-120 km south of Kargalyk on Tibet highway, steep south-east. slope of saddle of pass at entrance to juniper grove, June 4, 1959—Yun. and Yuan'; between Koshkulat village and Upal, No. 224, June 10, 1959—Lee et al.).

IIA. Junggar: ***Cis-Alt.*** (basin of Songinin-Gol river (upper course of Chern. Irtysh), in rock crevices, 25 km south-west of Dayan-Nur post, July 16, 1988—Kam., Bud., Gub.), ***Tien Shan*** (at head of Zanma river, 2100 m, June 18, 1877—Przew.; upper course of Khorgos river, 1500-1800 m, May [1]; between Sumbe pass and Kassan, 2100-2400 m, June 22—1878, A. Reg.; on lake Sairam, 2600-2700 m, July 16, 1878—Fet.; Borborogusun river, 2400 m, April 15; lower course of Taldy river, 1200 m, May 15; south. tributary of Karagol river, 2700 m, June 16; Aryslyn, 1800-2400 m, July 14—1879, A. Reg.; mittleres und vorderes Dschanart Tal, June 14-17, 1903—Merzb.; basin of Manas river, left bank of Ulan-Usu river valley at confluence with Koisu, midcourse, in shaded spruce forest, July 17, 1957—Yun. et al.; on bank of Danyu river, on slope, No. 2080, July 21; Datszymyao village, No. 1748, July 22—1957, Kuan; Danu-Gol river valley, 2600 m, in rock crevices, July 23, 1957—Yun. et al.).

IIIA. Qinghai: ***Nanshan*** (on south. slope of mountain range to south of Tetung river, on rocks, July 25, 1872—Przew.; on Yusup-Khatyma river, 2700-3000 m, July 24, 1880 —Przew.).

General distribution: Jung.-Tarb. (Jung. Ala Tau), Nor. and Cent. Tien Shan, East. Pam.; Fore Asia (Iran), Mid. Asia, West. Sib. (Altay), China (Nor.-West.), Himalayas (west., Kashmir).

22. Leptopyrum Reichb.

Consp. (1828) 192

1. **L. fumarioides** (L.) Reichb. Consp. (1828) 192; J. Drumm. et Hutch. in Kew Bull. (1920) 159; Kryl. Fl. Zap. Sib. 5 (1931) 1125; Schipcz. in Fl. SSSR, 7 (1937) 72; Kitag. Lin. Fl. Mansh. (1939) 220; Fl. Kazakhst. 4 (1961) 21; Fl. Sin. 27 (1979) 472; Grub. Opred. rast. Mong. [Key to Plants of Mongolia] (1982) 108; Claves pl. Xinjiang, 2 (1983) 289; Pl. Vasc. Helanshan (1986) 101; Y.Z. Zhao in Acta Sci. Nat. Univ. Intramong. 18, 2 (1987) 285; Fl. Intramong. ed. 2, 2 (1990) 446; Gub. Konsp. Fl. Vneshn. Mong. [Conspectus of Flora of Outer Mongolia] (1996) 50. —*Isopyrum fumarioides* L. Sp. pl. (1753) 557; Maxim. Fl. Tangut. 1 (1889) 17; id. Enum. pl. Mong. 1 (1889) 25.

—Ic.: Fl. Kazakhst. 4, Plate 2, fig. 4; Fl. Intramong. 2, tab. 108; Fl. Sin. 27, tab. 110, fig. 6-10; Grub. Plate 48, fig. 224; Pl. Vasc. Helanshan, tab. 13.

Described from Siberia.Type in London (Linn.).

On mountain slopes in subalpine belt, along banks of rivers and gorges, often as weed in meadow sections with damaged grass cover.

IA. Mongolia: ***Khobd.*** (Boro-Borgosun river, stream of Saksai river, July 2, 1903 —Gr.-Grzh.; south. macroslope of Sailyugem mountain range, 60 km nor.-west of Ulan-Khus settlement, July 9, 1988—Gub., Kam. et al. (MW), ***Mong. Alt.*** (Daingol lake lake, steppe slopes, July 5, 1906—Sap.; south. slope of Khan-Taishiri mountain range, on floor of Muryngin-Khataya gorge north of Khalyun somon, 1920 m, June 18, 1971 —Grub., Ulzij. et al.), ***Cent. Khalkha*** (? Bichikte river, unploughed pasture, June 1, 1883 —Klem.; 5 km east of upper bridge on Kerulen, on Ulan-Bator—Under-Khan road, valley of Usny-Gol river, steppe along mountain trails, July 22, 1949—Yun.; Khavtgoi-Tsagan mountain 60 km east of Under-Khan, south. slope, June 20, 1971—B. Dashnyam et al.), ***East. Mong.*** (near Kharkhonte railway station, sand, June 8, 1902—Litw.; around Khailar town, on ploughed land, June 8, 1951—S.H. Li et al.; Dariganga, narrow valley 10 km west of Dzasyn-Gol on border road, on garbage, Aug. 11, 1970—Grub. et al.; Gangin-Tsagan-Obo mountain, 23 km west of Dariganga, among rocks on summit, July 7, 1971—B. Dashnyam; 25 km east of Matad settlement, along slopes and floor of ravine, 800-850 m, June 25—1987—Kam., Ganbold et al. (MW), ***Gobi Alt.*** (Dundu-Saikhan mountains, west. slope, lower belt, July 4, 1909—Czet.), ***Alash. Gobi*** (Alashan mountain range, Yamata gorge, around kitchen gardens, in midbelt, June 13, 1908—Czet.).

IIA. Junggar: ***Tien Shan*** (on east. slope of a forest-free spur of Bogdo-Ola mountain range, on rock talus, June 17, 1897—Klem.).

IIIA. Qinghai: ***Nanshan*** (mountain range on southern side of Tetung river, forest belt, 2600 m, July 5, 1880—Przew.), ***Amdo*** (clifffs along Mudzhik river, 2700-2900 m, June 16, 1880—Przew.).

General distribution: West. Sib., East. Sib., Far East, Nor. Mong. Introduced and naturalised in Nor.-East. China, Korea, Japan, Europe and Nor. Amer.

23. **Paraquilegia** J. Drumm. et Hutch.

in Kew Bull. (1920) 154, 156, s. str.

1. Leaf blades trisected twice. Petals obovoid, roughly half as long as sepals. Seeds without wings, compact, short-hispid.................. ....1. **P. anemonoides** (Willd.) Engl. ex Ulbr.

\+ Leaf blades trisected thrice. Petals orbicular, 1/4 as long as sepals. Seeds broad-winged along suture, glabrous................................ ....2. **P. microphylla** (Royle) J. Drumm. et Hutch.

1. **P. anemonoides** (Willd.) Engl. ex Ulbr. in Feddes Repert. 12 (1922) 369; Pakhom. in Opred. rast. Sr. Azii [Key to Plants of Mid. Asia] 3 (1972) 148; Fl. Tadzh. 4 (1975) 21; Fl. Sin. 27 (1979) 483; Claves pl. Xinjiang, 2 (1983) 290; Pl. Vasc. Helanshan (1986) 101; Fl. Intramong. ed. 2, 2 (1990) 443. —*Aquilegia anemonoides* Willd. in Mag. Ges. Naturf. Freunde Berlin, 5 (1811) 401. —*Isopyrum grandiflorum* Fisch. ex DC. Prodr. 1 (1824) 48; Forbes

and Hemsl. Index Fl. Sin. 1 (1886) 18; Maxim. Fl. Tangut. 1 (1889) 17; Franch. in J. Bot. (Paris) 11 (1897) 191; Gr.-Grzh. Opis. put. Zap. Kit. [Report of Expedition in West. China] 3 (1907) 75; Pamp. Fl. Carac. (1930) 108; Kryl. Fl. Zap. Sib. 5 (1931) 1127. —*Paraquilegia grandiflora* (Fisch. ex DC.) J. Drumm. et Hutch. in Kew Bull. (1920) 156; Schipcz. in Fl. SSSR, 7 (1937) 76; Fl. Kirgiz. 6 (1955) 28; Fl. Kazakhst. 4 (1961) 22; Ikonnik. Opred. rast. Pamira [Key to Plants of Pamir] (1963) 113. —**Ic.:** Mag. Ges. Naturf. Freunde Berlin, 5, tab. 9, fig. 6; Royle, Ill. Bot. Himal. Mount. 2, tab. 11, fig. 3; Maxim. Fl. Tangut. tab. 8, fig. 10; Fl. Kazakhst. 4, Plate 2, fig. 2; Fl. Intramong. ed. 2, 2, tab. 179.

Described from Altay. Type in St.-Petersburg (LE).

In rock crevices, on talus, glacial moraine and in alpine belt, up to 4700 m above sea level.

IA. Mongolia: ***Alash. Gobi*** (Pl. Vasc. Helanshan, l.c.).

IIA. Junggar: ***Jung. Alat.*** (Kazan pass, 2700-3400 m, Aug. 11, 1878—A. Reg.; from Ven'tsyuan' [Arasan] 30 km westward, in water-divide, No. 2012, Aug. 25; 60 km nor. of Ulastai, No. 3990, Aug. 31—1957, Kuan), ***Tien Shan*** (pass to Kash river through Taldy upper course, 3000-3200 m, May 21; Irenkhabirga mountain range, Taldy river, 2700 m, May 26; between Dumbedan and Kumdaban, on nor. slope of Irenkhabirga mountain range, 2400-2700 m, May 28, nor. slope of Irenkhabirga mountain range, Kumdaban, 2700 m, May 29; Kara-Gol, 3000 m, June 14; south. tributary of Kara-Gol, pass to Nilki, 2400-2700 m, June 16; Arystyn, 3000-3400 m, Aug. 13—1879, A. Reg.; am Kukurtuk Pass (4400 m) etwa 200 m unter Passhöhe in dicken Polstern, June 26-July 2, 1903—Merzb.; Uch-Turfan, Airi gorge, under spruce, June 3, 1908—Divn.; Manas river basin, left bank valley of Ulan-Usu river at confluence with Dzhartas, upper forest boundary, July 18, 1957—Yun. et al.; Borgaty—Yakou, No. 1708, Aug. 3, 1957—Kuan).

IIIA. Qinghai: ***Nanshan*** (South. Tetung mountain range, in alpine belt, on rocks, July 13 and 25, 1872; Machan-Ula, alpine belt (3400-3700 m), under rocks, July 24, 1879; nor. slope of South. Tetung mountain range southward of Tetung river, on rocks, in alpine belt, frequent, 3000-3700 m, July 31, 1880—Przew.; An'si, Kha-Daban pass, May 3, 1890—Gr.-Grzh.; nor. slope of Humboldt mountain range, Blagodatnyi spring, on rocks, 3000-3400 m, June 3; nor. slope of Humboldt mountain range, Ulan-Bulak area, rocks, 4000 m, June 6; Yamatyn-Umru mountain range, 4000 m, rocks, June 13 —1894, Rob.; "South. Tetung mountain range, 3700 m, May 3, 1905—Gr.-Grzh." —Gr.-Grzh. l.c.).

IIIB. Tibet: ***Chang Tang*** (Nor. Tibet, 1884—Przew., "Zhitu"—Fl. Xizang. l.c.), ***South.*** ("Chzhada"—Fl. Xizang. l.c.).

General distribution: Jung.-Tarb., Nor. and Cent. Tien Shan, East. Pam.; Fore Asia, Mid. Asia (mountains), West. Sib. (Altay), East. Sib. (Sayans), China (Nor.-West.), Himalayas (west., east.).

2. **P. microphylla** (Royle) J. Drumm. et Hutch. in Kew Bull. (1920) 157; Schipcz. in Fl. SSSR, 7 (1937) 77; Fl. Kirgiz. 6 (1955) 29; Fl. Kazakhst. 4 (1961) 22; Fl. Sin. 27 (1979) 484; Grub. Opred. rast. Mong. [Key to Plants of Mongolia] (1982) 108; Fl. Xizang. 2 (1985) 160. —*Isopyrum microphyllum* Royle, Ill. Bot. Himal. Mount. 1 (1839) 54; Maxim. Enum. pl. Mong. 1 (1889) 26; Franch. in J. Bot. (Paris) 11 (1897) 192; Kryl. Fl. Zap. Sib. 5 (1931) 1128. —*I. grandiflorum* var. *microphyllum* (Royle) Finet et Gagnep. in Bull. Soc.

Bot. Fr. 51 (1904) 409. —*Paraquilegia anemonoides* auct. non Engl. ex Ulbr.: Gub. Konsp. Fl. Vneshn. Mong. [Conspectus of Flora of Outer Mongolia] (1996) 50. —**Ic.:** Royle, tab. 11, fig. 4; Maxim. tab. 8, fig. 11; Grub. Plate 48, fig. 223; Fl. Xizang. 2, tab. 18, fig. 1-3.

Described from Himalayas. Lectotype in London (KW). Isolectotype in St.-Petersburg (LE).

In rocky sites in alpine and subalpine mountain belts, up to 5000 m above sea level.

IA. Mongolia: ***Khobd.*** (Kharkira river, July 23, 1879—Pot.; valley of Turgen river, rock placers, 2300 m, July 6 and 8, 1973—Banzragch et al.; 20 km west-south-west of Ulangom, Mukhur-Ulyasty-gol gorge, in rock crevices on nor. slope, 2650 m, July 7, 1977—Karam. et al.), ***Mong. Alt.*** (Ak-su II river (=Belaya Kobdo), steep placers, July 22, 1909—Sap.; Khasagtu-Khairkhan, nor. slope of Tsagan-Irmyk-Ula, in upper Khunkerin-Ama river, larch forest, under rocks, 2500-2700 m, Aug. 23, 1972—Grub., Dariima et al.).

IIIA. Qinghai: ***Nanshan*** (on Tashitu river [2900 m], May 26, 1886—Pot.).

IIIB. Tibet: ***Weitzan*** (left bank of Yangtze river, on rocks, June 26; along Kon-chyun-chyu river, 4000-4300 m, June 30—1884, Przew.; Yangtze river basin, Donra area, on Dynechyu river, 4000 m, July 16, 1900—Lad.; "Bizhu"—Fl. Xizang. l.c.), ***South.*** ("Lhasa, Nan'mulin, Dinzhi"—Fl. Xizang. l.c.).

General distribution: Jung.-Tarb., Nor. and Cent. Tien Shan; Mid. Asia (Pamiro-Alay), West. Sib. (Altay), East. Sib. (Sayans), Far East, Nor. Mong. (Fore Hubs., Hang.), China (South-West.), Himalayas (west.).

24. **Aquilegia** L.

Sp. pl. (1753) 533

1. Flowers small, perianth 1-1.4 cm in diam., pink or lilac. Nectary petals without or with very thin and short (2-6 mm long and 1.5-2 mm thick) spur. Qinghai plant.................... 2. **A. ecalcarata** Maxim.

\+ Flowers considerably larger, of different colour shades. Nectary petals usually with spur.. 2.

2. Spurs broad-conical, recurved at tip, shorter than limb; even straightened spurs not longer than 14 mm; bent spurs 5-8 (11) mm long; in half-open flowers, very small, almost imperceptible (1-2 mm long). Flowers large, (4) 6-10 cm in diam., usually single (occasionally 2-3 each but then lateral flowers considerably smaller than terminal); sepals bright blue, turquoise blue or blue-violet, often turning pink when dry. Colour of petals same as of sepals or creamish. Altay plant...................... 3. **A. glandulosa** Fisch. ex Link.

\+ Spurs broad-, narrow- or cylindrical-conical, straight or recurved at tip; straightened spurs invariably longer than limb, (14) 17-23 mm long; bent spurs almost as long as limb or considerably longer, occasionally shorter.. 3.

3. Spurs long, twice or almost twice longer than sepals, thin, about 2 mm in cross-section, generally straight..4.
+ Spurs of medium length, thick, about 3-5 mm in cross-section; more or less hook-shaped at tip..5.
4. Flowers milk-white or light lilac, rarely pink, with white or yellow limb of petals. Stamens somewhat exserted from flower. Limb of petals narrower than sepal or equal to it...
....5. **A. lactiflora** Kar. et Kir.
+ Flowers greenish yellow, dingy blue or brownish red. Stamens mostly exsert perceptibly from flower. Limb of petals broader than sepals..7. **A. viridiflora** Pall.
5. Plant more or less pubescent; pedicels sometimes with glandular pubescence. Flowers monochromatic..6.
+ Plant glabrous. Flowers large, 5-6 cm in diam., 2-coloured, with violet-blue sepals and light blue, occasionally whitish petals.........
....6. **A. sibirica** Lam.
6. Flowers dark purple, about 4-5 cm in diam. Limb of petals slightly longer than spur.....................1. **A. atrovinosa** M. Pop. ex Gamajun.
+ Flowers violet-coloured, about 6 cm in diam. Limb of petal nearly as long as spur.....................4. **A. karelinii** (Baker) O. et B. Fedtsch.

1. **A. atrovinosa** M. Pop. ex Gamajun. in Not. Syst. (Alma Ata) 2 (1964) 12; Fl. Kazakhst. 4 (1961) 28, nom. nud.; Fl. Kirgiz. suppl. 1 (1967) 68; Pakhom. in Opred. rast. Sr. Azii [Key to Plants of Mid. Asia] 3 (1972) 152; Fl. Sin. 27 (1979) 500, p.p. —**Ic.:** Fl. Kazakhst. 4, Plate 3, fig. 6.

Described from West. Tien Shan (Talass Ala Tau). Type in Alma Ata (AA).

Along forest slopes, moist gorges and in alpine meadows.

IIA. Junggar: ***Cis-Alt.*** (in Koktogai region, No. 2154, Aug. 18, 1956—Ching), ***Jung. Alat.*** (Toli district, Albakzin, in forest, No. 2537, Aug. 6, 1957—Kuan), ***Tien Shan*** (in Savan region, No. 3602, Oct. 2, 1956—Ching; from Nyutsyuantsza 20 km southward, No. 220; Savan district, Datszymyao village, No. 1261, July 8; mountains south of Nyutsyuantsza, No. 655; Savan district, in gorge, No. 1645, July 18—1957, Kuan; upper Ili river, Tsanma waterdivide and Kunges, nor. slope toward Kunges, grassy spruce forest, Aug. 7, 1958—Yun. and Yuan').

General distribution: Nor. Tien Shan; Mid. Asia (West. Tien Shan, Pamiro-Alay).

2. **A. ecalcarata** Maxim. Fl. Tangut. (1889) 20; Finet et Gagnep. in Bull. Soc. Bot. Fr. 51 (1904) 41; Ulbr. in Notizbl. Bot. Gart. Mus. Berl. 10 (1929) 866; Hand.-Mazz. in Acta Horti Gotoburg, 13 (1939) 44; Munz in Gent. Herb. 7 (1946) 21; Fl. Sin. 27 (1979) 491; Borodina-Grabovskaya in Catalogue type specim. Centr. Asia (LE) (2000) 192. —*Semiaquilegia ecalcarata* (Maxim.) Sprague et Hutch. in Kew Bull. 1921 (1921) 175. —*S. ecalcarata* f. *semicalcarata* Schipcz. in Not. Syst. (Leningrad) 5 (1924) 54. —*Aquilegia*

ecalcarata f. *semicalcarata* (Schipcz.) Hand.-Mazz. l.c. 45. —**Ic.:** Maxim. tab. 8, fig. 12; Fl. Sin. 27, tab. 117, fig. 1-5.

Described from Qinghai (Nanshan). Lectotype and syntype in St.-Petersburg (LE). Plate VII, fig. 1. Map 2.

In coniferous forests.

IIIA. Qinghai: ***Nanshan*** (Jugum S a fl. Tetung, in silvis acerosis umbrosis [in coniferous mossy forests, mountain range along south. side of Tetung river], No. 160, June 25/July 7—lectotypus ! (Borodina-Grabovskaya), l.c. [same site] in subalpine belt, common among shrubs, July 13—1872; [same site] reg. silvaticum inferior, 7500 ft [2300 m], in silvis frequens, No. 698, July 25/Aug. 6, 1880, syntypus !—Przew.).

General distribution: China (Nor.-West., South-West.).

3. **A. glandulosa** Fisch. ex Link, Enum. Hort. Berol. 2 (1822) 84; Kryl. Fl. Zap. Sib. 5 (1931) 1133; Bulavkina in Fl. SSSR, 7 (1937) 95; Munz in Gent. Herb. 7 (1946) 38; Fl. Kazakhst. 4 (1961) 28; Pakhom. in Opred. rast. Sr. Azii [Key to Plants of Mid. Asia] 3 (1972) 152; Fl. Sin. 27 (1979) 501; Grub. Opred. rast. Mong. [Key to Plants of Mongolia] (1982) 109; Claves pl. Xinjiang, 2 (1983) 292; Gub. Konsp. Fl. Vneshn. Mong. [Conspectus of Flora of Outer Mongolia] (1996) 48. —**Ic.:** Fl. SSSR, 7, Plate 7, fig. 5; Fl. Kazakhst. 4, Plate 3, fig. 7; Fl. Sin. 27, tab. 117, fig. 7-10.

Described from Altay. Type in St.-Petersburg (LE).

In meadows, along banks of brooks, in placers, dwarf birch thickets in alpine belt; forest borders at upper boundary of forest belt.

IA. Mongolia: ***Khobd.*** (Gub. l.c.), ***Mong. Alt.*** (head of Belaya Kobdo river (Aksu), alpine tundra, June 29, 1906—Sap.; Kharagaitu-daba pass, in upper courses of Indertiin-Gol, alpine meadow, July 24, 1947—Yun.; Buyantu and Bulugun waterdivide, 2 km south of Akhuntyn-daba pass, on road to Kudzhurtu, 2840 m, upper course of Ulanchin-Gol, boulder pile along valley of brook within sedge-*Cobresia* waste land, July 2; upper Buyantu river, Chigirtei-gol, 21 km above lake, nor. slope of Chigirtei-ul, 2600-2800 m, young larch grove along creek valley with glacier brook, July 4—1971, Grub., Ulzij. et al.; Ulyastain-gola basin, valley of left tributary in upper reaches, June 28, 1973 —Golubk.; mountains along south. bank of Dayan-nur lake, July 26, 1977—Karam. et al.; Munkh-Khairkhan mountain range, 75 km south-west of Kobdo, near Ulan-Daba pass, on Olgoi-Nur lake, 3000 m, Aug. 11, 1982—Gub. (MW).

IIA. Junggar: *Cis-Alt.* (in Qinhe region, No. 1310, July 19; Burchum, No. 3276, Sept. 6 —1956, Ching; Altay mountains in Timulbakhan, 2300 m, No. 10753, July 18, 1959 —Lee et al.; basin of Songinin-Gol river (upper Chern. Irtysh) near boundary 25 km south-east of Dayan-Nur post, 1800-2100 m, July 16-17, 1988—Kam., Gub. et al. (MW), "Sinkiang"—Fl. Sin. l.c.).

General distribution: Jung.-Tarb., Nor. Tien Shan (Trans-Ili Ala Tau); West. Sib. (Altay), East. Sib., Nor. Mong. (Mong.-Daur.), China (Altay).

Note. Hybrid *A. glandulosa* x *A. sibirica* Lam., described as *A.* x *gubanovii* R. Kam. Bull. Soc. natur. Moscou, ser. biol. 96, 6 (1991) 114, is found in Mongolian Altay.

4. **A. karelinii** (Baker) O. et B. Fedtsch. in Trudy obshch. estestvoisp. Kazan. univ. 33, 3 (1899) 65; Bulavkina in Fl. SSSR, 7 (1937) 93, p.p.; Fl.

Kirgiz. 6 (1955) 34; Fl. Kazakhst. 4 (1961) 27, p.p.; Pakhom. in Opred. rast. Sr. Azii [Key to Plants of Mid. Asia] 3 (1972) 153; Claves pl. Xinjiang, 2 (1983) 292. —*A. vulgaris* L. var. *karelinii* Baker in Gard. Chron. new ser. 10 (1878) 76; Maxim. Enum. pl. Mong. 1 (1889) 27. —*A. vulgaris* auct. non L.: Forbes and Hemsl. Index Fl. Sin. 1 (1886) 18. —*A. atrovinosa* auct. non M. Pop. ex Gamajun.: Fl. Sin. 27 (1979) 500. —**Ic.:** Fl. SSSR, 7, Plate 7, fig. 11; Fl. Kirgiz. 6, Plate 1, fig. 3; Fl. Kazakhst. 4, Plate 3, fig. 2.

Described from Junggar Ala Tau. Type in St.-Petersburg (LE).

In moist gorges and forests on mountain slopes, mostly coniferous, sometimes deciduous forests.

IIA. Junggar: ***Cis-Alt.*** (on Kandagatai river, Sept. 15, 1876—Pot.), ***Jung. Alat.*** (mountains south of Sin'yuan' town (around Toli town), No. 1143, Aug. 22, 1957—Kuan), ***Tien Shan*** (on Sairam lake, June 1877—A. Reg.; same site, July 16 and 23, 1878—Fet.; on upper Kunges, 1200 m [June 24] 1877—Przew.; Talki, 1800-2400 m, July 16—1877; midcourse of Khorgos river, 1500-1800 m, April 16—1878, A. Reg.; Kutukchi (west of Kul'dzha), June 6, 1878—Fet.; Khonakhai mountains, 1500-2100 m, June 16, 1878—A. Reg.; south. slope of Kokkamyr mountain range, July; Bogdo mountain, 2100-2400 m, July—1878, Fet.; south. tributary of Kara-Gol river, 2400-2700 m, June 16; confluence of Aryslyn with Kash river, July 10; Aryslyn river, July 10 and 12; Kunges river, Aug.—1879, A. Reg.; mountains south of Nyutsyuan'tsza, No. 655, July 18, 1957—Kuan; 20 km south of Nyutsyuan'tsza, No. 220, July 8, 1959—Lee et al.), ***Jung. Gobi*** (east.: 36 km north of Barkul' lake, Sept. 26, 1957—Kuan).

General distribution: Jung.-Tarb., Nor. and Cent. Tien Shan; Mid. Asia (West. Tien Shan, Syr.-Dar.).

5. **A. lactiflora** Kar. et Kir. in Bull. Soc. natur. Moscou, 14 (1841) 374; Bulavkina in Fl. SSSR, 7 (1937) 92; Munz in Gent. Herb. 7 (1946) 98; Fl. Kazakhst. 4 (1961) 26; Pakhom. in Opred. rast. Sr. Azii [Key to Plants of Mid. Asia] 3 (1972) 154; Fl. Sin. 27 (1979) 497. —**Ic.:** Fl. SSSR, 7, Plate 7, fig. 6; Munz, fig. 25; Fl. Kazakhst. 4, Plate 3, fig. 4.

Described from Kazakhstan (Tarbagatai). Type in Moscow (MW). Isotype in St.-Petersburg (LE).

Along brooks, on rubble talus, at base of rocks, mostly in subalpine belt, in upper forest boundary.

IIA. Junggar: ***Jung. Alat.*** (in Toli district, No. 2697, Aug. 7, 1957—Kuan), ***Tien Shan*** (Manas river basin, left bank, valley of Ulan-Usu river at confluence with Dzhartas, upper forest boundary, on rubble talus, July 18, 1957—Yun. et al.; 15 km south of Nyutsyuan'tsza, on bank of Koisu river, No. 265, July 18; between Daban and Danyu, on slope, No. 2013, July 19—1957, Kuan), ***Tarb.*** ("Dachen"—Fl. Sin. l.c.).

General distribution: Fore Balkh., Jung.-Tarb., Nor. Tien Shan; Mid. Asia (mountains).

6. **A. sibirica** Lam. Encycl. 1 (1783) 150; Maxim. Enum. pl. Mong. 1 (1889) 26; Kryl. Fl. Zap. Sib. 5 (1931) 1134; Bulavkina in Fl. SSSR, 7 (1937) 96; Munz in Gent. Herb. 7 (1946) 31; Fl. Kazakhst. 4 (1961) 30; Fl. Sin. 27 (1979) 501; Grub. Opred. rast. Mong. [Key to Plants of Mongolia] (1982)

109; Claves pl. Xinjiang, 2 (1983) 292. —**Ic.:** Fl. SSSR, 7, Plate 7, fig. 3; Munz, fig. 6; Fl. Kazakhst. 4, Plate 3, fig. 3.

Described from Siberia. Type in Paris (P).

In larch and sparse mixed forests, forest borders, birch stands, rocky slopes in forest belt; in bouldery placers, on banks of brooks, bald peaks and small meadows in alpine belt.

IA. Mongolia: ***Khobd.*** (in Ulan-Daban pass, shaded and forest-covered gorge, June 23, 1879—Pot.; valley of Turgen' river, mossy larch grove, July 9; valley of Shur-Ulyastai-Gol river, larch grove, July 12—1973, Banzragch et al.), ***Mong. Alt.*** (Khan-Taishiri mountain range, south. slope in upper Shine-usu (north of Khamon somon), 2380-2450 m, larch forest on nor. slope under rocks of ridge, June 18, 1971—Grub., Ulzij. et al.; Uenchiin-Gol basin, Arshantyn-Gol valley right tributary of Khargaityn-Gol), 3 km from estuary, nor. slope of gorge, 2500-2600 m, larch forest, Aug. 14, 1979—Grub., Dariima et al.).

IIA. Junggar: ***Tien Shan*** ("[ed pedem m. Adzhi-Bogdo] in forests on northern base, June 12, 1877—Pot."—Maxim. l.c.; [Koshety-Daban], in forest, May 23, 1879—Przew.).

General distribution: Jung.-Tarb.; West. Sib., East. Sib., Nor. Mong. (Fore-Hubs., Hent., Hang., Mong.-Daur.).

Note. See note under *A. glandulosa* Fisch. ex Link.

7. **A. viridiflora** Pall. in Acta Ac. Sci. Petrop. (pro a. 1779) 3, 2 (1783) 260; Franch. Pl. David. 1 (1884) 21; Forbes and Hemsl. Index Fl. Sin. 1 (1886) 18; Maxim. Fl. Tangut. 1 (1889) 19; id. Enum. pl. Mong. 1 (1889) 26; Bulavkina in Fl. SSSR, 7 (1937) 91; Kitag. Lin. Fl. Mansh. (1939) 214; Munz in Gent. Herb. 7 (1946) 90; Fl. Sin. 27 (1979) 496; Grub. Opred. rast. Mong. [Key to Plants of Mongolia] (1982) 109; Y.Z. Zhao in Acta Sci. Nat. Univ. Intramong. 18, 2 (1987) 284; Fl. Intramong. ed. 2, 2 (1990) 439; Gub. Konsp. Fl. Vneshn. Mong. [Conspectus of Flora of Outer Mongolia] (1996) 48. —*A. atropurpurea* Willd. Enum. Hort. Berol. (1809) 577; Maxim. Enum. pl. Mong. 1 (1889) 26 (in adn. cum *A. viridiflora*) —*A. viridiflora* var. *atropurpurea* (Willd.) Finet et Gagnep. in Bull. Soc. Bot. Fr. 51 (1904) 412; Kitag. Lin. Fl. Mansh. (1939) 214. —*A. viridiflora* f. *atropurpurea* (Willd.) Kitag. in J. Jap. Bot. 34 (1959) 6; Fl. Sin. 27 (1979) 497; Pl. vasc. Helanshan (1986) 101; Fl. Intramong. ed. 2, l.c. 440. —**Ic.:** Pall. 1783, tab. 11, fig. 1; Fl. SSSR, 7, Plate 7, fig. 8; Munz, fig. 22; Fl. Intramong. 2, tab. 112, fig. 1-5; Grub. Plate 49, fig. 226; Fl. Sin. 27, tab. 118, fig. 5-8; Fl. Intramong. ed. 2, 2, tab. 176, fig. 3-5.

Described from East. Siberia (Dauria). Type in St.-Petersburg (LE). Isotype in London (BM).

In rock crevices, rocky and arid slopes of mountains.

IA. Mongolia: ***Mong. Alt.*** (Khara-Adzarga mountain range, valley of Sokhirsala river, on east. slope of Imertsik mountains, Aug. 21 and 22, 1930—Pob.; south. slope of Khan-Taishiri mountain range, Naran-Bulak gorge, north of Khalyun somon, June 19, 1971—Grub. et al.; south. slope of Adzhi-Bogdo mountain range, Ikhe-Gol river, 2500 m, Aug. 22, 1979—Grub. et al.; 35 km south-west of Khalyun settlement, Aug. 14, 1984—Gub. (MW), ***Cent. Khalkha, East. Mong.*** (Oulachan, en touffes sur les rochers ombrages, May 1864—David; Shara-Khada mountains, May 1871—Przew.; vicinity of Manchuria station, 1915—E. Nechaeva; same site, May 6, 1918—Kom.; Abdrant-Gol at border post, Aug. 20, 1970—Grub. et al.), ***Depr. Lakes*** (right bank of Khungul, Ikhe-Margats-Ul, south-east. branch of road to Urgamal somon, Aug. 18, 1972—Grub. et al.), ***Gobi-Alt.*** (Bain-Tsagan mountain range, in shaded gorge in Tsubulyur creek valley, Aug. 5; Dundu-Saikhan mountain range, Aug. 17; Dzun-Saikhan mountain range, Aug. 25—1931, Ik.-Gal.; Bain-Tsagan mountain range, mountain slopes from trail to upper belt, July and Aug. 1933—Khurlat; west. extremity of Dzun-Saikhan mountain range, gorge in juniper belt, June 18, 1945—Yun.; Dzun-Saikhan mountain range, Ëlyn-Ama gorge at upper end, July 21 and 22; Dundu-Saikhan mountain range, south. slope under main peak, 2825 m, July 23—1970, Grub. et al.; Gurban-Saikhan mountain range, about 100 km west-nor.-west of Dalan-Dzadagad town, along gorge slopes, 1800-2000 m, June 30 and 31, 1981—Gub. (MW), ***East. Gobi*** (Ikhe-Dzhargalant somon, southern fringe of Ikhe-Narat, May 31, 1941—Yun.; Khachig-Ula mountain range, July 13, 1972—Gur. and Rachk.; 200 km south of Sain-Shand town, Khutag-Ul mountains, 1250 m, July 10, 1982—Gub. (MW), ***Alash. Gobi*** (Alashan mountain range, Yamata gorge, west. slope, along river bed, May 5, 1908—Czet.).

IIIA. Qinghai: ***Nanshan*** (South. Tetung mountain range, in forests and along rocky slopes, July 11, 1872; same site, on rocks, in lower belt, May 29, 1873—Przew.; south. slope of Balekun mountain range, 3000 m, on rocks, in forest, June 6, 1880—Przew.; Kuen-Lun, nor. slope of Humboldt mountain range, Blagodatnyi spring, rocks, 3700 m, June 23, 1894—Rob.).

General distribution: East. Sib., Far East, Nor. Mong. (Hent., Hang., Mong.-Daur.), China (Dunbei, Nor., Nor.-West.).

Note. K.F. Ledebour and N.S. Turczaninov, as well as some contemporary investigators recognise 2 independent species, *A. viridiflora* and *A. atropurpurea*, based on differences in the colour of flowers (greenish yellow in the former and dingy blue or brownish red in the latter), forms of sepals (subacute in the former and blunt in the latter), length of spurs (very short in the latter) and length of stamens (in the latter, stamens perceptibly exsert from perianth). In "Flora Kitaya" [Flora of China] (Fl. Sin. l.c.), f. *atropurpurea* has been cited for East. Qinghai, Inner Mongolia, Nor. China and south. part of Dunbei.

* * *

The following species have also been cited for Central Asian territory:

A. moorcroftiana Wall. ex Royle, Ill. Bot. Himal. Mount. 1 (1839) 55.

IIIB. Tibet: ***South.*** (Chzhada);

A. yatabeana Kitag. in Rep. Inst. Sci. Res. Manchoukuo, 4, 4 (1936) 81.

IA. Mongolia: ***East. Gobi*** (Ulantsab).

Subfamily 5. THALICTROIDEAE Heintze

25. Thalictrum L.

Sp. pl. (1753) 545

1. Inflorescence racemose or flowers single. Small plants..................2.
\+ Inflorescence cymose—paniculate or corymbose............................3.

2. Inflorescence, simple raceme with single (very rarely 2 each), long, arcuate peduncles in axils of leafy bracts. Stem without leaves or with single leaf in lower 1/3; almost all leaves radical, bipinnately compound. Nuts 2-4 each..1. **Th. alpinum** L.
+ Flowers single. Inflorescence, spicate cluster with individual flowers in leaf axils. Plant with only cauline 3- or 4-pinnately compound leaves; leaves at stem base reduced to sheaths. Ovaries 4-5, only 2 (3) nuts developing fully from them..11. **Th. squamiferum** Lecoy.
3. Anther filaments clavately thickened above, as broad as anther or broader..4.
+ Anther filaments uniformly thin, filiform..6.
4. Anther filaments white, a few times broader than anthers in upper part. Plants 20-40 cm tall with leaves aggregated at base; cauline leaves 1-2, small..7. **Th. petaloideum** L.
+ Anther filaments yellow-green, violet or reddish, more narrow. Up to 1 m tall plant with leafy stems...5.
5. Nuts on very short stalk, globose. Inflorescence few-flowered corymbose panicle. Plant glabrous..2. **Th. baicalense** Turcz. ex Ledeb.
+ Nuts on well-manifest stalks equal to half or entire length of fruit, broad-ovoid. Inflorescence many-flowered loose oblong panicle. Plant with rare, white simple acicular hairs in nodes of stem and leaves as well as on underside of tender leaflets...8. **Th. przewalskii** Maxim.
6. Nuts flat or highly compressed. Leaflets small, elliptical or obovoid, entire..13. **Th. tenue** Franch.
+ Nuts ovoid, fusiform or subfusiform...7.
7. Entire plant, specially underside of leaflets, pubescent with fine glandular hairs. Plant with unpleasant odour...5. **Th. foetidum** L.
+ Glabrous plant..8.
8. Peduncles branched into 2 or 3 branchlets. Nuts usually 1-3, large, 5-7 mm long. Perianth segments thickish, serrated..12. **Th. squarrosum** Steph. ex Willd.
+ Nuts several but smaller...9.
9. Nuts 4 (6). Small, 2.5-10 cm tall plant...10.
+ Nuts 5-12. More large plant..11.
10. Nuts erect, upright, with small deltoid stylodium. Junggar plant. ....3. **Th. isopyroides** C.A. Mey.
+ Nuts highly curved, turned down, with highly recurved stylodium. Plant of Qinghai and Tibet......9. **Th. rutaefolium** Hook. f. et Thoms.
11. Inflorescence contracted, more or less corymbose...4. **Th. flavum** L.
+ Inflorescence elongated, more or less loose, broad or conical or oblong-ovoid..12.

12. Nut fusiform, 6-12 mm long, with 3-4 mm long perceptibly recurved stylodium..14. **Th. uncatum** Maxim.
+ Nut ovoid-elliptical with short deltoid stylodium........................13.
13. Inflorescence generally loose, broad, pyramidal (oblong-ovoid), without leaves or with leaves in lower part. Leaves declinate from stem, their blades (2)-4-pinnatipartite; lobes of latter order broad-obovoid, often subcircular, obovoid or oblong-obovoid, cordate or rounded-cuneate at base...6. **Th. minus** L.
+ Inflorescence compressed, elongated-conical (oblong), leafy. All leaves, except lowermost, caducous, appressed to stem, their blades bipinnatipartite; terminal lobes oblong-ovoid, cuneate at base........... ....10. **Th. simplex** L.

1. **Th. alpinum** L. Sp. pl. (1753) 545; Regel in Bull. Soc. natur. Moscou, 34 (1861) 15; Lecoyer in Bull. Soc. Bot. Belg. 24 (1885) 193; Forbes and Hemsl. Index Fl. Sin. 1 (1886) 8; Maxim. Fl. Tangut. 1 (1889) 6; id. Enum. pl. Mong. 1 (1889) 6; Hemsl. Fl. Tibet (1902) 162; Ostenf. and Pauls. in Hedin, S. Tibet, 6, 3 (1922) 82; Pamp. Fl. Carac. (1930) 113; Kryl. Fl. Zap. Sib. 5 (1931) 1212; Nevsk. in Fl. SSSR, 7 (1937) 519; Fl. Kirgiz. 6 (1955) 109; Fl. Kazakhst. 4 (1961) 123; Ikonnik. Opred. rast. Pamira [Key to Plants of Pamir] (1963) 122; Kovalevsk. in Opred. rast. Sr. Azii [Key to Plants of Mid. Asia] 3 (1972) 226; Fl. Tadzh. 4 (1975) 144; Fl. Sin. 27 (1979) 589; Grub. Opred. rast. Mong. [Key to Plants of Mongolia] (1982) 115; Claves pl. Xinjiang, 2 (1983) 296; Fl. Xizang. 2 (1985) 75; Pl. Vasc. Helanshan (1986) 109; Fl. Intramong. ed. 2, 2 (1990) 449; Gub. Konsp. Fl. Vneshn. Mong. [Conspectus of Flora of Outer Mongolia] (1996) 51. —**Ic.:** Fl. SSSR, 7, Plate 35, fig. 1; Fl. Kazakhst. 4, Plate 16, fig. 2; Fl. Sin. 27, tab. 121, fig. 21; tab. 119, 120, fig. 28; Grub. Plate 55, fig. 250; Fl. Intramong. ed. 2, 2, tab. 181, fig. 1-3.

Described from Nor. Europe. Type in London (Linn.).

On hummocky alpine marshy meadows and grass patches, sedge swamps and in marshy dwarf birch thickets and willow groves in upper courses of rivers, on melkozem-rubble slopes near moraine and glaciers in upper belt of mountains.

IA. Mongolia: ***Khobd.*** (on peak of Bairimen-Daban, 2400 m, June 20; Dzusylan gorge, July 13; valley of Kharkira river, in alpine belt, July 21; near head of Kharkira river, 2400 m, July 22—1879, Pot.; south. macroslope of Sailyugem mountain range, 70 km nor.-west of Ulan-Khus settlement, basin of Khara-Dzhamat-Gol river (left tributary of Uigoryn-Gol river), July 10, 1988—Gub., Bud., Dariima (MW), ***Mong. Alt.*** (south-west. bank of Daingol lake, July 29, 1908—Sap.; Kharagatiin-Daba, alpine meadow, No. 14998, July 23; Kharagaitu-Daba pass, in upper Indertiin-Gol, alpine meadow, No. 13638, 13565, July 24—1947, Yun.; Tsastu-Bogdo-Ula, in upper courses of Dzuilin-Gol, 2950-3000 m, cushion-like structure, June 24, 1971—Grub., Ulzij. et al.; Ikhe-Ulan-Daba pass, 2945 m, nor. slope, sedge tundra, on Must somon—Uenchi somon road, Aug. 13, 1979—Grub., Muld. and Dariima; pass from Dayan-Nur basin to Irtysh basin, 15-20 km south of Dayan-Nur settlement, alpine meadow and dwarf birch thickets, 2400-2500 m, July 15, 1988—Kam., Gub. et al. (MW), *Gobi Alt.* (Ikhe-Bogdo mountain range, upper mountain belt, May 29, 1945—Yun.; plateau-like crest of Ikhe-Bogdo mountain range, 3600 m, alpine carpet, July 29, 1948—Grub.), *Alash. Gobi* (Pl. Vasc. Helanshan, l.c.).

IB. Kashgar: *South.* (30 km south-east of Isagan, Cherchen district, No. 9408, June 4, 1959—Lee et al.).

IIA. Junggar: *Tarb.* (south. slope of Saur mountain range, valley of Karagaitu river, right bank Bain-Tsagan creek valley, alpine meadow, June 23, 1957—Yun.), *Tien Shan* (Ketmen' pass, 2400-2700 m, June [19]; Bogdo mountain, 3000 m, July 25—1878; Kumbel' pass, 2700-3000 m, May 30; on Borborogusun river, 2700 m, June 15—1879, A. Reg.; Pässe zwischen Kin-yu und Kurdai, July 3, 1907—Merzb.; 36 km south-east of Nyutsyuan'tsza, No. 131, July 19, 1957—Kuan; from Danyu in south, No. 406 and 424, July 18 and 21; from Nyutsyuan'tsza 30 km south along Ulan-Usu river bank, No. 282; between Daban and Danyu, on slope, No. 2053, July 19; from Nilki to Dzinkho through Yakou, on slope, No. 4031, Sept. 1—1957, Kuan; Turfan—Bogdoshan', No. 5691, June 17, 1958—Lee and Chu).

IIIA. Qinghai: *Nanshan* (Nanshan alps, 1879—Przew.; same site, 3400-3600 m, July 24, 1879—Pot.; nor. slope of Humboldt mountain range, 3700 m, alpine meadow, June 6, 1894—Rob.), *Amdo* (Mudzhik mountain range, alpine belt, 3000-3300 m, June 18, 1880—Przew.).

IIIB. Tibet: *Chang Tang* ("33°32′ N, 88°52′ E, 5100 m—Hedin"—Ostenf. and Pauls. in Hedin, l.c.; 30 km south-east of Isagan in Cherchen district, in alpine belt, 4190 m, No. 9408, June 4, 1959—Lee et al.; "Zhitu, Gaitsze"—Fl. Xizang. l.c.), *Weitzan* (Khunchunla, in meadow facing Yangtze river, June 21—Przew.: on meadowy descents of mountains in Burkhan-Budda mountain range, nor. slope, Nomokhun gorge, 4400 m, May 24, 1900; Huang He river basin, Russkoe lake and Serg-Chyu river, 4100 m, May 27, 1901—Lad.; "Bizhu, Sosyan'"—Fl. Xizang. l.c.), *South.* ("Chzhunba, Pulan', Chzhada, Tszyantsza, Dinzhi, Tszyanda"—Fl. Xizang. l.c.).

IIIC. Pamir (Kenkol river, Tom-Kara area, beyond Billuli pass, near brook, June 14; under Ucha pass, along arid clayey descent, June 16, 1909—Divn.; King-Tau mountain range, nor. slope, 3-4 km south-east of Kosh-Bulak settlement, 3500 m, juniper-*Cobresia* dwarf birch thicket, June 10, 1959—Yun. et al.).

General distribution: Jung.-Tarb., Nor. and Cent. Tien Shan, East. Pam.; Arct., Europe, Caucasus, Mid. Asia (West. Tien Shan and Pamiro-Alay), West. Sib., East. Sib., Far East, Nor. Mong. (Fore Hubs., Hent., Hang., Mong.-Daur.), China (Altay, Nor., Nor.-West., South-West.), Himalayas, Japan, Nor. Amer.

2. **Th. baicalense** Turcz. ex Ledeb. Fl. Ross. 1, 1 (1841) 7; Turcz. in Bull. Soc. natur. Moscou, 11 (1838) 85, nom. nud.; Regel in Bull. Soc. natur. Moscou, 34 (1861) 13; Lecoyer in Bull. Soc. Bot. Belg. 24 (1885) 164; Maxim. Fl. Tangut. 1 (1889) 5; Nevsk. in Fl. SSSR, 7 (1937) 515; Kitag. Lin. Fl. Mansh. (1939) 226; Fl. Sin. 27 (1979) 542; Fl. Intramong. ed. 2, 2 (1990) 451. —**Ic.:** Fl. SSSR, 7, Plate 35, fig. 7; Fl. Sin. 27, tab. 119; tab. 120, fig. 16; tab. 132; Fl. Intramong. ed. 2, 2, l.c. tab. 182, fig. 4-6.

Described from East. Siberia (Baikal coast). Type in St.-Petersburg (LE). On borders and under gently sloping forest.

IIIA. Qinghai: *Nanshan* (nor. slopes of South. Tetung mountain range, 1400 m, Aug. 9, 1880—Przew.; Chorten-tan temple, in spruce forests, 2100 m, Sept. 6, 1901—Lad.), *Amdo* (Mudzhik river, south of Guidui town, 2700 m, June 16, 1880—Przew.).

General distribution: East. Sib., Far East, Nor. Mong. [Hent., Cis.-Hing.), China (Dunbei, Nor., Nor.-West., South-West.), Korean peninsula (Nor.), Japan.

3. **Th. isopyroides** C.A. Mey. in Ledeb. Fl. alt. 2 (1830) 346; Regel in Bull. Soc. natur. Moscou, 34 (1861) 16; Lecoyer in Bull. Soc. Bot. Belg. 24 (1885) 196; Kryl. Fl. Zap. Sib. 5 (1931) 1214; Nevsk. in Fl. SSSR, 7 (1937) 523; Fl. Kirgiz. 6 (1955) 110; Fl. Kazakhst. 4 (1961) 124; Kovalevsk. in Opred. rast. Sr. Azii [Key to Plants of Mid. Asia] 3 (1972) 226; Fl. Tadzh. 4 (1975) 145; Fl. Sin. 27 (1979) 582; Claves pl. Xinjiang, 2 (1983) 295. —**Ic.**: Fl. SSSR, 7, Plate 35, fig. 6; Fl. Kazakhst. 4, Plate 16, fig. 5; Fl. Sin. 27, tab. 146, fig. 1-2.

Described from Southern Altay. Type in St.-Petersburg (LE).

In rocky wastelands, talus, rubble slopes from plains to lower belt of mountains.

IIA. Junggar: ***Tien Shan*** (Aktyube settlement, near Kul'dzha, 900-1100 m, May 13, 1877; Sarybulak river, north-west of Kul'dzha, 1200 m, April [10-12] 1878—A. Reg.; Montes Bogdo-Ola, in der Umgebung von Urumtschi, Aug. 26-29, 1908—Merzb.), ***Jung. Gobi*** (south.: nor. Bogdo-Ula foothills, 40 km north-east of Urumchi, on road to Fukan, April 26, 1959—Yun. and Yuan'), ***Zaisan*** (Maikapchagai mountain, rocky slopes, June 6, 1914—Schischk.).

General distribution: Aralo-Casp., Fore Balkh., Jung.-Tarb., Nor. Tien Shan, East. Pam.; Mediterr. (east.), Fore Asia, Caucasus, Mid. Asia, West. Sib. (Altay), China (Altay).

4. **Th. flavum** L. Sp. pl. (1753) 546; Regel in Bull. Soc. Nat. Moscou, 34 (1861) 47; Lecoyer in Bull. Soc. Bot. Belg. 24 (1885) 208; Kryl. Fl. Zap. Sib. 5 (1931) 1218; Nevsk. in Fl. SSSR, 7 (1937) 527; Fl. Kirgiz. 6 (1955) 111; Fl. Kazakhst. 4 (1961) 126; Kovalevsk. in Opred. rast. Sr. Azii [Key to Plants of Mid. Asia] 3 (1972) 227; Fl. Sin. 27 (1979) 586; Claves pl. Xinjiang, 2 (1983) 295. —**Ic.**: Lecoyer, tab. V, fig. 7; Fl. Kazakhst. 4, Plate 16, fig. 3.

Described from Nor. Europe. Type in London (Linn.).

In tugais, floodplains, scrubs.

IIA. Junggar: ***Cis-Alt.*** (in Shara-Sume region, No. 2657, Sept. 3, 1956—Ching; "Altay mountains"—Claves pl. Xinjiang, l.c.), ***Jung. Alat.*** (in Toli district, No. 1005, Aug. 6, 1957—Kuan).

General distribution: Fore Balkh., Jung.-Tarb., Nor. Tien Shan (M. Kemin river basin); Arct. (Europ.), Europe, Caucasus, West. Sib., East. Sib., Far East.

5. **Th. foetidum** L. Sp. pl. (1753) 545; Regel in Bull. Soc. natur. Moscou, 34 (1861) 31; Lecoyer in Bull. Soc. Bot. Belg. 24 (1885) 181; Maxim. Enum. pl. Mong. 1 (1889) 7; Pamp. Fl. Carac. (1930) 113; Kryl. Fl. Zap. Sib. 5 (1931) 1214; Nevsk. in Fl. SSSR, 7 (1937) 520; Kitag. Lin. Fl. Mansh. (1939) 227; Fl. Kirgiz. 6 (1955) 109; Fl. Kazakhst. 4 (1961) 123; Ikonnik. Opred. rast. Pamira [Key to Plants of Pamir] (1963) 122; Kovalevsk. in Opred. rast. Sr. Azii [Key to Plants of Mid. Asia] 3 (1972) 226; Fl. Tadzh. 4 (1975) 145; Fl. Sin. 27 (1979) 580; Grub. Opred. rast. Mong. [Key to Plants of Mongolia] (1982) 115; Claves pl. Xinjiang, 2 (1983) 294; Fl. Xizang. 2 (1985) 74; Pl. Vasc. Helanshan (1986) 109; Y.Z. Zhao in Acta Sci. Nat. Univ. Intramong. 18, 2 (1987) 28; Fl. Intramong. ed. 2, 2 (1990) 457; Gub. Konsp. Fl. Vneshn. Mong.

[Conspectus of Flora of Outer Mongolia] (1996) 51. —**Ic.**: Lecoyer, tab. IV, fig. 7; Fl. SSSR, 7, Plate 35, fig. 2; Fl. Kazakhst. 4, Plate 16, fig. 4; Fl. Intramong. 2, tab. 138, fig. 4-6; Fl. Sin. 27, tab. 119, 120, fig. 26; tab. 147, fig. 4-5; Fl. Xizang. 2, tab. 21, fig. 10-11; Fl. Intramong. ed. 2, 2, tab. 184, fig. 3-4.

Described from South. Europe. Type in London (Linn.).

On rubbly, rocky and stony steppe slopes, in arid sparse larch forests and their borders, *Cobresia* meadows, in upper belt of mountains.

IA. Mongolia: ***Khobd.*** (Khuren-Khairkhan-Ula town, 20 km west of Ulegei, July 13, 1988—Gub., Kam. et al.), ***Mong. Alt., Cent. Khalkha, East. Mong.*** (summit of Tsogt-Delger mountains, June 26, 1956—Dashnyam; Shiliin-Bogdo-Ula, on rubble slopes, Aug. 11, 1970—Grub., Ulzij. et al.; same site, July 8, 1971—Karam. et al.; same site, July 12, 1985—Gub.), ***Depr. Lakes, Gobi Alt., Alash. Gobi*** (Alashan mountain range, July 10, 1873—Przew.).

IB. Kashgar: *Nor.* (Karateke mountain range, on rocks in gorges, 1500-2400 m, June 7, 1889—Rob.; Uchturfan, Airi gorge in valley of brook, June 4, 1908—Divn.), ***East.*** (in Khalangao village in Khomote region, 2530 m, No. 7697, Aug. 11, 1958—Lee and Chu).

IC. Qaidam: ***Mount.*** (Dulankhit temple, 3400 m, Aug. 9, 1901—Lad.).

IIA. Junggar: ***Cis-Alt.*** (Qinhe, Chzhunkhaitsza, 2400 m, No. 1452, Aug. 7, 1956 —Ching), ***Tarb.*** (Saur mountain range, valley of Karagaitu river, Bain-Tsagan creek valley, subalpine meadow along valley on south. slope, June 23, 1957—Yun. et al.), ***Jung. Alat.*** (Toli-Otu, No. 943, Aug. 4; on slope of mountains in Toli district, No. 2759, Aug. 9 —1957, Kuan), ***Tien Shan*** (Talkibash mountains, 2100-2700 m, July 20; Sairam lake, July [20-24] 1877; Irenkhabirga mountain range, lower course of Taldy river, 1200 m, May 15; Naryngol, Tsagan-Usu, Dzhin village, 1800-2400 m, June 10; Irenkhabirga mountain range, Mengute, Aug. 2—1879, A. Reg.; Narin-Kirgut-Nor lake, on rocks in poplar forest, Sept. 21, 1893—Rob.; Oestliches Khaptu-Su Tal, Sept. 1-5, 1907; Südliches Kiukönik Tal, Larger am See Karakul and auf der Halbinsel am Gehänge, June 15, 1908—Merzb.; south-east of Urumchi, No. 177, July 14, 1956—Ching; on Danyu river, No. 2169, July 22; mountains east of Syata town (=Shaty), No. 1316, Aug. 11; Tekes district, Aksu region, Syate vicinity, No. 3362, Aug. 12—1957, Kuan; on Urumchi to Karashar road, valley terrace, 1900 m, No. 5925, July 21; Lotogou station, in Khetszin district, in Bagrashkul' region, on arid slope, No. 6285, Aug. 1; 2 km south-west of Bianbulak in Khotun Sumbul, No. 2650 and No. 6429, Aug. 10—1958, Lee and Chu; intermontane trough of Bol. Yuldus, residual mountain on right bank of Khaidyk river 3-5 km south-west of Bain-Bulak settlement, Aug. 10, 1958—Yun. and Yuan'), ***Jung. Gobi*** (east.: nor. slope of Baitag-Bogdo mountain range, 2400 m, steppe, Aug. 8, 1977 —Volk. and Rachk.; same site, forest and alpine belts of Budun-Khargaityn-Gol gorge, 15 km east of Baitag-Bogdo post, Aug. 1, 1988—Gub., Kam. et al. (MW).

IIIA. Qinghai: ***Nanshan*** (Kukunor lake, Kuisu island, nor. slope, 4000 m, Sept. 2, 1908—Czet.).

General distribution: Jung.-Tarb., Nor. and Cent. Tien Shan, East. Pam.; Europe, Mediterr., Balk.-Asia Minor, Fore Asia, Caucasus, Mid. Asia (Pamiro-Alay), West. Sib., East. Sib., Nor. Mong. (Fore Hubs., Hent., Hang., Mong.-Daur.), China (Dunbei, Nor., Nor.-West., South-West.), Himalayas, Japan.

6. **Th. minus** L. Sp. pl. (1753) 546; Regel in Bull. Soc. natur. Moscou, 34 (1861) 18; Lecoyer in Bull. Soc. Bot. Belg. 24 (1885) 199; Maxim. Fl. Tangut. 1 (1889) 5; id. Enum. pl. Mong. 1 (1889) 7; Pamp. Fl. Carac. (1930) 113; Kryl. Fl. Zap. Sib. 5 (1931) 1215; Nevsk. in Fl. SSSR, 7 (1937) 524; Fl. Kirgiz. 6 (1955) 110; Kovalevsk. in Opred. rast. Sr. Azii [Key to Plants of Mid. Asia]

3 (1972) 227; Fl. Sin. 27 (1979) 116; Claves pl. Xinjiang, 2 (1983) 295; Fl. desert Sin. 1 (1985) 503; Fl. Xizang. 2 (1985) 74; Y.Z. Zhao in Acta Sci. Nat. Univ. Intramong. 18, 2 (1987) 285; Fl. Intramong. ed. 2, 2 (1990) 460; Gub. Konsp. Fl. Vneshn. Mong. [Conspectus of Flora of Outer Mongolia] (1996) 51. —*Th. minus* var. *hypoleucum* (Sieb. et Zucc.) Miq. in Ann. Mus. Bot. Lugd.-Bat. 3 (1867) 3; Fl. Intramong. 2 (1978) 269; Fl. desert Sin. 1 (1985) 503; Pl. Vasc. Helanshan (1986) 109; Y.Z. Zhao, l.c. (1987) 285. —*Th. minus* var. *stipellatum* (C.A. Mey.) Tamura in Acta Phytotax. Geobot. (Kyoto) 15, 6 (1954) 87; Fl. Intramong. 2 (1978) 271; Fl. Sin. 27 (1979) 583. —*Th. minus* var. *minus*; Claves pl. Xinjiang, 2 (1983) 295. —**Ic.:** Lecoyer, tab. V, fig. 2-4; Fl. Intramong. 2, tab. 140, fig. 6, tab. 141, fig. 1-4 (var. *hypoleucum*), 5-6 (var. *stipellatum*); Fl. desert Sin. 1, tab. 184, fig. 12-14; Fl. Intramong. ed. 2, 2, tab. 186, fig. 1-2.

Described from Nor. Europe. Type in London (Linn.).

In larch forests, birch stands, forest meadows and borders, scrubs, meadow and rock slopes, solonchak meadows along river valleys, from plains to middle mountain belt.

IA. Mongolia: ***Khobd.*** (east of Ulan-Daban, June 12 and 22, 1879—Pot.), ***Mong. Alt., Cent. Khalkha*** (2 versts [1 verst = 1.067 km] from Bichikte river, June 1, 1893—Klem.; on Bichikte foothills, Aug. 31, 1925—Gus.; 5 km south of Tumen settlement, July 24, 1985—E. Alekseeva (MW), ***East. Mong.*** (Ourato, July 1866—David; Inshan', June 26, 1871—Przew.; vicinity of Manchuria station, July 5, 1922—B. Skvortzov; Delgermunkh, north of Choibolsan somon, east. base of mountains, Oct. 7, 1959—Dashnyam; Moluok-Els sand 5 km north of Narin settlement, July 3, 1978—Ogureeva (MW); Shiliin-Bogdo-Ula town, slope exposed northward, 1550 m, July 12, 1985—Gub. (MW), ***Depr. Lakes*** (Shargin-Gobi, between Ikhe-Gol and Tszak-Obo, Shargin-Gol river bank, Sept. 6, 1930 —Pob.; in Kobdo river valley, at turning to Kobdo-Ulangom road, Aug. 23, 1944—Yun.; basin of Tsagan-Nur lake, Aug. 9, 1982—Gub. (MW), ***Gobi Alt.*** (east fringe of Noin-Bogdo mountain range, near Debsyk khuduk collective, 2000-2200 m, in gorge, on arid river bed, Oct. 7, 1925—Glag.; Dzun-Saikhan town, Aug. 20, 1931—Ik.-Gal.; Dundu mountain range and Dzun-Saikhan, mountain slopes and gorge, July-Aug. 1933 —Simukova), ***Alash. Gobi*** (midportion of Alashan mountain range, in forest on west. and east. slopes, June 30 and July 3, 1873—Przew.; Alashan mountain range, Yamata gorge, west. slope, mid. and lower belts, May 4 and 5, 1908—Czet.).

IB. Kashgar: ***Nor.*** (between Aksu and Kucha at Yakka-Aryk village, in Muzart-Dar'ya floodplain, Aug. 13, 1929—Pop.; Temyn'chuan' village, in Karashar region, No. 6891, July 29, 1958—Lee and Chu), ***South.*** (Keriya river, along slopes, June 20, 1885 —Przew.).

IIA. Junggar: ***Cis-Alt.*** (Qinhe, No. 1240, Aug. 2; Koktogoi, No. 1958, Aug. 17; in Shara-Sume region, No. 2409, Aug. 26; Burchum, No. 3001, Sept. 11—1956, Ching; 20 km north-west of Shara-Sume, Juen 7, 1959—Yun. and Yuan'), ***Tarb.*** (east. valley of Chern. Irtysh, Karasu river, 13 versts from Zaisan post, July 13, 1876—Pot.; Saur mountain range, south. slope, valley of Karagaitu river, creek valley in Bain-Tsagan, June 23, 1957—Yun. et al.; Dachen, No. 1615 and 1627, Aug. 13, 1957—Kuan), ***Jung. Alat.*** (in Toli district, Albakzin mountain, No. 2668, Aug. 6, 1957—Kuan), ***Tien Shan, Jung. Gobi*** (nor.-west.: Savan district, Shamyn'tsza village, No. 956, June 17, 1957—Kuan; south.: in Kuitun' region No. 252 and 1115, June 29, 1957—Kuan; 3-4 km nor. of St. Kuitun' settlement, June 30, 1957—Yun. et al.; east.: floodplain of Bodonchi river, 2-3 km south of Bodonchin-Khure, July 18, 1947—Yun.; Bulugun somon, willow thicket and reeds at emergence of Damchzhigini-Usu spring, Sept. 16, 1948—Grub.; Bulgan-Gola basin, Ded-Nariin-sala 1 km from estuary, 1500-1600 m, Aug. 17, 1979—Grub. et al.; 35 km west of Bulgan settlement, on bank of Bulgan river, Aug. 14, 1982; south-west. fringe of Bulgan settlement, dampish sinkholes in steppe, July 29, 1988—Gub. (MW), ***Dzhark.*** (right bank of Ili river, west of Kul'dzha, May 4 and 14; Suidun, June 27 and July 4—1877, A. Reg.; same site, July 1878—Aliyakhun; Kutukchi, west of Kul'dzha, June 6, 1878—Fet.).

IIIA. Qinghai: ***Nanshan*** (Chzhan'e, 70 km south-east of Matisy mountain range, Aug. 12, 1958—Petr.), ***Amdo*** (upper course of Huang He river, 2100 m, in scrubs, near spring [Khagomi area], May 28 and June 22, 1880—Przew.).

General distribution: Jung.-Tarb., Nor. and Cent. Tien Shan; Arct., Europe, Balk.-Asia Minor, Fore Asia, Caucasus, Mid. Asia (Mount.), West. Sib., East. Sib. Far East, Nor. Mong., China (Nor., South-West.), Japan, Afr., Nor. Amer.

7. **Th. petaloideum** L. Sp. pl. ed. 2 (1762) 771; Regel in Bull. Soc. natur. Moscou, 34 (1861) 15; Franch. Pl. David. 1 (1884) 17; Lecoyer in Bull. Soc. Bot. Belg. 24 (1885) 165; Forbes and Hemsl. Index Fl. Sin. 1 (1886) 9; Maxim. Fl. Tangut. 1 (1889) 5; id. Enum. pl. Mong. 1 (1889) 5; Gr.-Grzh. Opis. put. Zap. Kit. [Report of Expedition in West. China] 3 (1907) 475; Kryl. Fl. Zap. Sib. 5 (1931) 1213; Nevsk. in Fl. SSSR, 7 (1937) 518; Kitag. Lin. Fl. Mansh. (1939) 227; Fl. Kirgiz. 6 (1955) 109; Fl. Kazakhst. 4 (1961) 122; Fl. Sin. 27 (1979) 544; Grub. Opred. rast. Mong. [Key to Plants of Mongolia] (1982) 115; Claves pl. Xinjiang, 2 (1983) 296; Fl. desert Sin. 1 (1985) 499; Fl. Intramong. ed. 2, 2 (1990) 453; Gub. Konsp. Fl. Vneshn. Mong. [Conspectus of Flora of Outer Mongolia] (1996) 51. —*Th. petaloideum* var. *supradecompositum* (Nakai) Kitag. Lin. Fl. Mansh. (1939) 227; Fl. Intramong. 2 (1978) 264; Fl. Sin. 27 (1979) 544; Fl. desert Sin. 1 (1985) 499; Fl. Intramong. ed. 2, l.c. 455. —**Ic.:** Fl. SSSR, 7, Plate 35, fig. 5; Fl. Kazakhst. 4, Plate 16, fig. 1; Fl. Sin. 27, tab. 121, fig. 16; Fl. Intramong. 2, tab. 137, fig. 1-6 (7—var. *suprodecompositum*); Grub. Plate 55, fig. 253; Fl. desert Sin. 1, tab. 183, fig. 6-10 (var. *supradecompositum*); Fl. Intramong. ed. 2, 2, tab. 183, fig. 1-6.

Described from Siberia. Type in London (Linn.).

In valley and coastal, steppified and steppe meadows, steppe rubble and rocky slopes, arid forest borders, larch forests, among coastal shrubs, riverine pebble beds.

IA. Mongolia: ***Khobd.*** (west. slope of Ulan-Daban, June 5, 1895—Klem.), ***Mong. Alt.*** (not far from Mogoi river, May 30 and June 3, 1893—Klem.), ***East. Mong.*** (la région occidentale de la chaine de l'Ourato, July 1866—David; Batukhan town—1870, Lom.; Inshan' mountains, July 1, 1871—Przew.; Kulun-buir-norsk plain, meadowy steppe between Aboer river and Borol'dzhitu, June 26; same site, Bilyutai town, July 8; Dzhalatu area, July 8—1899, Pot. and Sold.; vicinity of Manchuria station, 1915—E. Nechaeva; Choibolsan somon, Gurvan Bayan, east of Engershanda, June 4, 1958—Dashnyam; Shilin-Khoto town, steppe, 1959—Ivan.; Gangyn-Tsayun-Obo town, 23 km west of Dariganga, rock outcrops on south. slope, July 7; 43 km west of Dariganga, steppe on basalts, July 8—1971, Dashnyam, Karam. and Safronova), ***Gobi Alt.*** (Bago-Bogdo, steep canyon walls of lower mountain slopes and canyon bottoms, 1800-2100 m, 1925—Chaney), ***Ordos*** (on Ulan-Morin river, Aug. 23, 1884—Pot.).

IIIA. Qinghai: *Nanshan* (Kukunor lake, July 6; alpine belt between Nanshan and Donkir, on Rakogol river, 3000-3400 m, July 10—1880, Przew.; Xining mountains, Myndan'sha river, June 1, 1890—Gr.-Grzh.; Xining alps, 3000-3400 m, June 1905 —Gr.-Grzh."—Gr.-Grzh. l.c.; Kukunor lake, vicinity of Donkir, Aug. 11; same site, Kuisu island, 1200 m, Sept. 2—1908; Sanshantan village, Nor. Tetung mountain range, lower belt, July 22; Kukunor lake, Aug. 11—1909, Czet.; 33 km west of Xining, rocky slopes, Aug. 5, 1959—Petr.), ***Amdo*** (upper Huang He river, 2100 m, on pebble beds [Khagoma area], May 28; [Mudzhik river] June 1—1880, Przew.).

General distribution: Fore Balkh., Nor. (Ketmen' mountain range) and Cent. Tien Shan, West. Sib., East. Sib., Far East, Nor. Mong., China (Altay, Dunbei, Nor., Nor.-West.), Korean peninsula.

8. **Th. przewalskii** Maxim. in Bull. Ac. Sci. St. Pétersb. 23 (1877) 305; Lecoyer in Bull. Soc. Bot. Belg. 24 (1885) 157; Forbes and Hemsl. Index Fl. Sin. 1 (1886) 9; Maxim. Fl. Tangut. 1 (1889) 4; Fl. Sin. 27 (1979) 548; Fl. Intramong. ed. 2, 2 (1990) 453. —**Ic.:** Maxim. 1889, tab. 2, fig. 1-6; Fl. Sin. 27, tab. 135, fig. 1-4; Fl. Intramong. 2, tab. 136; Fl. Intramong. ed. 2, 2, tab. 182, fig. 1-3.

Described from Qinghai (Nanshan). Type in St.-Petersburg (LE).

In gently sloping forest, forest meadows and borders, meadowy slopes in forest and alpine belt of mountains.

IIIA. Qinghai: *Nanshan* (Regio silvestris declivit. borealis jugi a fl. Tetung meridium versus, ad silvarum margines [nor. slope of South. Tetung mountain range, along forest margin], No. 348, July 27/Aug. 8, 1872—Przew., holotypus !; mountains south of Tetung river, midbelt of forest, 2600 m, Aug. 3—1880, Przew.), ***Amdo*** (on Mudzhik river, June 28, 1880—Przew.).

General distribution: China (Nor., Nor.-West., South-West.).

9. **Th. rutaefolium** Hook. f. et Thoms. Fl. Ind. 1 (1855) 14; Lecoyer in Bull. Soc. Bot. Belg. 24 (1885) 195; Maxim. Fl. Tangut. 1 (1889) 5; Pamp. Fl. Carac. (1930) 113; Fl. Sin. 27 (1979) 559; Fl. Xizang. 2 (1985) 69. —**Ic.:** Lecoyer, tab. IV, fig. 14; Fl. Sin. 27, tab. 119, 120, fig. 20, tab. 140, fig. 1-3; Fl. Xizang. 2, tab. 21, fig. 4-6.

Described from West. Tibet and Himalayas (Sikkim). Type in Kew (K). Map 2.

In alpine and high-mountain meadows, mountain steppes.

IIIA. Qinghai: *Nanshan* (Humboldt mountain range, nor. slope, Ulan-Daban area, alpine meadow, 3800 m, June 24, 1894—Rob.; Mon'yuan', valley of Tatungkhe river, 2800 m, nor.-east. slope of mound, 1958; head of Ganshig river, tributary of Peishikhe, Aug. 18, 1958—Dolgushin).

IIIB. Tibet: ***Chang Tang*** ("Zhitu"—Fl. Xizang. l.c.), ***Weitzan*** (Amnen-kor mountain range, south. and nor. slopes, 4100 m, June 5 and 8; Dzhagyn-gol river, 4400 m, July 3 —1900; basin of Yantszytszyan (Goluboi), Makhmukh-Chyu brook, 4100 m, May 21, 1901—Lad.), ***South.*** ("Nan'mulin, Lhasa, Shigadze"—Fl. Xizang. l.c.).

General distribution: Himalayas (west., east.).

10. **Th. simplex** L. Mant. 1 (1767) 78; Regel in Bull. Soc. natur. Moscou, 34 (1861) 38; Franch. Pl. David. 1 (1884) 15; Lecoyer in Bull. Soc. Bot. Belg.

24 (1885) 204; Forbes and Hemsl. Index Fl. Sin. 1 (1886) 9; Maxim. Fl. Tangut. 1 (1889) 5; id. Enum. pl. Mong. 1 (1889) 8; Kryl. Fl. Zap. Sib. 5 (1931) 1217; Nevsk. in Fl. SSSR, 7 (1937) 526; Kitag. Lin. Fl. Mansh. (1939) 228; Fl. Kirgiz. 6 (1955) 110; Fl. Kazakhst. 4 (1961) 126; Kovalevsk. in Opred. rast. Sr. Azii [Key to Plants of Mid. Asia] 3 (1972) 227; Fl. Intramong. 2 (1978) 266; Fl. Sin. 27 (1979) 584; Grub. Opred. rast. Mong. [Key to Plants of Mongolia] (1982) 116; Claves pl. Xinjiang, 2 (1983) 295; Fl. desert Sin. 1 (1985) 501; Fl. Intramong. ed. 2, 2 (490) 458; Gub. Konsp. Fl. Vneshn. Mong. [Conspectus of Flora of Outer Mongolia] (1996) 51. —*Th. simplex* var. *brevipes* Hara in J. Fac. Univ. Tokyo, sect. 3, 4 (1952) 56; Fl. desert Sin. l.c. 502; Pl. Vasc. Helanshan (1986) 109; Fl. Intramong. ed. 2, l.c. 460. —**Ic.**: Lecoyer, tab. V, fig. 5; Fl. Kazakhst. 4, Plate 16, fig. 7; Fl. Intramong. 2, tab. 140, fig. 1-5, 6 (var. *brevipes*), Grub. Opred. rast. Mong. [Key to Plants of Mongolia], Plate 55, fig. 254; Fl. desert Sin. 1, tab. 184, fig. 6-8, 9-11 (var. *brevipes*); Fl. Intramong. ed. 2, 2, tab. 185, fig. 1, 2-4 (var. *brevipes*).

Described from Europe. Type in London (Linn.).

In coastal meadows and scrubs, bottomland deciduous forests, larch forests, birch stands, in forest meadows and borders, meadowy slopes.

IA. Mongolia: ***Khobd.*** (floodplain of Bukhu-Muren, 4-5 km north of Bukhu-Muren somon, July 31, 1945—Yun.), ***Mong. Alt.*** (Khara-Adzarga mountain range, valley of Shutyn-Gol river, deep rocky gorge, Aug. 27, 1930—Pob.), ***Cent. Khalkha*** (meadow of Uber-Dzhargalante river, Aug. 31; subalpine zone on left bank of Uber-Dzhargalante river, between Botoga and Agit mountains, Sept. 2; head of Kharukha river (Ara-Dzhargalante river), vicinity of Ulan-Khoda mountains, clayey slopes, Sept. 18—1925, Krasch. and Zam.; vicinity of Ikhe-tukhum-Nor lake, Bomkhona Dzhargalanta gorge [without precise date], June, 1926—Zam.), ***East. Mong.*** (Kuzupchi sand, facing Hekou town, Aug. 8, 1884—Pot.; Kulunbuir-norsk plain, Ulan-Dzhilgu area, July 3, 1899 —Pot. and Sold.; Khalkhin-Gol river, Symbur area, steppe, Sept. 4, 1928—Tug.; Shilin-Khoto town, steppe, 1959—Ivan.; Numutyn-Daba pass, 25 km south of Erdenotsagan somon, July 27, 1962—Dashnyam and Yun.; Shiliin-Bogdo-Ula, 1500-1600 m, July 10 and 12, 1985—Gub. (MW); Khalkh-Gol river valley, above Sumber settlement, 700-750 m, June 25, 1987—Kam., Gub., Ganbold et al.), ***Depr. Lakes*** (vicinity of Ulangom town, July 3 and Sept. 6, 1879—Pot.), ***Gobi-Alt.*** (Dzun-Saikhan town, narrow creek valley, along bank of brook, Aug. 20, 1931—Ik.-Gal.; Dzun-Saikhan mountain range, Elyn-Ama gorge, July 22, 1970—Grub. et al.), ***Alash. Gobi*** (Alashan mountain range, June 20-30, 1873—Przew.), ***Ordos*** (Ordos, valley of Huang He river, Aug. 10, 1871—Przew.).

IIA. Junggar: ***Cis-Alt.*** (Qinhe, No. 1657, Aug. 1; in Koktogoi region, No. 2213, Aug. 18 —1956, Ching), ***Tarb.*** (valley of Chern. Irtysh river, Mukur area, Aug. 11, 1876—Pot.), ***Tien Shan*** (near Khami, July 25 and Sept. 12, 1875—Pias.; south. bank of Sairam lake, 2100 m, July 20, 1877; along Agiaz river, 1800-2400 m, July [13], 1878—A. Reg.; in Shichan region, No. 1300, July 8; Danyu, No. 1413, July 16; south of Nyutsyuan'tsza, No. 545, July 16; Datszymyao, No. 445, July 22; Kizilzangi (south of Shichan), No. 804, July 23 (Shen-Tyan); 6 km south-west of Shichan in Savan district, No. 2282, July 26; from Chzhaosu in east for 10 km on road to Tekes and for 4 km in south, No. 799, Aug. 11; 20 km north of Ulastai, No. 3864, Aug. 28—1957, Kuan; Ketmen', east. part, Sarbushin pass saddle, on road to Kzyl-Kure from Ili, July 23, 1957—Yun. et al.; Ardyn-Daban pass, separating Mal. Yuldus trough from upper Kunges (Ili), nor.-west. slope to Kunges, subalpine meadow, Aug. 3, 1958—Yun. and Yuan'; on Narat mountain in Kunges, 2400 m, No. 9104, Aug. 7, 1958—Lee and Chu), ***Jung. Gobi*** (west.: Kuitun',

No. 253, June 29 (hybrid with *Th. minus*); 2 km north of Kuitun, No. 373, July 6—1957, Kuan; east.: east. macroslope of Baitag-Bogdo mountain range, in Budun-Kharkaityn-Gol floodplain, 1900 m, July 28, 1979—Gub. (MW), ***Zaisan*** (along Chern. Irtysh river, near Dyurbul'dzhin crossing, Aug. 24, 1876—Pot.), ***Dzhark.*** (Kul'dzha, 1876—Golike).

IIIA. Qinghai: ***Amdo*** (upper Huang He river, near Guidui town, 2100 m, near irrigation canals, July 1, 1880—Przew.).

General distribution: Aralo-Casp., Fore Balkh., Jung.-Tarb., Nor. and Cent. Tien Shan; Europe, Balk.-Asia Minor, Caucasus, Mid. Asia, West. Sib., East. Sib., Far East, Nor. Mong., China (Dunbei, Nor.), Korean peninsula, Japan.

11. **Th. squamiferum** Lecoy. in Bull. Soc. Bot. Belg. 16 (1880) 229 and ibid. 24 (1885) 184; Fl. Sin. 27 (1979) 588; Fl. Xizang. 2 (1985) 74. —**Ic.:** Lecoyer, 1885, tab. 4, fig. 8; Fl. Sin. 27, tab. 121, fig. 22; tab. 147, fig. 6-9.

Described from South. Tibet. Type in Brussels (BR).

On alpine slopes, about 5000 m.

IIIA. Qinghai ("Qinghai"—Fl. Xizang. l.c.).

IIIB. Tibet: ***South.*** ("Nan'mulin, Chzhada, Dyantszi, Chzhunba"—Fl. Xizang. l.c.).

General distribution: China (South-West.), Himalayas (west., east.).

12. **Th. squarrosum** Steph. ex Willd. Sp. pl. 2 (1799) 1299; Lecoyer in Bull. Soc. Bot. Belg. 24 (1885) 198; Forbes and Hemsl. Index Fl. Sin. 1 (1886) 3; Maxim. Enum. pl. Mong. 1 (1889) 8; Nevsk. in Fl. SSSR, 7 (1937) 523; Kitag. Lin. Fl. Mansh. (1939) 228; Fl. Sin. 27 (1979) 586; Grub. Opred. rast. Mong. [Key to Plants of Mongolia] (1982) 115; Fl. desert Sin. 1 (1985) 501; Fl. Intramong. ed. 2, 2 (1990) 457; Gub. Konsp. Fl. Vneshn. Mong. [Conspectus of Flora of Outer Mongolia] (1996) 51. —*Th. trigynum* Fisch. ex DC. Prodr. 1 (1824) 14; Franch. Pl. David. 1 (1884) 16. —**Ic.:** Lecoyer, tab. V, fig. 1; Fl. Intramong. 2, tab. 139; Fl. Sin. 27, tab. 148; Grub. Plate 55, fig. 249; Fl. desert Sin. 1, tab. 184, fig. 1-5; Fl. Intramong. ed. 2, tab. 184, fig. 1-2.

Described from Siberia. Type in Berlin (B).

On steppe rubble and rock slopes, in sand steppes, on rocks.

IA. Mongolia: ***Cent. Khalkha, East. Mong., East. Gobi*** (Alkha-Khoshuin-Gobi, Aug. 15, 1927—Zam.), ***Alash. Gobi*** (on way to Erdeni-Gol area from Bain-Barat, June 12, 1909—Czet.).

IIA. Junggar: ***Jung. Gobi*** (east.: 130 km south-east of Bulgan somon, Ikh-Khavtag-Ula mountain range, nor. macroslope, 2500 m, Aug. 12, 1977—Volk. and Rachk.).

General distribution: East. Sib., Far East, Nor. Mong., China (Dunbei, Nor.).

13. **Th. tenue** Franch. in Nouv. Arch. Mus. Hist. natur. (Paris), sér. 2, 5 (1883) 168; Franch. Pl. David. 1 (1884) 16; Lecoyer in Bull. Soc. Bot. Belg. 24 (1885) 175; Fl. Intramong. 2 (1978) 264; Fl. Sin. 27 (1979) 575; Fl. desert Sin. 1 (1985) 500; Pl. Vasc. Helanshan (1986) 109; Fl. Intramong. ed. 2, 2 (1990) 455. —**Ic.:** Franch. 1884, tab. 7; Fl. Sin. 27, tab. 140, fig. 9-13; Fl. desert Sin. 1, tab. 183, fig. 11-14; Fl. Intramong. ed. 2, 2, tab. 184, fig. 5-6.

Described from Nor. China. Type in Paris (P).

On steppified and semidesert rubble slopes of mountains, among xerophilous vegetation.

IA. Mongolia: ***East. Mong.*** ("Datsinshan' mountain range"—Fl. Intramong. l.c.), ***Alash. Gobi.*** (Pl. Vasc. Helanshan, l.c.).

General distribution: China (Nor.).

14. **Th. uncatum** Maxim. in Acta Horti Petrop. 11 (1890) 14; Fl. Sin. 27 (1979) 536; Fl. Xizang. 2 (1985) 66. —**Ic.:** Fl. Sin. 27, tab. 121, fig. 8; tab. 119, 120, fig. 18; tab. 130, fig. 1-4; Fl. Xizang. 2, tab. 19, fig. 1-3.

Described from Nor.-West. China. Type in St.-Petersburg (LE).

On meadow slopes, among shrubs in intermontane basins.

IIIA. Qinghai ("Qinghai"—Fl. Xizang. l.c.).

IIIB. Tibet: ***Weitzan*** ("Sosyan'"—Fl. Xizang. l.c.), ***South.*** ("Dzyanda"—Fl. Xizang. l.c.).

General distribution: China (Nor.-West.).

Family 48. BERBERIDACEAE Juss.

1. Shrubs; leaves entire, often serratidentate along margin; stipules in the form of spines, simple or trifid, sometimes multifid. Fruit, berry..1. **Berberis** L.
+ Perennial grasses with tubers or rhizome..2.
2. Perennial grasses with tubers; leaves ternate or pinnatisected; stipules lacking. Fruit, capsule..3.
+ Perennial grasses with long decumbent rhizome, long-petiolate, palmatilobate leaves. Fruit, berrylike. Each seed in proliferating fleshy placenta..........................4. **Podophyllum** L. (*P. emodi* Wall.).
3. Seeds with membranous aril; capsule not more than 8 mm in diam., not vesicular-intumescent, dehisces into regular lobes such that seed is visible.. ....2. **Gymnospermium** Spach (*G. altaicum* (Pall.) Spach).
+ Seeds without membranous aril; capsule 15-30 mm in diam., vesicular-intumescent, indehiscent.. .... 3. **Leontice** L. (*L. incerta* Pall.).

1. Berberis L.

Sp. pl. (1753) 330; id. Gen. pl. ed. 5 (1754) 153

1. Flowers single or 2-3 each in leaf axils..2.
+ Flowers in many-flowered racemes, paniculate inflorescences or gathered in bunches..7.
2. Style persistent on fruits..3.
+ Style lacking or barely perceptible on fruits..4.

3. Leaves broad, with 15-40 spines along margin..4. **B. circumserrata** Schneid. var. **occidentalis** Ahrendt.
+ Leaves narrow, with 1-12 spines along margin or entire..6. **B. diaphana** Maxim.
4. Flowers single..5.
+ Flowers 2-3 each in leaf axils..6.
5. Flowers 12-14 mm in diam.............1. **B. angulosa** Wall. ex Hook. f.
+ Flowers 6-7 mm in diam..19. **B. sibirica** Pall.
6. Inflorescence 4-9 mm long; pedicels 3-5 mm long; fruit 4-6 mm long..20. **B. ulicina** Hook. f. et Thoms.
+ Inflorescence 10-15 mm long; pedicels 2-3 mm long; fruit 8 mm long..12. **B. kashgarica** Rupr.
7. Leaves entire..8.
+ Leaves with spines..17.
8. Style persistent on fruits..9.
+ Style lacking on fruits..12.
9. Flowers gathered in bunches..10.
+ Flowers gathered in racemes..3. **B. caroli** Schneid.
10. Leaves lustrous underneath..............................18. **B. sabulicola** Ying.
+ Leaves dull underneath..11.
11. Leaves with distinct nervation.................8. **B. hemsleyana** Ahrendt.
+ Leaves glabrous, nervation not manifest......14. **B. obovatifolia** Ying.
12. Berries black..9. **B. heteropoda** Schrenk.
+ Berries red..13.
13. Ovary single..15. **B. poiretii** Schneid.
+ Ovaries 2-3 or more..14.
14. Leaves narrow, 4-8 (12) mm broad.................22. **B. vernae** Schneid.
+ Leaves very broad, 13-25 (30) mm broad..15.
15. Inflorescence 4-8-flowered.............................2. **B. boschanii** Schneid.
+ Inflorescence 15-30-flowered..16.
16. Berries globose, fruit-stems 5-6 mm long......13. **B. nummularia** Bge.
+ Berries oblong, fruit-stems 7-15 mm long......10. **B. integerrima** Bge.
17. Flowers gathered in racemes..18.
+ Flowers gathered in bunches..20.
18. Fruit-stems 1 mm long....................................17. **B. purdomii** Schneid.
+ Fruit-stems 4-8 mm long..19.
19. Leaves with gray bloom underneath; young stems dark red; fruits 8 (8.5) mm long, 5 mm in diam.................11. **B. kansuensis** Schneid.
+ Leaves without bloom underneath; young stems pale yellow; fruits 7 mm long, 5 mm in diam...........................5. **B. dasystachya** Maxim.
20. Leaves ever-green..21.
+ Leaves caducous..7. **B. dubia** Schneid.

21. Flowers 9-10 mm in diam..
....21. **B. umbellata** Wall. var. **brianii** Ahrendt.
\+ Flowers 14 mm in diam..................................16. **B. pulangensis** Ying.

1. **B. angulosa** Wall. ex Hook. f. Fl. Brit. Ind. 1 (1872) 111; Fl. Xizang. 2 (1985) 146; Ahrendt in J. Linn. Soc. (London) Bot. 57 (1961) 113. —**Ic.**: Fl. Xizang. 2, tab. 44, fig. 1-7.

Described from Himalayas (Nepal). Type in London (K).

On rocky slopes, 3400-3500 m alt.

IIIB. Tibet: ***Chang Tang*** ("Dansyun"—Fl. Xizang. l.c.).

General distribution: Himalayas (east.).

2. **B. boschanii** Schneid. in Sarg. Pl. Wils. 1 (1913) 367; Ahrendt in J. Linn. Soc. (London) Bot. 57 (1961) 157.

Described from South-West. China. Type in London (K).

On rocky slopes, among rocks, in coniferous forests.

IIIA. Qinghai: ***Nanshan*** (Kokonor Region: Roko gorges, No. 13261, Sept.; Richthofen range and adjacent region in Hung-Shui-Kou on rocky slopes, 3200-3500 m, No. 13300, Oct., 1925—Rock), ***Amdo*** (Radja and Yellow River gorges, rocky slopes back of Radja, 3600 m, No. 13920, May 20; along streambed in outskirts of spruce forest in Dachso canyon, north of Radja, 3500 m, No. 14073, June 2, 1926—Rock).

General distribution: China (South-West.).

3. **B. caroli** Schneid. in Bull. Herb. Boiss. sér. 2, 5 (1905) 459 and sér 2, 8 (1908) 202; id. in Österr. Bot. Zeitschr. 66 (1916) 218; Ahrendt in J. Linn. Soc. (London) Bot. 57 (1961) 176; Fl. Intramong. 2 (1978) 275; Fl. desert Sin. 1 (1985) 516. —*B. integerrima* var. *stenophylla* Maxim. Fl. Tangut. 1 (1889) 29.

Described from Mongolia (Ordos). Type in St.-Petersburg (LE). Plate VIII, fig. 1.

On arid rocky slopes and riverine valleys.

IA. Mongolia: ***East. Mong.*** ("Datsinshan', Ulashan'"—Fl. Intramong. l.c.); ***Ordos*** ([valley of Ulan-Morin river, Aug. 9] 1884—Pot., holotypus !).

IIIA. Qinghai: ***Nanshan*** (On Tetung-gol river, Aug. 10, 1880—Przew.; Xining mountains, 1908—Czet.), ***Amdo*** (camp on Churmyn river, May 17, 1880—Przew., before entering Huang He river at uppermost point of its flow, May 18, 1880—Przew.; Guidatin town, along irrigation canals, Oct. 20, 1908—Czet.; Ba valley, in dry arid loess ravines of lower valley, No. 14362, June 1926—Rock).

General distribution: endemic (?).

4. **B. circumserrata** Schneid. var. **occidentalis** Ahrendt in J. Linn. Soc. (London) Bot. 57 (1961) 122; Hao in Bot. Jahrb. 68 (1938) 599.

Described from Qinghai (Nanshan). Type in London (K).

On mountain slopes along riverine valleys.

IIIA. Qinghai: *Nanshan* ("Kokonor region, Roko gorge, No. 13272, Sept. 1925—Rock" —Ahrendt, l.c.; "Kokonor, Nan-Schan, 3000 m, No. 1296, Nov. 14, 1930"—Hao, l.c.).
General distribution: China (South-West.).

5. **B. dasystachya** Maxim. in Bull. Ac. Sci. St. Pétersb. 23 (1877) 308; Hemsl. in J. Linn. Soc. (London) Bot. 23 (1886) 31; Forbes and Hemsl. Index Fl. Sin. 1 (1886) 31; Maxim. Fl. Tangut. 1 (1889) 30; Fedde in Bot. Jahrb. 36 (1905) 43; Schneid. in Bull. Herb. Boiss. sér. 2, 5 (1905) 664 and sér. 2, 8 (1908) 262; id. in Österr. Bot. Zeitschr. 67 (1918) 287; Ahrendt in J. Linn. Soc. (London) Bot. 57 (1961) 184; [Drevesn. rast. Tsinkhaya—Woody Plants of Qinghai] (1987) 209; Mikhailova in Catalogue type specim. Centr. Asia (LE) (2000) 60.

Described from Qinghai. Lectotype and syntypes in St.-Petersburg (LE).

In forests, midbelt of mountains, along riverine valleys.

IIIA. Qinghai: *Nanshan* (Jugum a fl. Tetung S versus silvis frondosis, in rupibus sylvatis frequens, solo humoso, No. 62, May 19/31, 1873—Przew., lectotypus (Mikhailova, l.c.) et 3 isolectotypi; Jugum a fl. Tetung versus, ad No. 62, June 1873, Przew., syntypus; Tetung river, Aug. 5, 1880—Przew.; along banks of Datung-khe river, 2300-2700 m, Sept. 2, 1901—Lad.).
General distribution: China (Nor.-West.).

6. **B. diaphana** Maxim. in Bull. Ac. Sci. St. Pétersb. 23 (1877) 309; id. Fl. Tangut. 1 (1889) 32; Forbes and Hemsl. Index Fl. Sin. 1 (1886) 31; Schneid. in Bull. Herb. Boiss. sér. 2, 5 (1905) 398 and sér. 2, 8 (1908) 195; id. in Österr. Bot. Zeitschr. 66 (1916) 321; Ahrendt in J. Linn. Soc. (London) Bot. 57 (1961) 123; [Drevesn. rast. Tsinkhaya—Woody Plants of Qinghai] (1987) 216; Mikhailova in Catalogue type specim. Centr. Asia (LE) (2000) 60. —**Ic.:** Maxim. (1889) tab. 8, fig. 1-7.

Described from Qinghai. Type in St.-Petersburg (LE).

On mountain slopes, in undergrowth of spruce forests, along river banks.

IIIA. Qinghai: *Nanshan* (Jugum S a fl. Tetung versus, solo humoso in vallibus silvestris...frequens, No. 177, June 27/July 9—lectotypus ! (Mikhailova, l.c.); jugum a fl. Tetung S versus, regio sylvestris, ad No. 177, Aug. 25/Sept. 6—1872 Przew., 2 syntypi !, Tetung river valley, 1800 m, Aug. 5 and Aug. 14, 1880—Przew.; mountains west of Dzhamba river valley, March 15, 1885—Pot.; south. slope of South Kukunor mountain range, 3600-4000 m, June 3, 1895—Rob.; bank of Daitung-khe (Tetung) river, 1700 m, Oct. 2, 1901—Lad.; Kokonor Region: Lalaku gorge, 3600 m, No. 13273, Sept.; Lienhoa Shan, 3000 m, No. 13474, Oct. 1925—Rock), *Amdo* (Mudzhik mountains, June 18, 1880—Przew., alpine region, between Radja and Jupar range; near Radja, among conglomerate, 3300 m, No. 14100, June 6; Radja and Yellow River gorges; outskirts of spruce forest on slopes, 3300 m, No. 14100, June 10, 1926—Rock).
IIIB. Tibet: *Weitzan* (Basin of Mekong river, Chok-chyu river, 4000 m, Aug. 31, 1900—Lad.).
General distribution: China (South-West.).

7. **B. dubia** Schneid. in Bull. Herb. Boiss. sér. 2, 5 (1905) 663 and sér. 2, 8 (1908) 259; id. in Österr. Bot. Zeitschr. 67 (1918) 228; Ahrendt in J. Linn.

Soc. (London) Bot. 57 (1961) 164; Mikhailova in Catalogue type specim. Centr. Asia (LE) (2000) 61. —*B. sinensis* var. *crataegina* Regel in Acta Horti Petrop. 2 (1873) 417; Maxim. Enum. pl. Mong. 1 (1889) 33; id. Fl. Tangut. 1 (1889) 30.

Described from Nor.-West. China (Gansu), Qinghai and Mongolia (Alashan mountain range), lectotype and syntypes in St.-Petersburg (LE).

Along riverine valleys, in rock crevices.

IA. Mongolia: ***East. Mong.*** (on Suma-khada mountains, July 9, 1871—Przew.), ***Alash. Gobi*** (camp in Alashan mountain range, return into Dyn'yuan'in town, July 9, 1872; mont. Alaschan, pars media decliv. occidentalis, in fomibus raro, July 2/14, 1873, syntypus !—Przew.; Yamata gorge, on rock crevices at summit, May 7, 1908 —Czet.).

IIIA. Qinghai: ***Nanshan*** (at Cheibsen temple, base of South. Tetungsk mountain range, July 9, 1872; on way to Chertynton temple, May 19, 1973—Przew.; valle fl. Karyn, in abruptis limosis siccis, May 4, 1885—Pot., syntypus !; on Teshitu river, 2900 m, May 26, 1886—Pot.), ***Amdo*** (up to emergence in Huang He river at the highest point of its flow, May 8; in gorge of Churmyn river, May 17, 1880—Przew.; Lachistan pass, between Xining-gola and Huang He valleys, Karym river and road up to Kadzhir river, 3967 m, May 4, 1881—Przew.).

IIIB. Tibet: ***Weitzan*** (on Yantszy-Tszyan river, June 24, 1884—Przew.).

General distribution: Nor. Mong., China (Nor., Nor.-West.).

Note. Specimen from Nor.-West. China (valle fl. Labran, May 21, 1885—Pot. (Mikhailova, l.c.) was chosen as lectotype.

8. **B. hemsleyana** Ahrendt in J. Linn. Soc. (London) Bot. 57 (1961) 213; Fl. Xizang. 2 (1985) 137. —**Ic.:** Fl. Xizang. 2, tab. 42, fig. 3-5.

Described from south-east. Tibet (Sikan). Type in London (BM).

On mountain slopes.

IIIB. Tibet: ***South.*** ("Shigatsze, Zhen'bu, Lankatsza, Lhasa, Tsyuishui, Tszyan'tszy"—Fl. Xizang. l.c.).

General distribution: China (South-West.: Sikan).

9. **B. heteropoda** Schrenk in Fisch. et Mey. Enum. pl. nov. 1 (1841) 102; Maxim. Enum. pl. Mong. 1 (1889) 33; Schneid. in Bull. Herb. Boiss. sér. 2, 5 (1905) 457; id. Handb. Laubholz Kunde, 1 (1906) 295; Kryl. Fl. Zap. Sib. 5 (1931) 1227; B. Fedtsch. in Fl. SSSR, 7 (1937) 555; Fl. Kazakhst. 4 (1961) 138; Ahrendt in J. Linn. Soc. (London) Bot. 57 (1961) 227; Opred. rast. Sr. Azii [Key to Plants of Mid. Asia] 3 (1972) 234; Claves pl. Xinjiang, 2 (1983) 327; Fl. desert Sin. 1 (1985) 514.

Described from East. Kazakhstan (Jung. Ala Tau). Type in St.-Petersburg (LE).

On rocky and clayey slopes, river terraces.

IB. Kashgar: ***Nor.*** (nor. slope of Kara-Teke mountain range, in recesses in gorges and on rocks, July 17, 1889—Rob.), ***West.*** (King-Tau mountain range, 3-4 km south-east of Kosh-Kulak, lower part of forest belt, June 10, 1959—Yun. et al.).

IIA. Junggar: ***Cis-Alt.*** (south. slope of Chin. Altay, Bugotar area, June 11; Kurtu river, June 19, 1903—Gr.-Grzh.; valley of Kairta river, Aug. 2, 1906—Sap.), ***Jung. Alat.*** (Argaty, north of Borotala, 1600-2000 m, Sept. 5, 1880—A. Reg.; ascent to Kuzyun' pass, rocky site, Aug. 2, 1908—B. Fedtsch.; south-west. fringe of Mainli mountain range, 10-15 km south-south-east of Karaganda pass on road to Ebi-Nur lake and Junggar gateways, mountain steppe belt, Aug. 14; south. slope of Argatu-gol river valley at the site of its interruption by fringe of mountain range, between Borotala settlement and Khundelyun, along fringe of poplar floodplain forest at transition to desert steppe, Aug. 17, 1957—Yun. et al.), ***Tien Shan*** (on Ulastai river, Aug. 8, 1875 —Larionov; on Kunges river, Sept. 6, 1876—Przew.; Talki gorge, 2800-2900 m, July 8 and July 25; Dzhagastai gorge, 1600-2300 m, Aug. 9—1877; Chapchal gorge and pass, 2600 m, June 16, 1878—A. Reg.; Urtak-Sary river west of Sairam, July 19, 1878—Fet.; Shara-Tugutal [Tekes river] pass, Sept. 29, 1878; Kash river, 900-1300 m, Aug. 16, 1879—A. Reg.; Turgen—Aksu, June 3, 1886—Krasn.; Tien Shan, Aug. 29, 1898—Klem.; valley of Dzhanart river, June 14-17, 1903—Merzb.; high terrace of Chendokhoze river on right, 2000 m, May 26, 1954—Mois.; Tien Shan foothills in Savan district, Oct. 2, 1956—Ching; 7 km south-south-east of Urumchi, on road to Turfan, valley of Urumchinki river at Yanervo settlement, floodplain forest, June 1, 1957—Yun. et al.; south. slope of Boro-Khoro mountain range, descent to Ili valley, 6 km south of Nov. Ortai settlement, on highway to Kul'dzha from Sairam-nura, Aug. 19, 1957—Yun. et al.; Ili river at confluence of Kash river in it 2 km east of Yamatu settlement of road to Sarbushin and Kzyl-Kure, Aug. 21; Ili valley, Tekes river floodplain, 1 km east of Kobo settlement of road to Kalmak-Kure, floodplain forest, Aug. 23, 1957—Yun. et al., valley of Muzart river, lowermost right-turn of Lengar before exit of Muzart in Baisk basin, scrubs on lower fringe of forest, Sept. 12, 1958—Yun. et al.), ***Jung. Gobi*** (west.: Kuitun river basin, right bank, Bain-gol creek valley, south of Tushantszy settlement, midportion of valley, steppe belt along valley floor, June 29, 1957—Yun. et al.).

General distribution: Jung. Tarb.; Mid. Asia.

10. **B. integerrima** Bge. in Del. Sem. Horti Dorp. (1843) 4; id. in Mém. Ac. Sci. St. Pétersb. 7 (1854) 189; Maxim. Fl. Tangut. 1 (1889) 29; id. Enum. pl. Mong. 1 (1889) 33; Deasy, Tibet, Chin. Turk. (1901) 399; Danguy in Bull. Mus. Hist. natur. (Paris) 17 (1911) 269; B. Fedtsch. in Fl. SSSR, 7 (1937) 556; Fl. Kazakhst. 4 (1961) 140; Ahrendt in J. Linn. Soc. (London) Bot. 57 (1961) 178; Opred. rast. Sr. Azii [Key to Plants of Mid. Asia] 3 (1972) 235; Claves pl. Xinjiang, 2 (1983) 328.

Described from Mid. Asia (Zeravshan). Type in St.-Petersburg (LE).

On rocky-rubbly slopes, banks of rivers, in lower and middle belts of mountains.

IA. Mongolia: ***Khesi*** (Satszau-Yuan'tszy area, Dankhe river, June 26, 1895—Rob.).

IB. Kashgar: ***Nor.*** (Uital river gorge, June 2; Yarkend-Dar'ya, June 5—1889, Rob.), ***West.*** (major road between Ming-Yul and Kong-Dunugan, Aug. 17, 1913—Knorr.), ***South.*** (Keriya oasis, 1885—Przew.).

IIA. Junggar: ***Tien Shan*** (Kash river, 1000 m; Tekes river valley—1878, A. Reg.; Khaidyk-Gol river, Aug. 12, 1893—Rob.; Muzart river, 1902 and 1903; Kunges river, May 1-5, 1908—Merzb.; Ili river valley, 42 km east of bridge on Kash river, on road to Ziekta, Aug. 29, 1957—Yun. et al.), ***Dzhark.*** (valley of Ili river, right bank, sand between Suidun and Ili, Aug. 13, 1957—Yun. et al.).

General distribution: Mid. Asia (Pamiro-Alay), Fore Asia.

11. **B. kansuensis** Schneid. in Österr. Bot. Zeitschr. 67 (1918) 288; Walker in Contribs. U.S. Nat. Herb. 28 (1941) 619, Rehd. and Wils. in J. Arn. Arb. 9 (1928) 48; Ahrendt in J. Linn. Soc. (London) Bot. 57 (1961) 182; [Drevesn. rast. Tsinkhaya—Woody Plants of Qinghai] (1987) 209.

Described from Nor.-West. China (Gansu). Type in London (K).

On rocky slopes.

IIIA. Qinghai: ***Nanshan*** ("Shui Mo Kou, near Lien Cheng, 1923—K.C. Ching" —Walker, l.c.; "Kemang-ssu, 2950 m, No. 13290; N.E. of Tankar, 2950 m, No. 13291, Oct. 1925—Rock"—Ahrendt, l.c.; "Komang-ssu, 2950 m, No. 13292, Oct. 1925—Rock" —Rehd. and Wilson, l.c.).

General distribution: China (Nor.-West.).

Note: This species is highly variable in plant height. On this basis, two varieties have been distinguished: shrubs 1-1.8 m tall—var. *kansuensis* Ahrendt, l.c.; and shrubs 4-5 m tall—var. *procera* Ahrendt, l.c.

12. **B. kashgarica** Rupr. in Mém. Ac. Sci. St. Pétersb., sér. 7, 14, 4 (1869) 38; Krasn. in Zap. Russ. Geogr. Obshch. 19 (1889) 333; Maxim. Fl. Tangut. 1 (1889) 31; Schneid. in Bull. Herb. Boiss., sér. 2, 5 (1905) 400 and sér. 2, 8 (1908) 195; B. Fedtsch. in Fl. SSSR, 7 (1937) 555; Fl. Kirgiz. 6 (1955) 166; Fl. Kazakhst. 4 (1961) 138; Ahrendt in J. Linn. Soc. (London) Bot. 57 (1961) 228; Ikonnik. Opred. rast. Pamira [Key to Plants of Pamir] (1963) 123; Opred. rast. Sr. Azii [Key to Plants of Mid. Asia] 3 (1972) 234; Claves pl. Xinjiang, 2 (1983) 326. —**Ic.:** Maxim. tab. 23, fig. 1-2; Fl. Tadzh. 4 (1975) Plate 32, fig. 1-5.

Described from Sinkiang (Tien Shan). Type in St.-Petersburg (LE).

On rocky, arid slopes in middle and upper mountain belts, on pebble beds and among boulders along riverine valleys.

IB. Kashgar: ***Nor.*** (15 km nor. of Kucha in Bai, on gorge terrace, 2000 m, Sept. 6, 1958—Lee and Chu; south. slope of Baisk basin, point of interruption by Muzart river, Sept. 6; south. slope of Muzart river valley in the upper course of river, between Yangimalle settlement and Kurgan, 2100 m, steppe belt, rocky slope, Sept. 12, 1958 —Yun. et al.; on Bilushtum-Upal road, 20 km from Bilushtum, 2500-3000 m, June 15, 1959—Lee et al.), ***West.*** (Kenkol river gorge, July 29, 1913—Knorr.; Kizyl-su river basin, around Ulugchat, July 2, 1929—Pop.; valley of Tiznaf river (Kholosstan-dar'ya) 15 km before Kyude on Tibetan road to Ak-këz pass and Karganlyk, June 4, 1959—Yun. et al.; Sinkiang—Tibet highway at Kyude settlement, 3000 m, June 5, 1959—Lee and Chu; 20 km from Upal on road to Kokshaltau, June 9, 1959—Lee and Chu), ***South.*** (in Taktal region of Chira district, in valley, May 20, 1959—Lee and Chu; Sandzhu river valley, between Akar and Kanka, May 28, 1959—Yun. et al.; in upper course of Sandzhu river, in valley, 2500 m, May 29, 1959—Lee and Chu).

IIA. Junggar: ***Tien Shan*** (valley of Dzhanart river, June 14-17; Inilshchek-Tal, Aug. 17-25, 1903—Merzb.).

IIIB. Tibet: ***Chang Tang*** (Sinkiang—Tibet highway at Maichata settlement, 3700 m, June 3, 1959—Lee and Chu; 17 km south of and beyond Polur, on road to Khoneilyak, south. slopes in alpine belt, May 12, 1959—Yun. et al.).

General distribution: Cent. Tien Shan; Mid. Asia (Pamiro-Alay).

13. **B. nummularia** Bge. in Del. Sem. Horti Dorpat. (1843) 4; Schneid. in Bull. Herb. Boiss. sér. 2, 5 (1905) 459; B. Fedtsch. in Fl. SSSR, 7 (1937) 559; Fl. Kazakhst. 4 (1961) 142; Claves pl. Xinjiang, 2 (1983) 327. —*B. densiflora* var. *nummularia* Boiss. Fl. or. 1 (1867) 102; Schneid. l.c. 459. —*B. nummularia* var. *schrenkiana* Schneid. l.c. 460. —*B. iliensis* M. Pop. in Index Sem. Horti Almaatensis, 3 (1936) 3.

Described from Pamir. Type in St.-Petersburg (LE). Plate VIII, fig. 2.

IB. Kashgar: ***Nor.*** (Zigdeidag in Kel'pin, 2000 m, No. 7412, Sept. 9, 1958; from Talake in Aksu, No. 8916, Sept. 26, 1958—Lee and Chu), ***West.*** (basin of Yarkend-Dar'ya river, Aug. 10, 1890—Grombch.), ***South.*** (Keriya, oasis, June 5, 1885—Przew.).

IIA. Junggar: ***Tien Shan*** (Tien Shan, Aug. 26, 1876—Przew.; Tekes river, Sept. 19, 1878—A. Reg.; Bedel' pass, Aug. 1, 1886—Krasn.; Khaidyng-gol river, Tsagan-usu tributary, Aug. 12, 1893—Rob.; from bridge on Kash river 28 km east, No. 1187, Aug. 19, 1957—Lee and Chu), ***Dzhark.*** (Kul'dzha, Aug. 18, 1880—A. Reg.; valley of Ili river, right bank, sand between Suidun and Ili 6 km nor.-east of border post Santokhodze on road to Suidun, No. 1732, Aug. 31, 1957—Yun. et al.).

General distribution: Mid. Asia, Fore Asia (Iran).

14. **B. obovatifolia** Ying in Fl. Xizang. 2 (1985) 146.

Described from South. Tibet. Type in Beijing (PE).

On mountain slopes.

IIIB. Tibet: ***South.*** ("Lhasa, alt. 3900 m, May 25, 1975, Qinghai—Xizang Complex Exp. 75, No. 55—typus"—Fl. Xizang. l.c.).

General distribution: endemic (?).

15. **B. poiretii** Schneid. in Mitt. Deutsch. Dendr. (1906) 180; id. in Bull. Herb. Boiss. sér. 2, 8 (1908) 258; id. in Österr. Bot. Zeitschr. 67 (1918) 228; B. Fedtsch. in Fl. SSSR, 7 (1937) 558; Ahrendt in J. Linn. Soc. (London) Bot. 57 (1961) 174; Fl. Intramong. 2 (1978) 275; Fl. desert Sin. 1 (1985) 515. —*B. sinensis* var. *angustifolia* Regel in Acta Horti Petrop. 2 (1873) 416; [Drevesn. rast. Tsinkhaya—Woody Plants of Quinghai] (1987) 207. —**Ic.:** Fl. desert Sin. 1, tab. 189, fig. 6-8.

Described from Nor. China. Type ?

On mountain slopes, along sandy banks of rivers.

IA. Mongolia: ***East. Mong.*** ("Shilingol, Datsinshan'"—Fl. Intramong, l.c.).

General distribution: China (Nor., Nor.-West.).

16. **B. pulangensis** Ying in Fl. Xizang. 2 (1985) 133.

Described from South. Tibet. Type in Beijing (PE).

On mountain slopes, in midbelt.

IIIB. Tibet: ***South.*** ("Pulan', 76-8494, Ching"—Fl. Xizang. l.c.).

General distribution: endemic (?).

17. **B. purdomii** Schneid. in Sarg. Pl. Wils. 1 (1913) 372; Walker in Contribs. U.S. Nat. Herb. 28 (1941) 619; Ahrendt in J. Linn. Soc. (London)

Bot. 57 (1961) 175; [Drevesn. rast. Tsinkhaya—Woody Plants of Qinghai] (1987) 201.

Described from Nor.-West. China (Shen'si). Type in London (K).

On mountain slopes.

IA. Mongolia: ***Alash. Gobi*** ("Holanshan, Hsi Yeh Kou, No. 167, R.C. Ching" —Walker, l.c.).

General distribution: China (Nor.-West.).

18. **B. sabulicola** Ying in Fl. Xizang. 2 (1985) 133.

Described from South. Tibet. Type in Beijing (PE).

On rubbly slopes in midbelt of mountains.

IIIB. Tibet: ***South.*** ("Lhasa, alt. 3810 m, July 19, 1976, Qinghai—Xizang Complex Exp. 76, No. 8494—typus"—Fl. Xizang. l.c.).

General distribution: endemic ?

19. **B. sibirica** Pall. Reise, 2 (1773) 737; Maxim. Enum. Pl. Mong. 1 (1889) 32; Schneid. in Bull. Herb. Boiss. 2, sér. 5 (1905) 396, Kryl. Fl. Zap. Sib. 5 (1931) 1226; B. Fedtsch. in Fl. SSSR, 7 (1937) 554; Rehd. Man. Cult. Trees and Shrubs, ed. 2 (1940) 232; Fl. Kazakhst. 4 (1961) 138; Ahrendt in J. Linn. Soc. (London) Bot. 57 (1961) 131; Opred. rast. Sr. Azii [Key to Plants of Mid. Asia] 3 (1972) 234; Fl. Intramong. 2 (1978) 272; Grub. Opred. rast. Mong. [Key to Plants of Mongolia] (1982) 116; Claves pl. Xinjiang, 2 (1983) 326. —**Ic.:** Fl. SSSR, 7, Plate 37, fig. 1.

Described from Altay. Type in St.-Petersburg (LE).

On rocks, rock placers, granite outliers.

IA. Mongolia: ***Khobd.*** (transit to Burgassutai river in Uryuk-Nor lake basin, 26 versts [1 verst = 1.067 km]—from Kut river toward Bairimen-daban pass leading from basin of Achit-Nor lake into basin of Uryuk-Nor lake, 2575 m, July 8, 1879—Pot.), ***Mong. Alt.*** (Onkattu lake, Chingistei sentry, alp. meadow, June 25, 1906—Sap.; upper Kobdo river, valley of Khoton-Gol, left bank, east. slope of Modot-Tologoi mountain, near winter road, on rocks, 2400 m, July 8, 1971—Grub., Ulzij., Dariima; 50 km south of Ulangom, Mort-Ula massif, larch grove, 1900 m, June 20, 1978; 22 km south-south-west of Altan-tsugts, Tokte-Ula, 2450 m, July 19, 1979—Karam. et al.; valley of Uenchiin-gol 8 km below Khargaityn-gol estuary, on granite rocks in gorge, Aug. 15, 1979 —Grub., Muld. and Dariima), ***Depr. Lakes*** (Dzun-Gobi somon, Ugumur mountain on west. bank of Bayan-nur, upper rocky portion of slope, July 15, 1973; 34 km south of ? Umun-Gobi, mound, rock crevices, Aug. 3, 1977—Karam.), ***Gobi-Alt.*** (Baga-Bogdo, Altay mountains 2500 m, No. 257, 1925—Chaney; Dzun-Saikhan mountains, on rocks, along mountain slope, Aug. 23, 1931—Ik.-Gal., Dundu- and Dzun-Saikhan mountain range, mountain slopes and gorges from trail to upper belt, July-Aug. 1933—Simukova; Khongor-Obo-somon, Dzun-Saikhan mountain range, June 19, 1945—Yun.).

IIA. Junggar: ***Cis-Alt.*** (slope of Chin. Altay, Bor-Balgasun river, July 2, 1903 —Gr.-Grzh.), ***Tarb.*** (Saur mountain range, south. slope, valley of Karagaitu river at its emergence from mountains on to trail, south. rocky slope, June 23, 1957—Yun. et al.), ***Jung. Alat.*** (Toli district, Myaoergou village, on rubble slope, Aug. 4, 1957—Kuan; Dzhair mountain range, 1-1.5 km north of Otu settlement on road to Chuguchak from Aktam, mountain-steppe belt, on granite pillow lavas on left bank of a small brook, Aug. 4, 1957—Yun. et al.).

General distribution: West. Sib. (Altay), East. Sib., Mid. Asia, Nor. Mong., China (Dunbei).

20. **B. ulicina** Hook. f. et Thoms. Fl. Brit. Ind. 1 (1872) 111; Henderson and Hume, Lahore to Jarkand (1878) 309; Hemsl. in J. Linn. Soc. (London) Bot. 30 (1894) 123; Persson in Bot. Not. (Lund) (1938) 284; Ahrendt in J. Linn. Soc. (London) Bot. 57 (1961) 228.

Described from Kashgar. Type in London (K).

On arid mountain slopes, in middle and upper belts of mountains (2700-5100 m).

IB. Kashgar: ***West.*** ("Common in Jarkand, from 2700 to 3900 m"—Henderson and Hume, l.c.).

IIIB. Tibet: ***Chang Tang*** ("Tibet, Kuen-lun, plants at about 5100 m, H.P. Picot" —Hemsl. l.c.).

General distribution: Himalayas (west., Kashmir).

21. **B. umbellata** Wall. var. **brianii** Ahrendt in J. Linn. Soc. (London) Bot. 80 (1942) 94 and ibid. 57 (1961) 104; Hook. f. Fl. Brit. Ind. 1 (1872) 110, p.p.; Hao in Bot. Jahrb. 68 (1938) 599.

Described from specimens cultivated from seeds imported from India. Types in Oslo (O) and London (BM).

On mountain slopes.

IIIA. Qinghai: ***Nanshan*** ("Kokonor: Sining, Schang-wu-chuang, 2900 m, No. 824, Aug. 3; auf dem südkokonorischen Gebirge von 3800 m, No. 1261, Sept. 12; auf den Abhangen des östlichen Nan-Schan, No. 1296, Sept. 14, 1930, Hao"—Hao, l.c.).

General distribution: Himalayas (east.).

22. **B. vernae** Schneid. in Sarg. Pl. Wils. 1 (1913) 372; id. in Bull. Herb. Boiss. 2 sér. 8 (1908) 440; id. in Österr. Bot. Zeitschr. 66 (1916) 218; Rehd. and Wils. in J. Arn. Arb. 9 (1928) 44; Walker in Contribs. U.S. Nat. Herb. 28 (1941) 620; Ahrendt in J. Linn. Soc. (London) Bot. 57 (1961) 175; Fl. Intramong. 2 (1978) 272; Fl. desert Sin. 1 (1985) 516. —*B. caroli* var. *hoanghensis* Schneid. in Bull. Herb. Boiss. 2 sér. 5 (1905) 459; [Drevesn. rast. Tsinkhaya—Woody Plants of Qinghai] (1987) 209. —**Ic.:** Fl. Intramong. 2, tab. 142, fig. 111.

Described from Nor.-West. China.Type in London (K).

In mountain forests and riverine valleys.

IIIA. Qinghai: ***Nanshan*** (between Hsining and Tankar, 2700 m, No. 13256, Sept. 10; foot of Loyni Shan, near Sin Cheng, NE Tankar, No. 13289, Oct.; Richthofen range, adjacent region to Liyuanku, No. 13315, Nov.—1925, Rock).

General distribution: China (South-West.).

* * *

B. nullinervis Ying in F. Xizang. 2 (1985) 141 ("Namling, alt. 4200-4300 m") has been cited for South. Tibet.

2. **Gymnospermium** Spach

in Hist. natur. vég. 8 (1839) 66

1. **G. altaicum** (Pall.) Spach in Hist. natur. vég. 8 (1839) 67; Opred. rast. Sr. Azii [Key to Plants of Mid. Asia] 3 (1972) 231; Claves pl. Xinjiang, 2 (1983) 328. —*Leontice altaica* Pall. in Acta Ac. Sci. Petropol. (pro a. 1779) [3] pars 2 (1783) 255; Kryl. Fl. Zap. Sib. 5 (1931) 1225; B. Fedtsch. in Fl. SSSR, 7 (1937) 549; Fl. Kazakhst. 4 (1961) 134.

Described from West. Siberia (Altay). Type not preserved.

On clayey-rocky mountain slopes, loessial mounds, from foothills to upper belt of mountains.

IIA. Junggar: ***Tarb.*** ("Dachen"—Claves pl. Xinjiang, l.c.), ***Jung. Alat.*** ("Ala Tau" —ibid.).

General distribution: Jung.-Tarb.; Mid. Asia (Tien Shan), West. Sib. (Altay).

3. **Leontice** L.

Sp. pl. (1753) 312; id. Gen. pl. ed. 5 (1754) 147

1. **L. incerta** Pall. Reise, 3 (1776) 726; Kryl. Fl. Zap. Sib. 5 (1931) 1225; B. Fedtsch. in Fl. SSSR, 7 (1937) 545; Fl. Kazakhst. 4 (1961) 133; Opred. rast. Sr. Azii [Key to Plants of Mid. Asia] 3 (1972) 232; Claves pl. Xinjiang, 2 (1983) 328; Fl. desert Sin. 1 (1985) 518. —*L. vesicaria* Pall. in Acta Ac. Sci. Petropol. (pro a. 1779 [3] pars 2 (1783) 257.

Described from West. Kazakhstan (Inderskoe lake). Type not preserved. Plate VIII, fig. 3.

On clayey, rubbly and rocky slopes and trails of mounds and mountains, on sandy and loessial soils in lower and middle belts of mountains.

IIA. Junggar: ***Jung. Gobi*** (south.: "Khutubi, Manas"—Claves pl. Xinjiang, l.c.), ***Dzhark.*** (Bayandai, 2000 ft, May 5, 1878; Kul'dzha, 3000-6000 ft, May 1, 1879—A. Reg.).

General distribution: Aralo-Casp., Fore Balkh., Jung.-Tarb.; Mid. Asia.

4. **Podophyllum** L.

Sp. pl. (1753) 505; id. Gen. pl. ed. 5 (1754) 223

1. **P. emodi** Wall. ex Hook. f. et Thoms. Fl. Ind. 1 (1855) 232; Fl. Brit. Ind. 1 (1872) 112; Walker in Contribs. U.S. Nat. Herb. 28 (1941) 620.

Described from Himalayas. Type in London (K). Plate VIII, fig. 4.

In shaded deciduous forests.

IIIA. Qinghai: ***Nanshan*** (in forests on South. Tetung mountain range, rarely on rivers, Aug. 9, 1872; midbelt of South. Tetung mountain range, in valleys, rare, April 23, 1873; same site, in forests, frequent, Aug. 3, 1880—Przew.; Xining mountains,

Cha-dorsi gorge, N 129, May 13, 1890—Gr.-Grzh.; "Tu Er Ping, in woods, No. 370, June 1923, Ching"—Walker, l.c.).

General distribution: China (Nor.-West., South-West.), Himalayas.

Family 49. MENISPERMACEAE Juss.

1. Menispermum L.

Sp. pl. (1753) 340

1. **M. dahuricum** DC. Syst. 1 (1817) 540; id. Prodr. 1 (1824) 102; Ledeb. Fl. Ross. 1, 1 (1841) 78; Turcz. Fl. baic.-dahur. 1 (1842) 90; Maxim. Prim. Fl. amur. (1859) 30; Kom. in Fl. SSSR, 7 (1937) 563; Grub. Konsp. fl. MNR [Conspectus of Flora of Mongolian People's Republic] (1955) 146; id. Opred. rast. Mong. [Key to Plants of Mongolia] (1982) 116; Fl. Intramong. 2 (1978) 277; Fl. desert Sin. 1 (1985) 518; Fl. Intramong. ed. 2, 2 (1990) 583; Fl. Sin. 30, 1 (1996) 39. —**Ic.:** Fl. SSSR, 7, Plate 38, fig. 11-17; Fl. Intramong. 2, tab. 145, Fl. desert Sin. 1, tab. 191, fig. 1-4; Fl. Intramong. ed. 2, 2, tab. 237; Fl. Sin. 30, 1, tab. 8, fig. 10-12.

Described from East. Siberia (Dauria). Type in Geneva (G).

IA. Mongolia: ***East. Mong.*** (Kalgan gorge, Aug. 23, 1898—Zabolotnyi; "Datsinshan"—Fl. Intramong. l.c.).

General distribution: East. Sib. (Ang.-Sayan: Angara river; Dauria), Far East (south), Nor. Mong., China (Dunbei, Nor., Nor.-West., South-West.: nor.), Korean peninsula, Japan.

Plate I

1 — *Paeonia veitchii* Lynch; 2 — *Callianthemum cachemirianum* Cambess.; 3 — *Hepatica falconeri* (Thoms.) Steward; 4 — *Anemone obtusiloba* D. Don

Plate II

1 — *Halerpestes salsuginosa* (Pall. ex Georgi) Greene; 2 — *Trollius pumilus* D. Don; 3 — *T. dschungaricus* Regel; 4 — *Caltha scaposa* Hook. f. et Thoms.

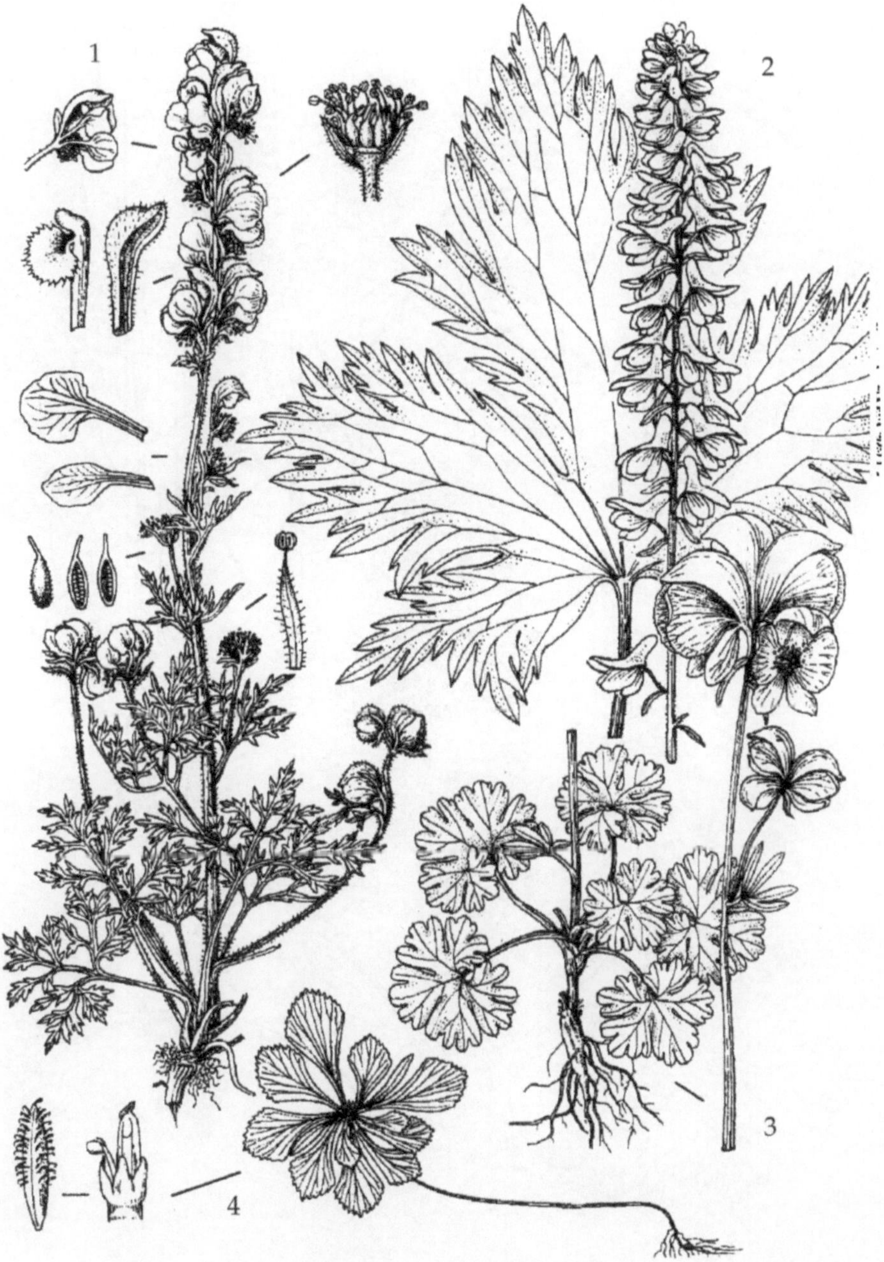

Plate III

1 — *Aconitum gymnandrum* Maxim.; 2 — *A. apetalum* (Huth) B. Fedtsch.; 3 — *A. tanguticum* (Maxim.) Stapf; 4 — *Circaeaster agrestis* Maxim.

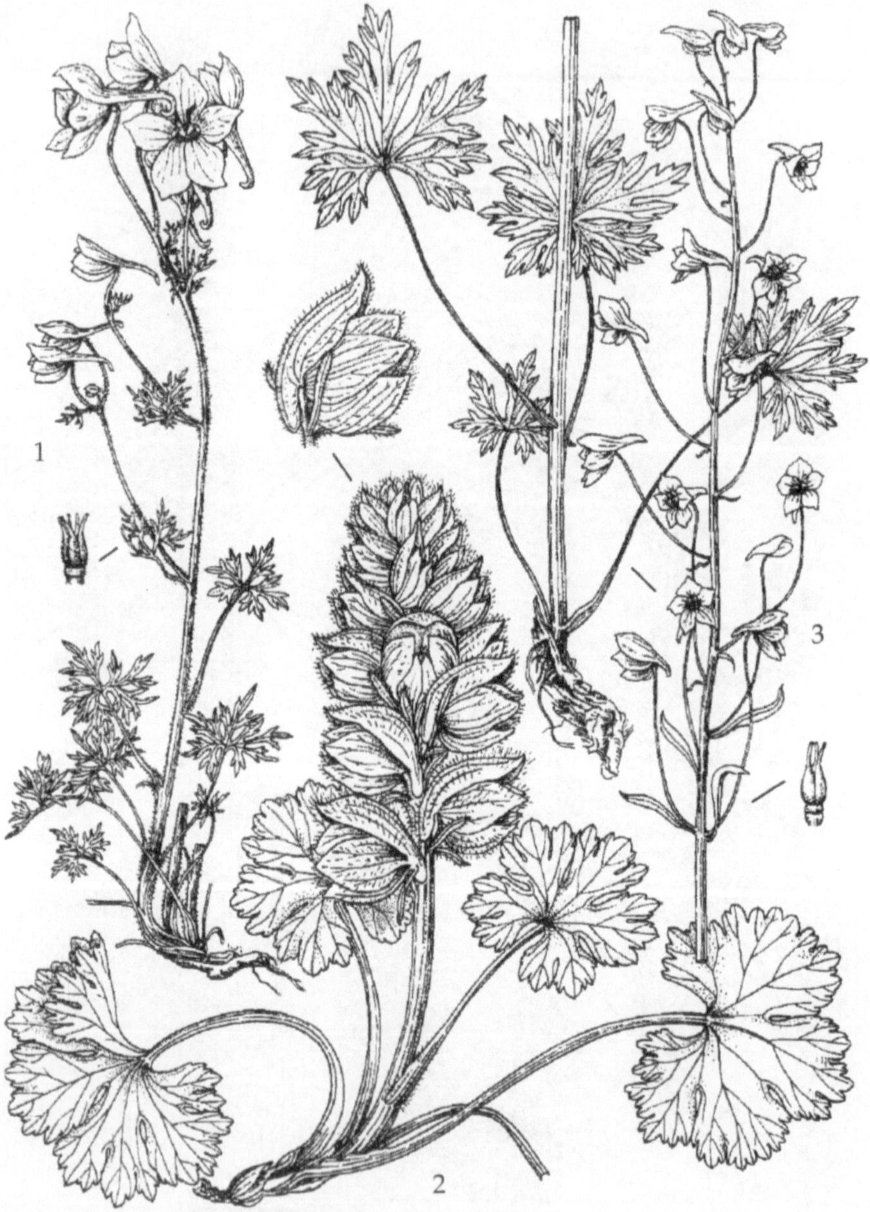

Plate IV
1— *Delphinium pylzowii* Maxim.; 2 — *D. densiflorum* Duthie ex Huth;
3 — *D. sparsiflorum* Maxim.

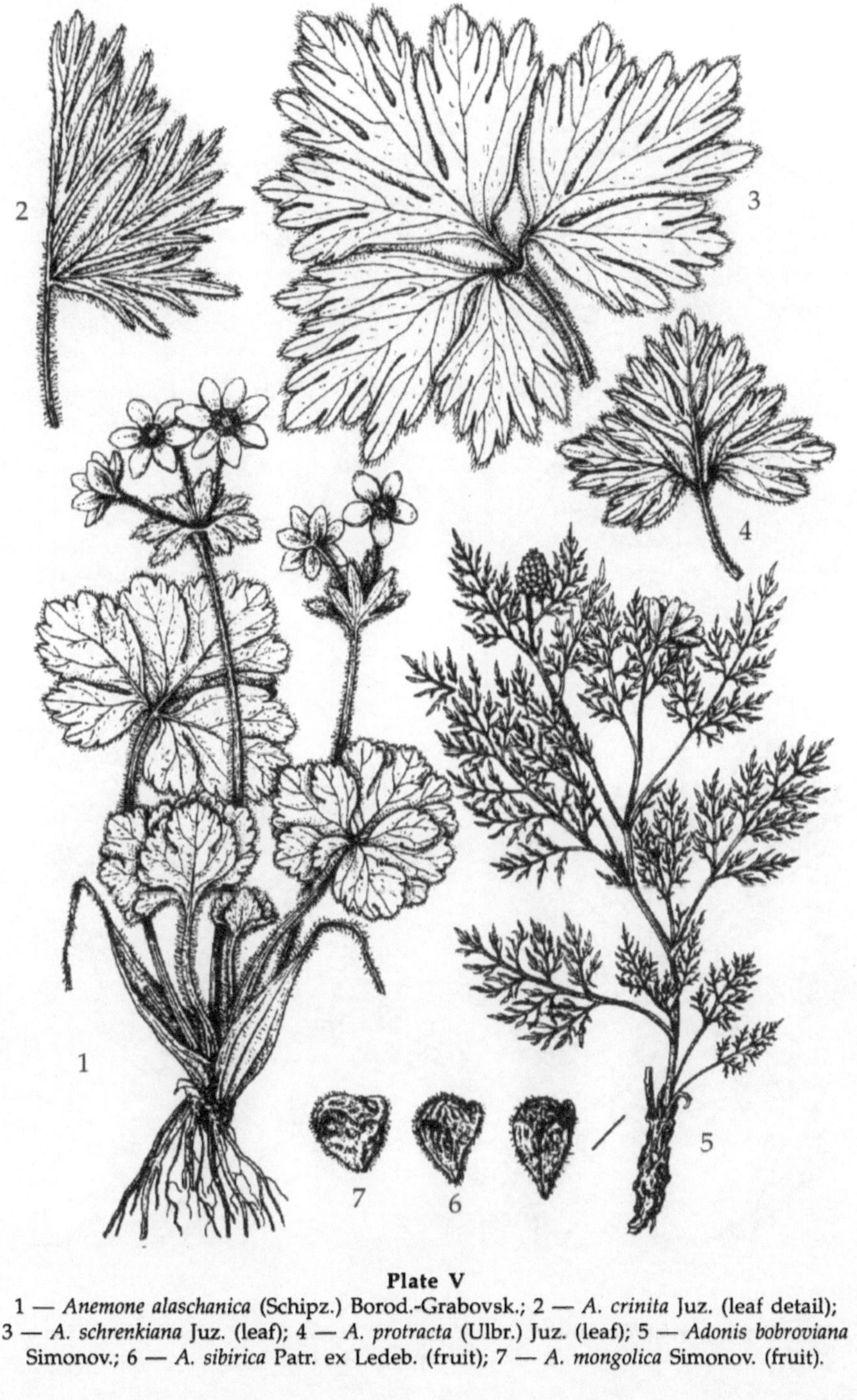

Plate V

1 — *Anemone alaschanica* (Schipz.) Borod.-Grabovsk.; 2 — *A. crinita* Juz. (leaf detail); 3 — *A. schrenkiana* Juz. (leaf); 4 — *A. protracta* (Ulbr.) Juz. (leaf); 5 — *Adonis bobroviana* Simonov.; 6 — *A. sibirica* Patr. ex Ledeb. (fruit); 7 — *A. mongolica* Simonov. (fruit).

Plate VI

1 — *Clematis aethusifolia* Turcz.; 2 — *C. fruticosa* Turcz.; 3 — *C. aspleniifolia* Schrenk; 4 — *C. nannophylla* Maxim.

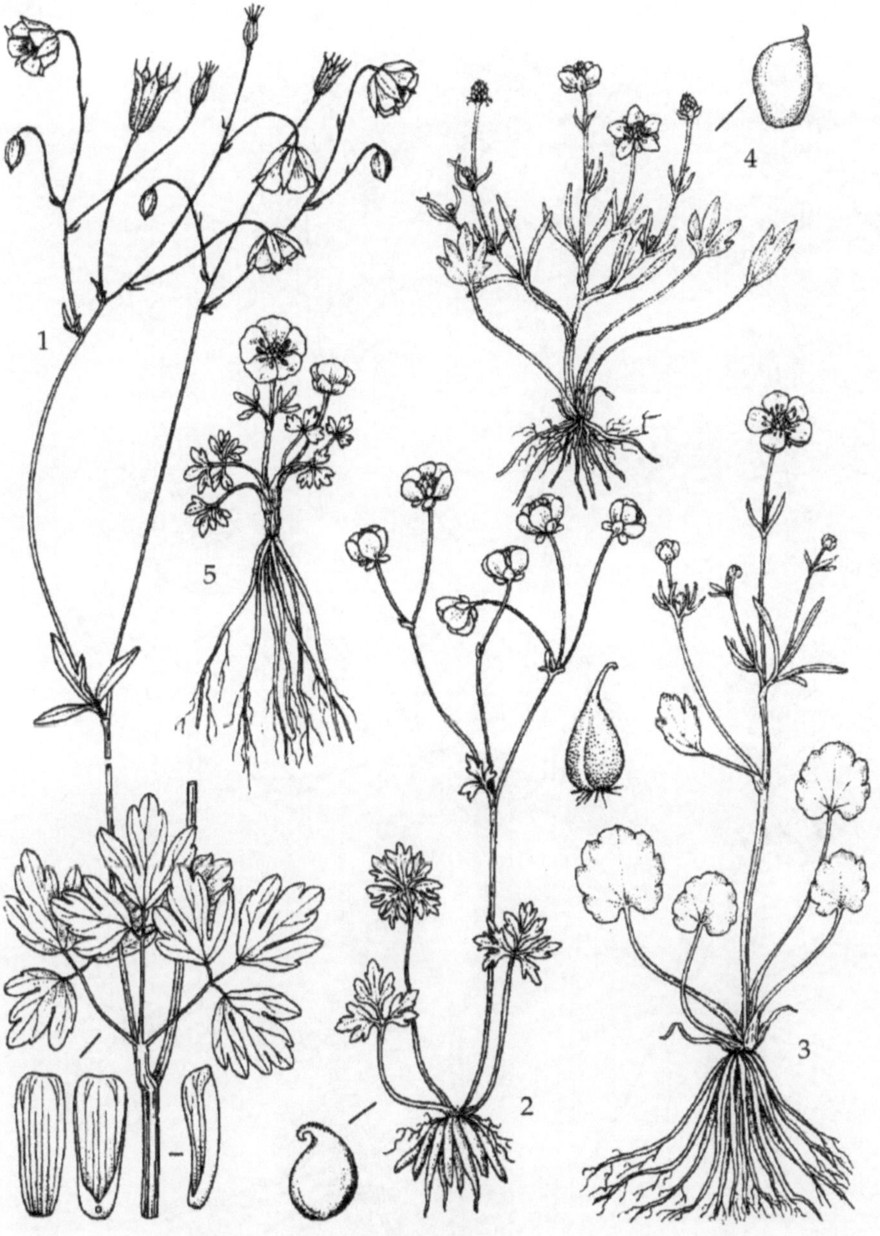

Plate VII

1 — *Aquilegia ecalcarata* Maxim.; 2 — *Ranunculus gobicus* Maxim.; 3 — *R. indivisus* (Maxim.) Hand.-Mazz.; 4 — *R. popovii* Ovcz.; 5 —*R. tibetanus* (Tolm. et Jurtz.) Borod.-Grabovsk.

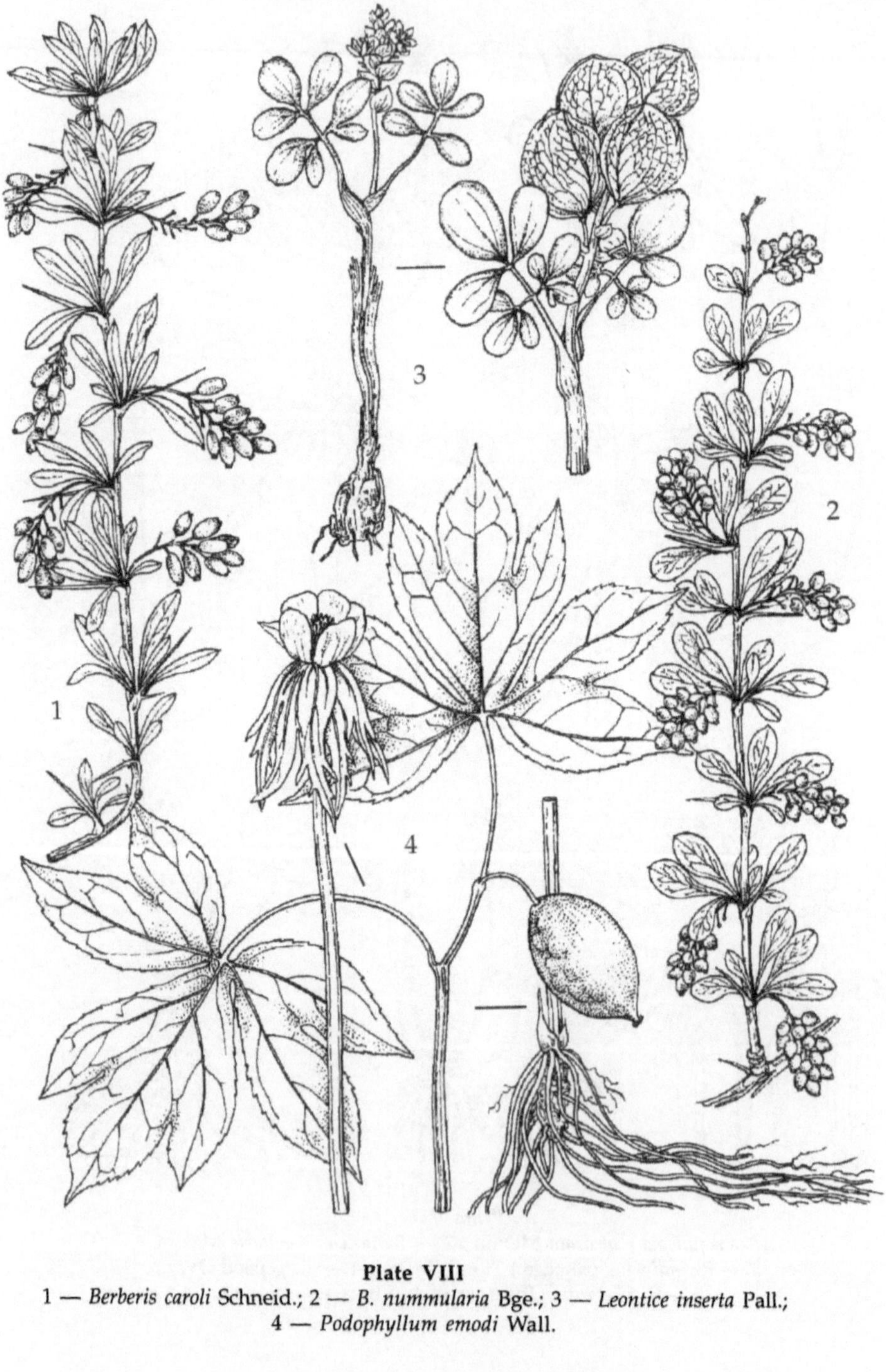

Plate VIII

1 — *Berberis caroli* Schneid.; 2 — *B. nummularia* Bge.; 3 — *Leontice inserta* Pall.; 4 — *Podophyllum emodi* Wall.

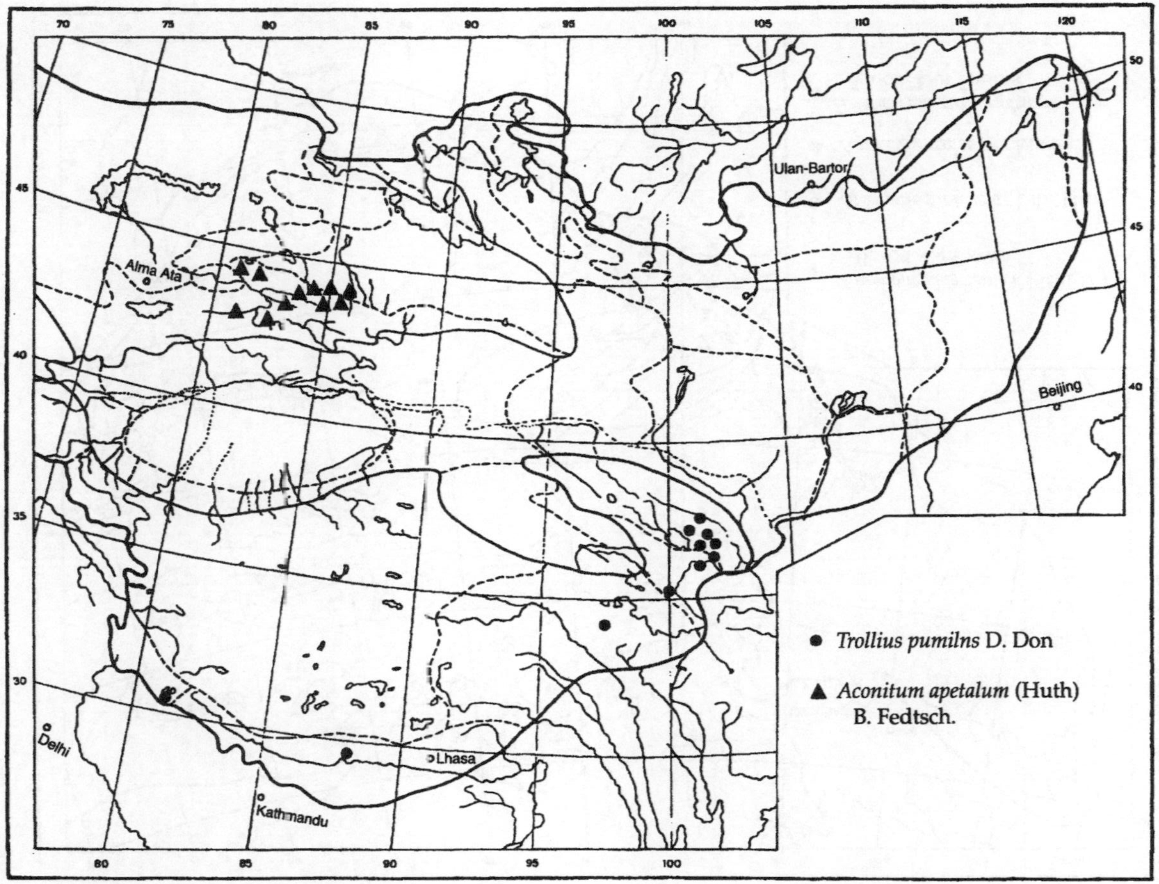
Alma Ata
Ulan-Bartor
Beijing
Delhi
Kathmandu
Lhasa
70
75
80
85
90
95
100
105
110
115
120
50
45
40
35
30
● Trollius pumilns D. Don
▲ Aconitum apetalum (Huth)
B. Fedtsch.

Map 1

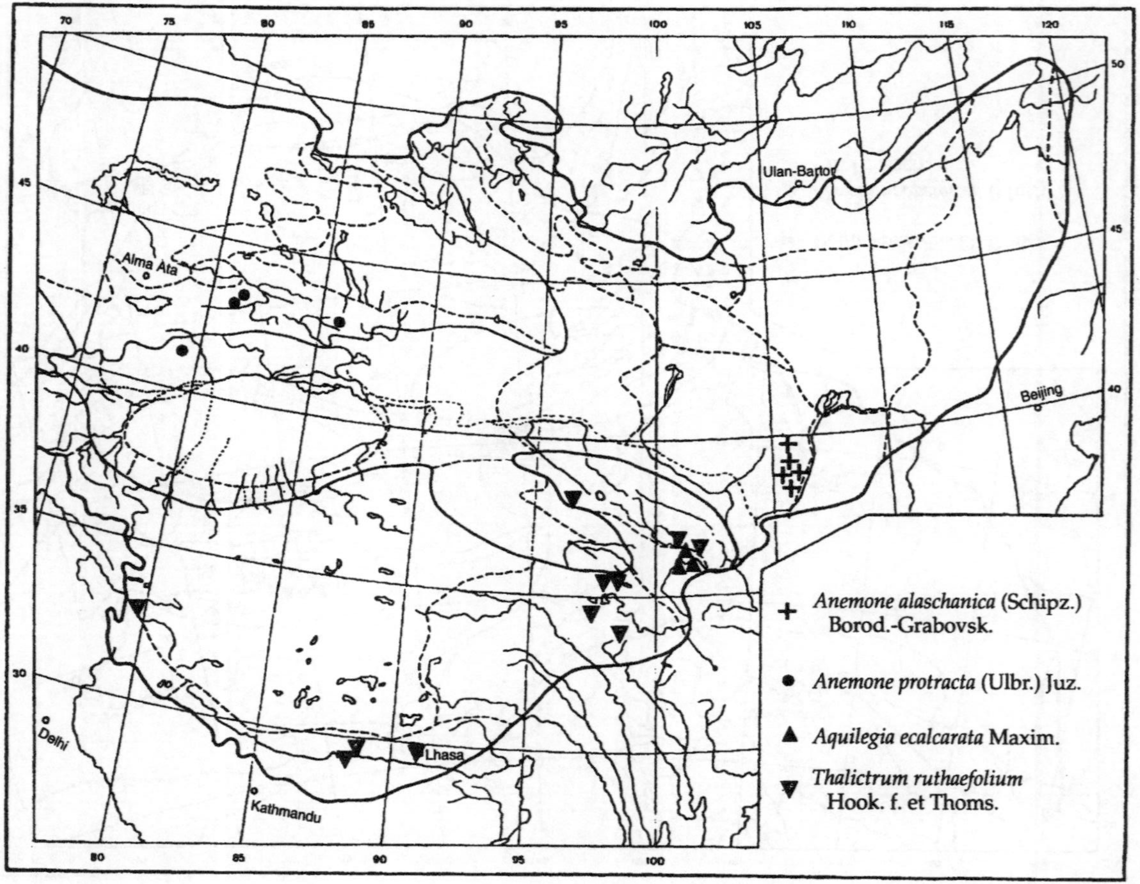
Alma Ata
Ulan-Bartor
Beijing
Delhi
Kathmandu
Lhasa
70
75
80
85
90
95
100
105
110
115
120
50
45
40
35
30
+ Anemone alaschanica (Schipz.) Borod.-Grabovsk.
● Anemone protracta (Ulbr.) Juz.
▲ Aquilegia ecalcarata Maxim.
▼ Thalictrum ruthaefolium Hook. f. et Thoms.

Map 2

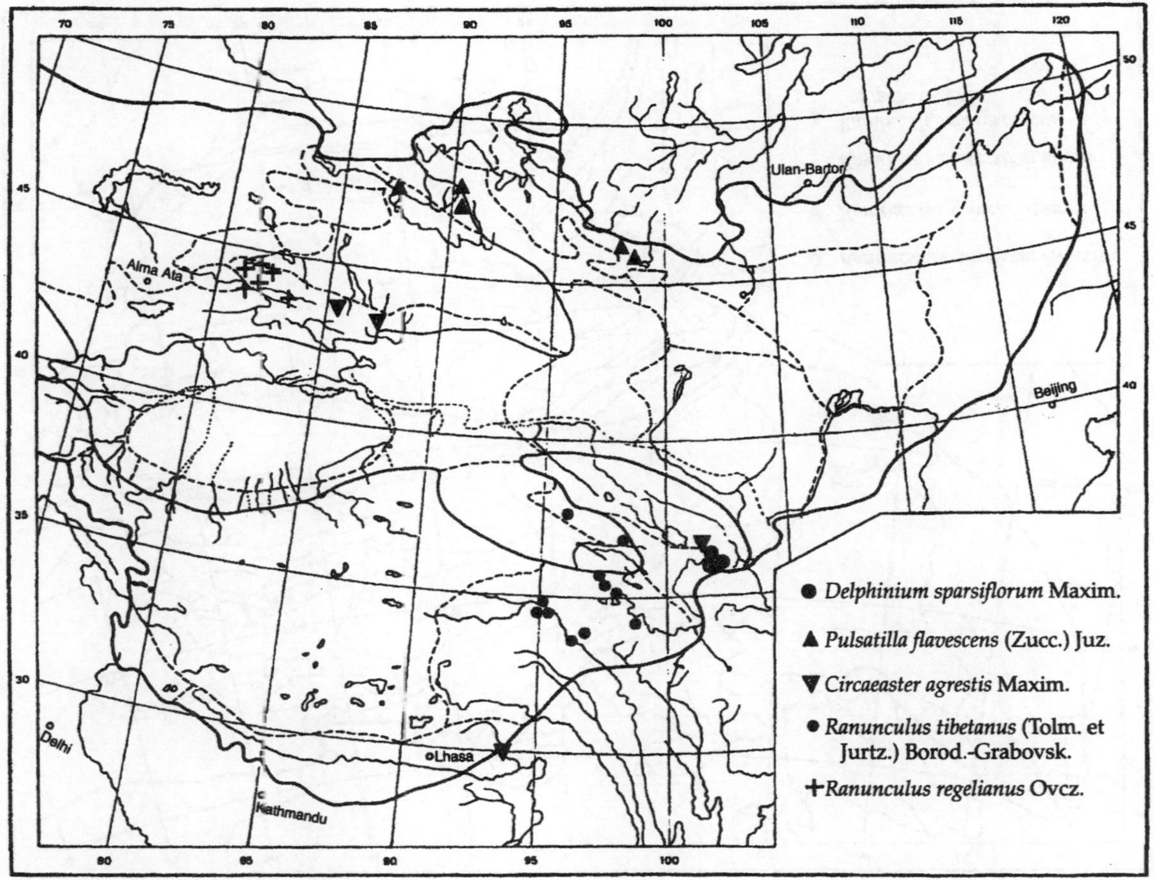
Ulan-Bator
Alma Ata
Beijing
Lhasa
Delhi
Kathmandu
● Delphinium sparsiflorum Maxim.
▲ Pulsatilla flavescens (Zucc.) Juz.
▼ Circaeaster agrestis Maxim.
● Ranunculus tibetanus (Tolm. et Jurtz.) Borod.-Grabovsk.
+ Ranunculus regelianus Ovcz.

Map 3

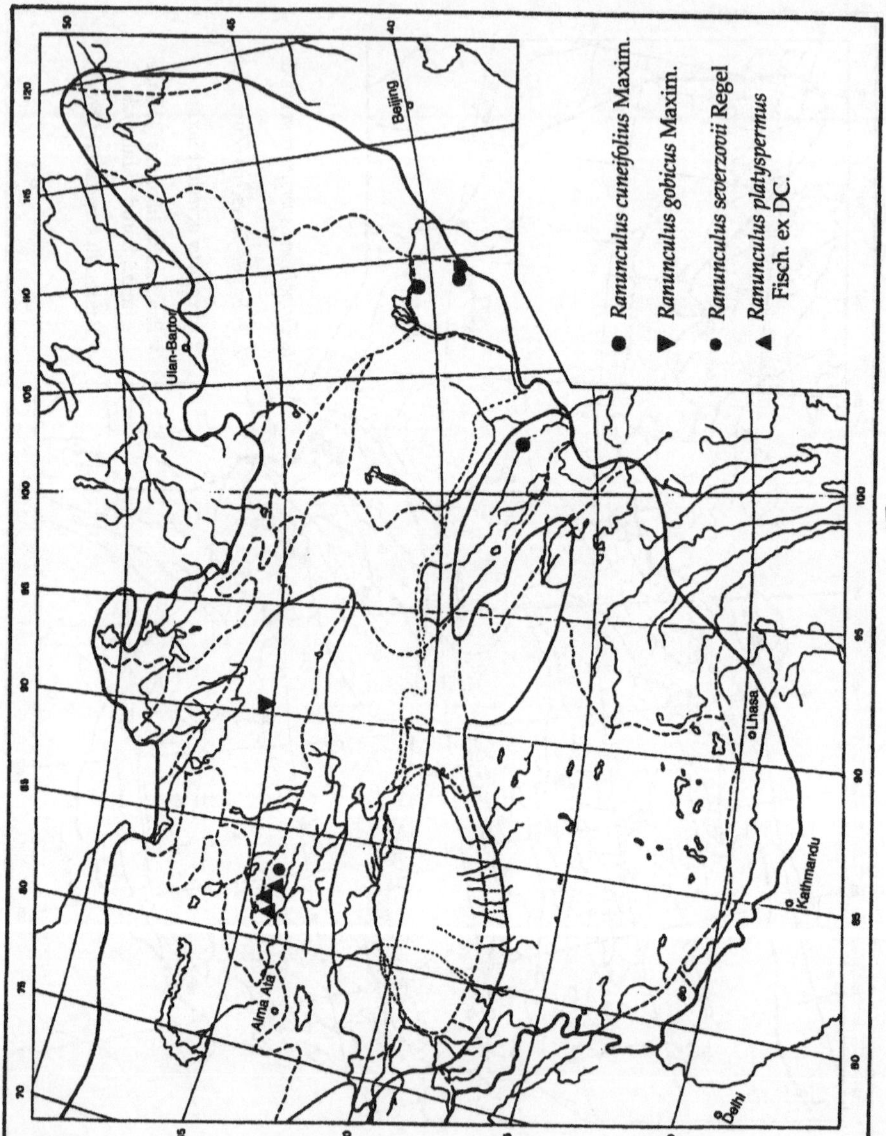
Ulan-Bator
Beijing
Alma Ata
Lhasa
Kathmandu
Delhi
Ranunculus cuneifolius Maxim.
Ranunculus gobicus Maxim.
Ranunculus severzovii Regel
Ranunculus platyspermus Fisch. ex DC.

Map 4

INDEX OF LATIN NAMES OF PLANTS

INDEX OF PLANT ILLUSTRATIONS

INDEX OF PLANT DISTRIBUTION RANGES